James Chapman

Travels in the Interior of South Africa

Vol. 1

James Chapman

Travels in the Interior of South Africa
Vol. 1

ISBN/EAN: 9783744760874

Printed in Europe, USA, Canada, Australia, Japan

Cover: Foto ©Andreas Hilbeck / pixelio.de

More available books at **www.hansebooks.com**

TRAVELS

IN THE

INTERIOR OF SOUTH AFRICA,

COMPRISING

Fifteen Years' Hunting and Trading;

WITH JOURNEYS ACROSS THE CONTINENT FROM NATAL TO
WALVISCH BAY, AND VISITS TO LAKE NGAMI
AND THE VICTORIA FALLS.

BY

JAMES CHAPMAN, F.R.G.S.

ILLUSTRATED WITH MAPS AND NUMEROUS ENGRAVINGS.

IN TWO VOLUMES.

VOL. I.

LONDON:
BELL & DALDY, YORK STREET, COVENT GARDEN;
EDWARD STANFORD, 6, CHARING CROSS.
MDCCCLXVIII.

PREFACE.

THE narratives of Travel comprehended within these volumes extend, as will be seen, over a lengthened period, and cover a wide area of the African continent. With, indeed, a single great and striking exception—whose name is on everyone's lips, and whose fate engages at the present moment the hopeful expectations of civilized mankind in every quarter of the globe—no traveller can pretend to an acquaintance with the interior of Southern Africa, from the shores of the Atlantic to those of the Indian Ocean, and from the Cape to the Zambesi, so prolonged in duration, and so varied in range, as the Author of these volumes. If Mr. Chapman's records of personal experience, acquired in the pursuits of the hunter and trader, lay for the most part no claim to the merits that belong to original investigation or discovery, they at least embody the results of prolonged and intelligent observation, directed, during many years, towards regions, many of which have hitherto been seldom visited by civilized man, and towards native races whose social life presents a highly interesting problem of inquiry to all.

It is due alike to the Author and the reader to state that this work appears before the public without the advantage of personal superintendence on the part of the former—still resident in the country within which so many years of his previous life have been passed. And justice to the present Editor requires a statement of the divided responsibility between himself and the late Mr. Thomas Forester, to whom the laborious task of revising Mr. Chapman's voluminous manuscripts had been originally intrusted by the publishers. Mr. Forester's labours, directed mainly to such curtailment of various redundancies in the Author's narrative as were absolutely necessary in order to bring it within the required limit of two volumes, were themselves incomplete when interrupted by the hand of death. In completing the task thus left unfinished by his predecessor, the Editor has, necessarily, been guided by the plan which Mr. Forester had marked out, and upon which he had worked. Beyond such verbal alterations as were naturally desirable in the case of journals transcribed from notes made while in actual course of travel, the only liberties taken with Mr. Chapman's MSS. have been those of abridgment and omission. Care has been taken, however, to preserve, with scrupulous fidelity, whatever constitutes an addition, however seemingly slight, to the facts of geographical science, or to the truths of natural history—the latter more especially the favourite object of the Author's regard, and his contributions to which will, it is believed, be found to form one of the most attractive features of his work.

The Author desires the expression of his gratitude to the Right Honourable Sir George Grey, K.C.B. (formerly Governor of the Cape Colony), for aid and patronage generously extended to him throughout his undertakings; and also to the following gentlemen: John Sanderson, Esq., for his " Notes on the Botany of Natal;" Charles John Andersson, Esq., for notes on the birds found between the West Coast and the Victoria Falls; the Rev. J. C. Brown, Professor of Botany and Colonial Botanist; Dr. Bleek, Author of the paper on the Languages of South Africa; the Rev. Messrs. Rath and Kolbe, on the Damara Languages; E. L. Layard, Esq., Curator of the South African Museum, &c.; Henry Hall, Esq., late of the Royal Engineers' Office, Cape Town; Charles Bell, Esq., Surveyor-General, of Cape Town, for drawings and sketches used in illustrating these volumes; as well as to other kind friends not specially named, but whose aid and advice is thankfully acknowledged.

W. H.

LONDON, *December*, 1867.

ERRATA TO VOL. I.

Page 23 for "*Oriotragus Sattatrix*" read "*Oreotragus Saltatrix.*"
" 186 for "*Anser Melanogoster*" read "*Anser melanogaster.*"
" 193 }
" 299 } for "*Juncus serratus*" read "*Juncus serratus.*"
" 208 for "*Pyrecephalus adspersus*" read "*Pyxicephalus adspersus.*"
" 230 for "*Aigocerus*" read "*Aegocerus.*"
" 416 for "*Cucumberi*" read "species of *Cucumis.*"
" 453 for "*Galaga Moholi*" read "*Galago Maholi.*"

CONTENTS OF VOL. I.

CHAPTER I.

Approach to Port Natal—Town of D'Urban—Pietermaritzburg—Progress of the Colony—The Author accepts an appointment under the Colonial Government—Resigns it to embark on a Commercial Enterprise in the Interior—Crosses the Drakenberg—Abundance of Game—A Flight of Locusts—Bushmen—The Trans-Vaal—Establishment on the Mooi River—Difficulties with the Boers—Their Relations with the Native Tribes 1

CHAPTER II.

Preparations for a Journey into the Interior—Start from the Mooi River—Matebe and Mabotsa Missionary Stations—Hunting on the Notuani River—The Author kills his first Elephant—Mode of extracting the Tusks—A Boa-Constrictor and a variety of Snakes—Curious Cell of a large Spider—Arrival at Sekomi's Town—Manners and Customs of the Bamañwatos: their Language 20

CHAPTER III.

Rain-makers and Doctors—Arrival at Sekomi's of Griqua Traders—Ceremonies of the "Boyali"—Bamañwato Superstitions—Proceed on the Journey—Fall in with Giraffes—Palmyra Trees—The Kalahari Desert—Bushmen—Salt Pans of Shogotsa—The Bakurutsie—Chapo's Village and Marsh 42

CHAPTER IV.

A Bushman Village—Cross the Botletlie River—The Ntwetwe Salt-pan—A magnificent Baobab Tree—Reach Thumtha, or Goroge's Post—A superior Tribe of Bushmen—The Makalakas—Trek Northwards—A Bushman Smoking-bout—Elephant-hunting in the Madinasana Desert—A dangerous Encounter 60

CHAPTER V.

Turn back, Water failing—Travel North-east—Goroge's Post—A Bushman Dance—Attack of Fever—Return to Sekomi's—Native Politics—A fearful Tragedy averted—Proceed to Sechelli's Town—The Chief Sechelli: His People massacred by the Boers—Arrival at Kolobeng—The sack of Dr. Livingstone's House—Leave Kolobeng—Arrival in the Trans-Vaal—The Boers: their Position and Prospects 87

CHAPTER VI.

Another Expedition—Outfit—Start from the Trans-Vaal—Rhinoster-Kop—Junction of the Valsch and Vaal Rivers—Koranna Villages—Mahura's Town—Alarms of an Invasion of the Boers—Motieto—Mr. Moffat's Missionary Station—Arrival at Kuruman—Threatening Aspect of Affairs 125

CHAPTER VII.

Leave Kuruman—Threatened Attack—The Bawankitze—Trial by Ordeal—At Sechelli's and Sekomi's again—Massacre of Boer Hunters—Proceed towards Lake Ngami—Chapo's Marsh—Perils of the Desert—Water fails—The Passage effected—The Chenamba Hills—Hunting Adventures—Native Cure for Madness 135

CHAPTER VIII.

Travel North for the Chobé—The Command River—The Grass on Fire—Moreymi's Town—Sebetoane, Chief of the Makololo: his History and Character—Dr. Livingstone: his Intercourse with the Natives—The Makololos and their Country: Trade with—The Tsetse Fly—The Chief Sekelètu—The Author hears from Dr. Livingstone, and endeavours to meet him 161

CHAPTER IX.

Travel South—Mababe Flats—Interior Watercourses—The Tamalakan: embark on it—Native Roguery—Crocodiles in the Tamalakan—The Botletlie River—Makobas—The Chief Lechulatèbe: Trade with him—Visit to Lake Ngami—Lechulatèbe's Town—Journey to Chapo's: Disasters on the Way—Journey to Scchelli's Town—Shock of an Earthquake—Reach Kuruman   182

CHAPTER X.

A new Expedition—Fauresmith—Cession of the Sovereignty—At Phillipolis—Back to Kuruman—Rev. Mr. Moffat—A Stalactite Cavern—Reach Sechelli's—The Question of Polygamy—Arrival at Sekomi's—Plans of Procedure—The Author again sets out for the Lake—Reach Chapo's—Buffalo-hunting—Crocodiles in the Botletlie River—A rare Antelope Shot—Arrival at Lake Ngami   215

CHAPTER XI.

Return to Chapo's—Start for the North-eastern Interior—Alarms by Natives—The Ntwetwe Salt-pan—Encounter with Lions—Abundance of Game—Reach the Shua River—An Elephant-hunt—Makarikari Salt Lake—Confluence of Rivers—Borders of Moselikatze's Country—Dealings with some petty Chiefs—A Night Encounter with Makalakas — Bushmen massacred — The Simonani River 232

CHAPTER XII.

The Great Salt Lake—Sable Antelopes—Ostriches—Vast Herds of Bucks and Buffaloes—Effects of a Snow-storm—The Bakalihari—The Lotlotlani River—Reach Sekomi's Town again—Charms and Wizards—A Court of Justice—Bamañwato Customs—Native Music—Leave Sekomi's—Native Love of Battle—Start North for the old Hunting-fields 260

CHAPTER XIII.

Journey Northward from Sekomi's—Altered Plans—Reach Kamakama—Wild Grape-vines—Beginning of a New Year—Camp at the Chenamba Hills—Mr. Edwards departs for Linyanti—Almond

Trees—Food in the Desert—Move Eastward in search of Game—Hunting Elephants—Native Fruits—Various Caterpillars—The Epidemic Fever—Curious Cisterns of Rain-water—Increasing Difficulties 277

CHAPTER XIV.

Dr. Livingstone and Sekelètu—South African Missionaries—A Bushman Revel—Boa-constrictors—At Chapo's—Potato Culture—Hunting on the Botletlie River—Return to Lake Ngami—Transactions with Lechulatèbe—An Unhealthy Season—Plague of Snakes and Mosquitoes—Sufferings from Famine amongst the Native Population—Up the Teougé River—A Visit to Lebèbè—The Bakuba Nation 288

CHAPTER XV.

Preparations for Journey Westward—Professor Wahlberg—Departure from the Lake—The Western Desert—Increasing Scarcity of Water and Game—Elephant's Kloof—The Damaras—The Namaqua Chief, Lamert—The Nosop River—Jonker Africaner—Mission Station of Barmen—Otjimbengue—The Kuisup River—Missionary Life in South Africa—Walvisch Bay—The Voyage to Cape Town .. 317

CHAPTER XVI.

Preliminaries—A New Expedition—From Cape Town to Walvisch Bay—Fisheries—Discouraging Prospect—Oesip—Geological Features—The Schwagoup River—The Berg-Damaras—The Chobé River—Intense Heat—Otjimbengue—Cattle-stealing—Missionary Labours 365

CHAPTER XVII.

At Otjimbengue—Loss among Cattle—Return to Walvisch Bay—Bed of the Schwagoup—A Buffalo-hunt—Stories of African Adventure—Lion-killing—M. Gérard and his Exploits—Wreck of the Canute—Walvisch Bay again, and return to Otjimbengue—Meteors—Shocks of Earthquake—Journey farther Inland—Change in the Aspect of the Country—Hot Springs of Barmen—the Schwagoup—Jaager Africaner 384

CHAPTER XVIII.

Journey towards the Lake—Mineral Springs of Windhoek—Difficulties of Progress—The Namaqua Chief Jan Jonker and his Family—Elephant's River—Drooge Vely—Adullam—Further Impediments—Edible Roots—Gobabies, or Elephant's Fountain—Amraal—A Novel Mode of Duelling—Lions and their Victims—Amraal again—Want of constituted Authority — Commerce with the Cape — Namaqua Habits, Manners, and Superstitions 407

CHAPTER XIX.

Start in search of Baines—An amicable Exchange of Wives—Meet Baines — Cattle-sickness—Peculiarities of Hottentot Language—Water-tubers—Riet Fontein—Gert, the Hottentot—Eland-shooting—Gnathais—Difficulties of watering the Cattle—Our Horses stolen—Ghanze — Whirlwinds — Thounce — Advance to Koobie—Poison-Grubs—Native Plants—Living on Ostrich Eggs—Plan of future Proceedings—Leave Koobye 430

LIST OF ILLUSTRATIONS TO VOL. I.

	PAGE
TOILET OF A BECHUANA BELLE (from a Drawing by CHARLES BELL, Esq.) .. .. *Frontispiece*	
NATAL FROM THE SEA	2
KILLING LOCUSTS (from a Sketch by CHARLES BELL, Esq.)	7
NATIVE WOMEN CULTIVATING THEIR GARDENS (from a Sketch by CHARLES BELL, Esq.)	36
KORANNA PACK-OXEN (from a Drawing by CHARLES BELL, Esq.) ..	128
FIRE IN THE FOREST (from a Drawing by CHARLES BELL, Esq.) ..	163
LECHULATÈBE'S TOWN	206
NATIVE WOMEN PREPARING WINTER STORES (from a Sketch by CHARLES BELL, Esq.)	268
DISTANT VIEW OF WALVISCH BAY	346
WELWETSCHIA MIRABILIS	377
THE NAMAQUA KRAAL (from a Drawing by CHARLES BELL, Esq.) ..	426

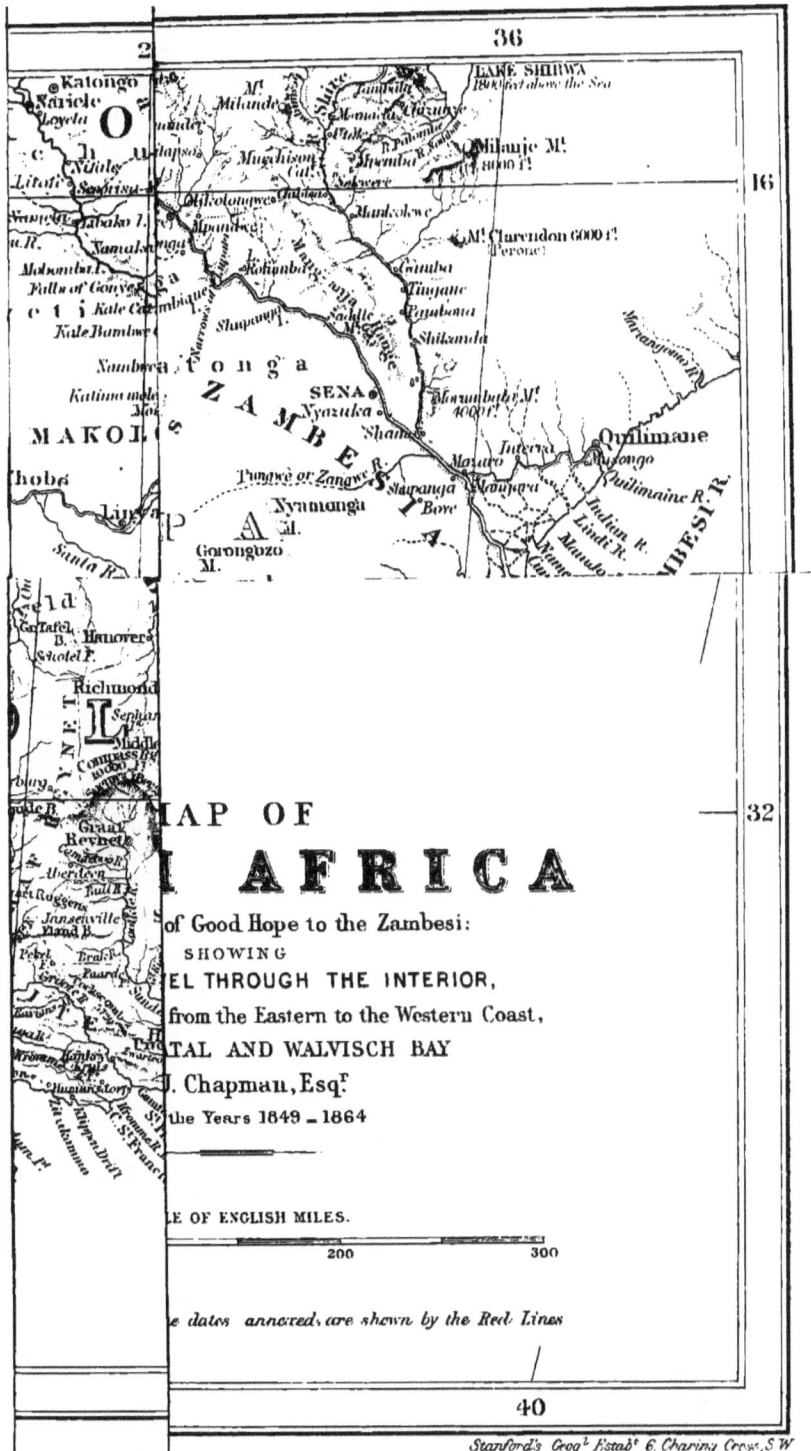

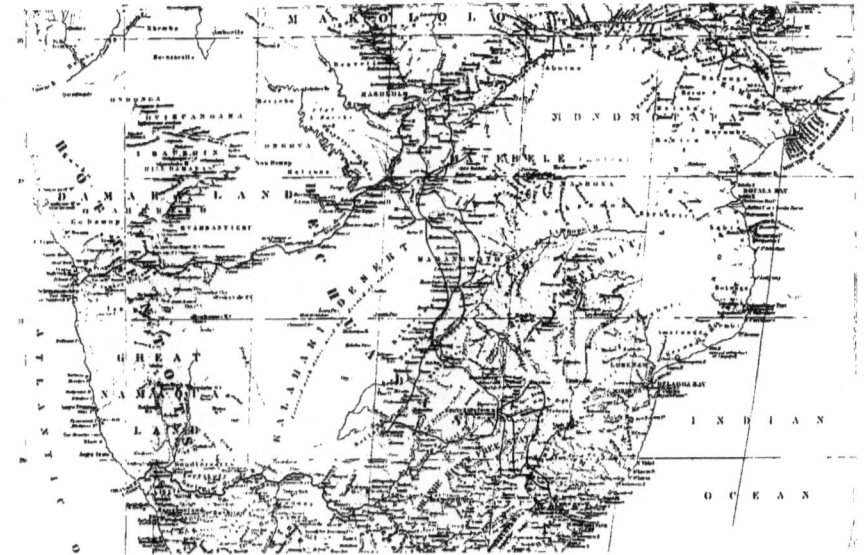

CHAPMAN'S TRAVELS.

CHAPTER I.

Approach to Port Natal—Town of D'Urban—Pietermaritzburg—Progress of the Colony—The Author accepts an Appointment under the Colonial Government—Resigns it to embark on a Commercial Enterprise in the Interior—Crosses the Drakenberg—Abundance of Game—A Flight of Locusts — Bushmen — The Trans-Vaal — Establishment on the Mooi River—Difficulties with the Boers—Their Relations with the Native Tribes.

NOTHING can be more refreshing to the eyes of a traveller, after the monotony of a long voyage, than the sight of the beautiful and verdant hills gently sloping to the sea, and forming the coast-line of Natal; the wooded valleys intersecting which indicate the course of innumerable * streams flowing into the Indian Ocean.

On rounding the bluff of Natal, endless forests, in which every shade and hue of colour richly commingle, break upon the view, and the voyager already hears in imagination the songs of the feathery tribes, or the shrill cry of the elephant and other beasts of chase, which he knows by repute to abound in them. Having crossed the bar, sometimes a

* Two hundred streams flow into the ocean between Port Natal and the eastern frontier of the British possessions in South Africa.

dangerous impediment to the approach of sailing vessels, the Bay of Natal is entered by a very narrow passage, scarcely fifty yards broad. In the port seven or eight vessels, of from 100 to 400 tons burthen, are generally to be seen moored within a stone's throw of the custom-house, and the work and bustle entailed by the operations of loading and discharging cargoes give life and activity to the scene, while the many sailing or pleasure boats on a breezy afternoon,

NATAL FROM THE SEA.

skimming the surface like swans in the distance, perhaps still more enliven the view.

Having lived at Natal for four years before my return there in 1849, I had seen the colony rising with such rapidity as to appear to me quite astonishing. Instead of the thirty or forty mud and wattle-and-daub houses which formed the town of D'Urban when I left it in 1845, scattered for the most part over a large area of unoccupied and uncultivated ground, I now found the houses so crowded into new streets and avenues, that I could not again recognise old and familiar places, nor did I know my way through the town. Wooden, iron, brick, and stone houses and stores had now superseded the former mud dwellings, and the town, I was informed, contained at that time a population of upwards of 1000 souls, mostly English.

Natal having made this rapid progress, and possessing every advantage of climate and extraordinary fertility of soil, already bids fair to become one of the most important colonies under the British crown. Sugar, coffee, indigo, cotton, arrowroot, tea, ginger, and many other tropical productions, are already, or soon will be, exported, and no doubt in large quantities. One comfort the English emigrant may expect in Natal is, that he will meet there a society almost purely English.

I propose, without further details, to conduct my readers along the high road, already thickly studded with English cottages, road-side inns, and smiling farms, with their broad cotton-fields and other plantations, amidst a variety of the most beautiful and diversified scenery, to the town of Pietermaritzburg, 56 miles distant from D'Urban. It was at Pietermaritzburg, the capital of Natal, that I made preparations for my journey into the interior.

When I first visited Pietermaritzburg, it was a mere village of about thirty or forty houses, and in 1849 it was reputed to contain a population of from 2000 to 3000 souls. The town, laid out on a larger area than D'Urban, stands on a fine plain surrounded entirely by different streams of water, the largest of which is crossed by a substantial bridge before we reach the place. From a distance the houses have a very picturesque appearance, the red-tiled roofs contrasting well with the bright green of the gardens surrounding them. The streets are broad, with a fine stream of water gurgling on either side. The town is well protected by a fort erected on a small conical hill commanding the place.

I had returned to Natal with the intention of settling there, and returning to my former avocations in a mercantile establishment; but having been favoured with the offer from the Lieutenant-Governor, Martin West, Esq., of an appointment as chief clerk in the diplomatic department, I accepted the office. Being accustomed, however, to more varied and active employment, better suited to my natural bent, and receiving from friends very tempting proposals to embark in a mercantile enterprise beyond the colonial limits, I soon begged permission of his Excellency to resign the appointment he had so kindly conferred upon me.

In one short month my preparations were completed. Wagons, oxen, and merchandise were purchased; servants procured; and I started on the journey of which a detailed account will be found in the following chapters, with the design of settling in either the Orange River Sovereignty, now the Free State, the Trans-Vaal country, or wherever else I might find it most eligible to establish myself.

It was on a beautiful day in the month of October, 1849, the season of spring in those latitudes, and just before the setting in of the periodical thunder-storms peculiar to Natal, that we started from Pietermaritzburg for a town or village far in the interior, then but little known to Europeans.

Of the locality of Potschefstroom we had but a vague idea, namely, that it was somewhere "ober de Berg" (meaning the Drakenberg mountains), on the river Mooi, or Pretty, a stream flowing into the Vaal, one of the largest and most distant sources of the great Orange river. Potschefstroom, the metropolis of a republic recently established, had been only occasionally visited, and that seldom by Englishmen, not one of whom was allowed to settle there. The communications between this town and Natal had now been for some time interrupted; and it was for the purpose of opening friendly relations with the country, as well as of encouraging trade and indulging a youthful passion for exploring unfrequented routes, that I set out for the capital of the Vaal River State.

The proposed journey appeared to me somewhat adventurous; it was the longest I had yet accomplished, nor were many of its incidents devoid of interest; but, with so wide a field before us as the South African continent to be crossed and re-crossed from sea to sea, neither time nor space can well be allowed for details which are of far less importance than those connected with the vast interior we are about to explore. It would be needless to detain the reader on the threshold of the enterprise with any description of those parts of Natal through which I passed, so various and valuable are the sources from which ample information regarding that colony may now be obtained. The most striking natural objects on that part of my route were the lofty range of the Drakenberg, the pass through which with our heavily-laden wagons was, perhaps, the most toilsome part of our undertaking; and the well-known falls of the Umgani river, one of the most picturesque scenes in south-east Africa.

All the rivers rising on the north-western side of the Drakenberg are affluents of the Vaal river, which, as already mentioned, flows into the great Orange river. When the summer rains fall, these rivers, fed by the mountain torrents, swelled from a thousand sources, become flooded, and great danger attends the crossing them. The country through which we "trekked," till we crossed the frontier on the Vaal, being the north-eastern portion of what was then termed, as being under British rule, the Sovereignty, but has since become the Orange River Free State, is well watered, and contains vast plains, at present overspread with innumerable troops of antelopes, but well suited for pastoral purposes, and partially occupied at wide distances by Dutch farmers. The road lies for more than a hundred miles along the widest of these plains, which is distinguished by a hillock, called the Vegt-Kop, rising above the vast level.

The plains abounded in all sorts of game; lions and other carnivorous animals frequently beset us. But what to us was most remarkable consisted in the myriads of quaggas, blesboks, springboks, and wildebeest or black gnus, which continually crossed our path. I could have formed no conception of the reality of the accounts I had often heard of the vast quantity of game to be met with here. Indeed, at certain times of the day the plains, for miles around, had somewhat the appearance of a living ocean, the tumultuous waves being formed by the various herds crossing and re-crossing each other in every direction.

Besides these vast numbers of all kinds of animals, we saw that great scourge of the country, a host of unfledged locusts, rolling and tumbling over each other in small waves. I walked in them to the depth of seven inches! They cannot fly before they arrive at a certain age, and are, if anything, more destructive than those that are winged, devouring everything that grows, even down to the roots. In this stage the Dutch call them "voetgangers," or infantry. They generally travel towards the east, hopping along at the rate of two miles an hour, and when they fly they are generally carried away with the wind towards the westward, alighting wherever they are attracted by the corn-fields or the pasturage. I have myself known them to devour, in two hours, a field of corn, from which the proprietor expected to reap 300 muids.* The locusts lay their eggs by inserting the extremity of the abdomen into the ground, turning round and round, with the aid of their legs and wings, until the whole of the ab-

* A muid is equal to about two bushels.

domen, as far as the thorax, is buried: the upper part of the body is then wrenched off, and the abdomen, full of eggs or larvæ, remains buried until a shower of rain falls sufficient to produce a crop of grass in their immediate vicinity, sufficing for the sustenance of the larvæ after they are hatched. Providence has admirably provided some check to their ravages in the locust-birds, the invariable attendants on the pests from which they derive their name. From a peculiarity in the structure of their digestive organs, these birds are able to

KILLING LOCUSTS.

feast continually on the locusts, destroying myriads. It was amusing to watch the motions of these useful birds, as they whirled about in dense clouds, assuming in their flights all kinds of fantastic shapes and appearances. They are considered very good eating.

I observed that the locust is eaten with avidity by horses, oxen, goats, sheep, dogs, cats, as well as by the elephant and other wild animals; and last, though not least, it constitutes an important article of food for man himself, all the natives of South Africa considering locusts as great delicacies. Horses sometimes eat them to excess, and die of constipation. Nothing stays the progress of these destructive insects. They march in myriads over the bodies of their own species, sometimes to the depth of more than a foot, and on arriving at a stream they plunge in, and the foremost, being drowned, form a bridge for the rest to cross over in safety; many are thus destroyed, but the great bulk of the column effect their passage.

The Dutch farmers are averse from settling in the north-eastern parts of this country in consequence of the bold and frequent depredations of the Bushmen, whose habits of plunder enable them to secure their booty in the fastnesses of the Drakenberg before the unfortunate farmer is on the trail of his lost sheep or cattle; when it happens that their pursuers are too quick for them, they will destroy as much of it as they can before making their escape, or, if overtaken, will fight desperately for their lives, using their only weapon, the deadly poisonous arrow. I have known instances of 400 and 500 head of cattle having been captured by them, and, though they were pursued by horsemen twelve hours after they had stolen the cattle, they managed to escape with their booty. Indeed it is seldom that the robbers are overtaken, as they drive their herds night and day at a fearful rate, and kill on the spot any animals that are knocked up from fatigue, rather than let them fall alive into the hands of their pursuers.

The Trans-Vaal, the territory of the then recently-formed Dutch republic, comprises an extent of about 50,000 square miles, over which were scattered, at the time of my visit, between 2500 and 3000 Dutch families, mostly in lone farms. There were three villages separated by a distance of between 200 and 300 miles from each other—namely, Origsberg, Zoutpansberg, and Potschefstroom, or Mooi (Pretty river)

Dorp, sometimes called Vryberg. Other villages have been more recently built. Potschefstroom, the most important, is situated on the Mooi river, within a few miles of its junction with the Vaal, in lat. 26° 43', and long. 27° 40' E., and is the capital of the Trans-Vaal State. The site of the town has been admirably chosen, in a sanitary as well as a commercial point of view, being in the healthiest part of the Trans-Vaal, and only 13 miles from the nearest bend of the Vaal river. The advantages for irrigation are so many that the plain for 15 or 20 miles could be watered by merely extending the present watercourse. The Boers, being a strictly pastoral people, live on their farms, the inhabitants of Potschefstroom being chiefly the officials, a few traders, and the rest mechanics. Of the officials of the village, the "landroost," or magistrate, is the principal. Next to him are the hemeraaden, the field-cornet, a jailor, and other subordinates.

At the period of my first visit to Potschefstroom there was no church or school of any kind, but a church has since been built: the members of different families sometimes met on Sundays at the house of the landroost to sing psalms, or hear a sermon read.

On my arrival, my wagons having been drawn up in the market-place, along with those of other traders, I got an introduction to the landroost, Hermanns Lombard. Upon making known to this magistrate the object of my visit, he informed me, to my infinite surprise, that I could not be permitted to establish myself permanently in the country—the consequence, he affirmed, of having allowed individuals like myself to settle among the Boers in Natal having been the expulsion of his countrymen, and that the like result must be expected from permitting Englishmen to settle in the Trans-Vaal. The landroost, therefore, strongly advised me to abandon my project.

I returned to my wagons rather discouraged, after receiving this communication, and towards the evening sent the landroost two cases of cognac, with a letter which had been entrusted to me by a friend at Natal. The next day I received a visit from the old gentleman, whose mood being now changed, I took advantage of his good-humour to obtain from him a favourable reply to my application for liberty to hire a house for a couple of months, or until such time as I had disposed of my property; which license was afterwards indefinitely extended.

The exports from the Trans-Vaal consist chiefly of ivory and cattle. Butter, wheat, and a variety of miscellaneous articles, are brought to market, but they are not generally exported. Wool, in small quantities, has of late been introduced into the market here. This, with dried fruits, wine, and brandy, will probably form hereafter a considerable item in the exports from the Trans-Vaal. In barter, or exchange, the medium through which most of the traffic takes place, groceries, with cotton and fustian goods, are in great demand. Orleans and alpaca cloths, voerschits, and other articles of clothing, with a considerable quantity of hardware in the shape of pots and pans, brass and copper kettles, &c., generally meet the requirements of the market, profits on these goods averaging nearly cent. per cent. to the importers. Still the success of the traders in procuring ivory in return for imports is rather precarious, and the expenses and risks with cattle, by losses from disease, straying, and robbers, are so great as to diminish considerably the amount of looked-for gain.

The climate of the Trans-Vaal country is generally very salubrious, and has proved especially beneficial to persons suffering from asthma and other pulmonary diseases. The rains commence in the beginning of summer, about October, and last to the end of March. They are frequently accompanied with violent thunder-storms, showering large hailstones, which sometimes cut down every blade of corn to the ground, stripping the houses of their plastering, and the fruit-trees of every particle of bark and leaf. Nothing can present a more desolate scene than the appearance of the country the day after one of these storms. However, one of such severity occurred only once during my four years' stay.

The thermometer, during the greatest heat of the day, generally between 3 and 4 o'clock P.M., seldom exceeds 100° of Fahrenheit in the shade, but more frequently ranges at about 80° to 85° or 90°. Thunder-storms prevail during the whole of the summer months. These storms are more frequent and intense in the low wooded districts, where the atmosphere is less pure. The summer nights are generally very pleasant, or, if the heat is at all oppressive, it may be attributed to the low and confined houses in which we sleep. In the winter mornings the Boers are glad to see the hoar-frost lying thick on the ground, and the small pools of stagnant water encrusted with ice. They then begin to breathe freely, not having much to fear from horse-sickness; for though the sun shines brightly every day, it is generally cold and fresh until about 9 o'clock, when the frost is dissolved, and after sunset the fireside is the most comfortable place. During the whole winter the atmosphere is very clear, and at mid-day the sun's heat is as intense as in summer. No rain ever falls in the winter, and light particles of snow only fall once in three or four years.

For the growth of wheat, oats, barley, all descriptions of vegetables and fruits, the whole Trans-Vaal country is well adapted. The Magallisberg and all the warmer districts are particularly suited for the vine, which they are beginning to cultivate rather extensively, and wine may some day be the principal export of that part of the country. The Sweet Reed (*Holchus saccharatus*) is also grown, and converted into brandy, wine, and vinegar. The most fruitful parts, such as

Magallisberg and the Mariqua, are, however, rather more unhealthy, being subject occasionally, at intervals of ten years or so, to an epidemic fever; this, however, does not seem much to affect the rising generation, who, being reared in a hardy way, appear to become quite acclimatised. These warmer districts are considered more healthy for cattle, but not for sheep and horses, which seem to thrive better on some parts of the Vaal river country, where the grass is not so luxuriant — those, indeed, which are also esteemed most healthy for man. I have no doubt that at some future period, when the country becomes grazed down, and our breed of European sheep are thoroughly acclimatised, they will thrive as well as the native sheep; and as to horses, I expect time will work a change for the better, as it did at Natal. When I was there, the Boers, for six or seven months of the year, dared not keep their horses (excepting those that are acclimatised within their own country), but were obliged to send them into what was British territory, now the Free State, till the dangerous season was past.

During my stay at the Mooi river I collected a variety of living specimens in the zoological department of natural history: at this period my menagerie comprised four young lions, two panthers, two meer-kaatges (a kind of weasel), two springboks, one young gnu, three zebras, a baboon, some monkeys, a porcupine, an armadillo, a jackal, several hedgehogs, and a number of paroquets. I have many a time been surprised to see how animals of such opposite dispositions would become reconciled to each other. I have frequently seen these young lions huddle together in one mass in the shade with a cat and her kittens, a litter of pointer pups, not forgetting the weasels, and a hen with her chickens perched on the top. This group formed a singular, though, in my eye, a very pleasing picture.

Besides the animals above enumerated, I had sent me from

the Mariqua three small lemurs. These pretty little animals, with their furry hides and fiery eyes, bounded from the floor on to the beams of the ceiling with the grace and ease of a bird. They are most active at night, and they build nests like birds in the branches of trees, but in their general habits they partake more of the monkey than any other animal, living on gum, berries, &c. The jackal proved very mischievous, going out on regular foraging expeditions against my neighbours' poultry, but in some way or other possessed the sagacity not to interfere with mine. He was as tame as any dog, and I have often wondered at his escaping from the curs of the town during his extended expeditions. A monkey which I purchased met with a melancholy fate, dying of grief after the departure of his former owner, and refusing to eat any of the dainty things I offered him, amongst which were sweetmeats, for which monkeys generally have a ravenous appetite.

I spent nearly two years in the Trans-Vaal country, compelled by circumstances to live in a community the society of which was far from congenial to my disposition and habits, and suffering annoyances and persecutions which rendered my position almost intolerable. Being too much involved in business to leave the country till my affairs were arranged, I had sought relief, from time to time, by embracing opportunities of indulging my passionate love of sport and the chase, in the stirring enjoyment of which, in the free air of the veldt and the forest, I found some respite from the cares and trials that beset me, returning to my task with spirits refreshed and my frame reinvigorated.

On one occasion I indulged my bent by a wider excursion into the Mariqua, a pleasant and romantic district lying on the west of the Trans-Vaal, my destination being the farm of Mr. J. W. Vilgoen, field-cornet of the district, and a good friend, who afterwards became my companion in a journey far into the interior. My friend being at the time absent on a hunting expedition to the Lake Ngami (the first Boer who attempted the enterprise), I extended my excursion, waiting his return, still farther west, to the high lands about the Molopo river, a healthy and fruitful part of the country. During these excursions I had very good sport, and was received with much kindness and hospitality at the places of the numerous Dutch farmers which lay scattered at intervals throughout my route. I found the inhabitants of part of the Mariqua district in a high state of alarm, expecting an attack from Mahura, a petty chief of some importance in the neighbourhood of Lattakoo. The Boers, with their families, cattle, and moveable property, were, as is frequent on hostilities occurring, collected for mutual defence in a sort of fortified camp, or *lager*, consisting of a square inclosed by four rude stone walls, within which their wagons and tents, consisting of at least two hundred, were drawn up or pitched in rows. I visited this lager, and found the people there cooped up in a miserable plight. Their cattle also were suffering severely, and many perishing from being huddled together in the narrow range of pasture surrounding the camp. However, the alarm soon afterwards subsiding, the Boers deemed it safe to break up the lager and disperse to their farms. They suffered heavy losses not only in their cattle, but from the circumstance that their corn-fields were ripening just at this time.

The country in the neighbourhood of the Trans-Vaal State, and, as will be seen hereafter, far into the interior, was at this time liable to continual disturbances from the hostile feeling subsisting between the Boers and the native tribes, and the outrages and alarms thence originating. There were wrongs to be revenged, and acts of atrocious violence perpetrated, both on the one side and on the other, the natives being, I regret to say, the most injured, and that most unjustly. The Boers from time to time organized against them commandos, as they are termed, being levies in arms of all the able-bodied men, under the command of the field-cornet of the district. It was easy work for these men, well-mounted, inured to hardships in their hunting expeditions, and expert in the use of fire-arms, to carry devastation wherever they went. The cattle were swept off, villages burnt, the inhabitants massacred, and, what was perhaps the worst feature in the case, the women and children, and often the men, were dragged away to become forced labourers—in fact, slaves—on the Dutchmen's farms. Against such attacks the natives could offer little resistance; but they retaliated, when opportunity offered, by waylaying and murdering small parties of the Boers, and more frequently by lifting their cattle. The root of the evil now lay in the assumption of the Trans-Vaal Boers that all the country from the Orange river north, to an extent unlimited, and from sea to sea, belongs to them, and only waits the occupation which their roving propensities and the increasing demands of pasture for their cattle will in course of time necessitate.

The lands they at present occupy they profess to have taken from Moselikatze; but it is well known that not one tenth part of it was ever claimed by that great chief, or any chieftains subordinate to his paramount authority. In point of fact, at the time the Boers intruded into these Trans-Vaal territories they were held by independent chiefs and tribes, who inherited the lands with the titles of their fathers; and in consequence of that intrusion, the Boers are found in continual struggles with some tribe or other which has asserted its independence. By a law passed in the volksraad the natives are subjected to a species of taxation, which is to be paid in labour; and this enactment, in the way it is often arbitrarily enforced, is a continual source of oppression and grievance to the natives. The Boers also purchase many native children, who, with those captured in their wars with the tribes, remain in a condition of slavery until released by death. I have had many of these unfortunate beings offered me, either in exchange for a horse, a quantity of merchandise, or in liquidation of a debt, and have often been tempted to purchase one or other to redeem it for charity's sake; but on the other hand there was something so repulsive to my feelings in the very idea of such a transaction, that I was compelled to refrain from doing the good I had intended. Two of these wretched little creatures were sold and resold, and afterwards redeemed by an agent of Messrs. Young & Co., of Natal. They were of that particular tribe of diminutive Bushmen with peppercorn heads, mostly dwelling in holes and caves. They were sent to England, where they were made a show of. For aught that I know they may still be alive, and their craniums and skeletons will one day probably become an interesting study for anatomists and phrenologists.

But not only are children thus acquired: men and women, of any age, taken by illegitimate means, are sold, or exchanged for cattle and goods. It hence not unfrequently happens that the unfortunate natives, when they have hard masters, taking with them their wives and children, will seek to escape from their thraldom by flight. Being well aware that the law offers them no protection, and as they cannot live in an inhabited country without great risk of being discovered and brought back—with the sure experience, in such a case, of everlasting revenge and persecution—they take refuge in the mountains and deserts, living on such wild produce as is at their disposal, until driven by the pangs of hunger to the alternative of cattle-stealing.

The Boers, with very few exceptions, take no interest in the spiritual welfare of these unhappy beings, nor do they even allow them to be present at their worship. It would be considered sacrilege for them to enter a church; and the idea of a coloured man obtaining entrance into heaven appears to a Boer monstrous and absurd.

Another outrage, which greatly exasperated the native tribes, was taking their cattle from them on some frivolous pretext, and sometimes by perfectly illegitimate means. The consequence is, that they rob in return; but, being the weaker party, are generally made to suffer in the end. I have known a single Boer to turn out twenty head of fine large cattle from the herd of a petty chief, and make them his own, under pretence of the cattle having trespassed on his lands; the Boer himself being at the time not even armed with any authority from a landroost or veld-cornet, although, according to law, cattle found trespassing are to be impounded, and the damage done assessed and defrayed. But it is deemed quite unnecessary to resort to this mode of proceeding when dealing with natives. In consequence of some occurrence of this sort, the Boers had once a skirmish with a chief in Magallisberg, called Zambok, who was severely thrashed by the field-cornet in the presence of his own people, on which occasion Pretorius himself was nearly shot. Another fight took place between the Boers and a powerful chief to the eastward, named Sequaxatie, who, when the Boers afterwards employed a native as an envoy to sue for peace, sent the messenger back with his hands and ears cut off.

Among other chiefs dwelling in the neighbourhood of the Boers was one Linchoè, who, on seeing me for the first time, declared me at once to be an Englishman, and, having confidence in them, he paid me two or three visits while I was in his village, to make his complaints. He said that he found it very hard to be compelled to live with his people under the oppression of the Boers, and that he could brook it no longer. He had been invited by the chief of a tribe beyond their

VOL. I.   C assumed jurisdiction to bring his tribe and come and dwell with him, and enjoy the freedom their fathers had. Linchoè and his people began moving towards the place of refuge, but before he had reached it he was overtaken by a commando of Boers, and, being brought back, was condemned to forfeit nearly the whole of his cattle, to satisfy the Boers sent to compel his return. After this affair, being unencumbered with cattle, he succeeded in making his escape, but in the end the poor fellow was shot by his implacable persecutors.

The younger Boers, in general, are not averse to war with the tribes, the horrors of which are disregarded in anticipation of the plunder they expect to secure, as well as their share in the native children taken. These commandos are generally ruinous to the people attacked, it being the policy of the Boers to destroy or impoverish the natives, as by such means only are the tribes, in their opinion, to be humbled and reduced to submission. Hendrick Potgeiter, the commandant of the eastern country, was one of the most arbitrary in his dealings with the natives, and exceedingly cruel in the warfare against them: it is a well-known fact that on one occasion, when a portion of the tribe he had conquered sought refuge in a cave, the whole were burnt and smothered in it.

The Boers are very averse to Englishmen having intercourse or trading with the natives within, or even beyond, their territory, over whom they also assume authority. Horses, guns, and ammunition are articles the sale of which is strictly prohibited; but although they keep an exceedingly vigilant look-out that Englishmen do not supply the natives with them, they are not averse, on occasions, to drive a bargain in these contraband articles themselves; and, to my certain knowledge, great quantities of guns and ammunition are frequently exchanged by the Boers with the natives for ivory, cattle, and other produce. This of course was a contraband traffic, the policy of the state being to cut off all supplies of arms and ammunition, the possession of which might enable the native tribes to offer effectual resistance to the exactions and encroachments systematically practised by the Boers. It was seen in the sequel with what avidity these articles had been sought by the chiefs of the interior, and how those who could arm even a small portion of their followers with muskets were not only able to maintain their independence, but were placed in a position to menace more warlike tribes, their neighbours.

CHAPTER II.

Preparations for a Journey into the Interior—Start from the Mooi River—Matebe and Mabotsa Missionary Stations—Hunting on the Notuani River—The Author kills his first Elephant—Mode of extracting the Tusks—A Boa-Constrictor and a variety of Snakes—Curious Cell of a large Spider—Arrival at Sekomi's Town — Manners and Customs of the Bamaũwatos: their Language.

THE experiences I had gained in my hunting expeditions, though on a comparatively limited field, joined with the strong desire I felt to rival the exploits of the Boers in hunting the elephant, of which I had heard them boast so much during our sport in company while in the Mariqua country, gave fresh zest to the intention I had long entertained of taking the earliest possible opportunity of embarking on a wider field of enterprise. This, added to the renewal of the annoyances to which I was exposed while resident on the Mooi river, led to the determination of at once carrying my designs into execution. I therefore settled my affairs in the Trans-Vaal as quickly as possible, arranging that the accounts for goods sold on credit, and still owing, might be collected during my absence.

In anticipation of the intended journey, I had, in the spring of 1852, collected a fine stud of fourteen horses, all of which, with only one exception, were swept off by the prevailing distemper, of which nearly every horse in the country died. At the same time some epidemic, resembling in every respect the lung-sickness, was also rife among cattle, of which some eighty head perished on a single farm. The breaking out of these distempers was accompanied by unusually violent rains, before which the heat of the weather had been most intense, insomuch that the morning's milk became sour by 10 o'clock, and the myriads of flies swarming in all quarters became so troublesome as to be a perfect plague.

I lost no time in recruiting my stud of horses, though, as may be imagined, with considerable difficulty and expense; and on the 15th of May, 1852, the unhealthy season being passed, I was fully prepared and equipped for an expedition of six months. I took with me three wagons laden with provisions, consisting of 6 muids of meal and wheat, 2 bags each of rice and coffee, 4 of sugar, some chests of tea, and a small stock of French brandy; about 250 lbs. of gunpowder, 800 lbs. of lead, 100 lbs. of tin, a few thousand percussion caps, and a bag of gun-flints. My companions were an experienced hunter, Jan Vilgoen, field-cornet in the Trans-Vaal, an intimate friend, and two young Boers, Jan de Waal and Peter Greyling. I was also accompanied by Otto Wirsing, a young gentleman lately arrived from Germany, anxious to see the country and distinguish himself in the field. My servants were three Hottentot wagon-drivers, Abraham, David, and Joachim; the latter, being a half-bred Mozambique and Malay from the Cape, acted also in the capacity of cook and steward. The party also included several other natives. My stock of cattle consisted of six spans or teams of fourteen oxen each, and a stud of twenty-five horses, two of which died at starting.

The first day of our journey we visited Matebe, the missionary station of the Rev. Walter Inglis, and the residence of a tribe of the Bahurutsi numbering about 300 men, under their chief Moilo. We made an ineffectual attempt to bargain with this chief for some Kaffir corn for our horses, but were gratuitously provided with some, and also with a good supply of barley by Mr. Inglis. I shall never forget the many tokens of kindness we received from him and his good lady. The chief Moilo, though amiable enough, was, like most of his race, a hard bargainer, and, finding that he was disposed to take advantage of our necessities, we declined treating with him. Most of his clan were away, working for the Boers under the system of compulsory labour before alluded to. A few were busy smelting iron by the primitive process in use amongst the natives, and converting it into hoes and picks, which they bartered with the neighbouring tribes for sheep or cattle. Copper, as well as iron, is found in the neighbourhood, and converted with great ingenuity into a very fine wire, which is worked up into armlets. Their principal occupation, in addition, seems to consist in tending their flocks and herds. They also make karosses (mantles) of jackal and panther skins, neatly fashioned and sewn, some of which they barter with the traders for cattle, &c. The women till the ground, and were now reaping the crops.

Leaving the farm of Mr. Inglis, and then traversing a narrow pass, we entered the picturesque valley of the Bakatla, where we found the Rev. Mr. and Mrs. Edwards, at the mission station of Mabotsa. By their kindness our wagons were laden with abundance of fresh butter and vegetables; and, proceeding on our way, we were joined by a party of hangers-on called Baastards (half-bred), with one wagon. On the 16th we "trekked" along the valley of the Notuani river, through a rugged and hilly country, studded with various kinds of mimosas, aloes, euphorbias, and cacti. Here we startled herds of pallahs, quaggas, and tsèsèbis (or bastard hartebeests), shooting one of each during the day. We encamped rather early in the evening on the banks of the Notuani, for the purpose of making a strong kraal or fold of thorny branches, to protect our oxen from the lions, of which this place is a noted resort. The proximity of those animals was sufficiently evidenced by the roaring which (mingled with the howling of wolves and jackals) assailed our ears during the ensuing night; while our dogs, thirty-seven in number, kept up an incessant barking, and one of the oxen, which had remained outside the kraal, was carried off.

The Notuani river is here nothing more than a dry "nullah," the water losing itself in the sandy channel long before it reaches this spot. Its bed, however, still contained a few very small pools of water, all swarming with fish, which lay so thick that many had already died from lack of their natural element.

Continuing our course down the Notuani, and meeting with an abundance of sport to our hearts' content, we killed several buffaloes and four rhinoceroses. Besides our every-day game, we started on one occasion a family of water-bucks or "kringgat-bok" (*Kobus ellipsiprymnus*), three lions, a pack of hyenas, and a number of ostriches, with several families of "pallahs" (*Antilope melampus*), and a few straggling spring-boks.

Encamping rather early one evening near the base of two rugged conical hills, the soft ground about which was full of the impressions of elephants' feet of the preceding year, I ascended one of the hills and killed two klipspringers (*Oriotragus sattatrix*). This animal, of the size of a small goat, is sure-footed, and bounds with amazing agility over immense chasms and precipices: the points of its hoofs are worn quite blunt. When pressed by dogs, it takes to the highest neighbouring acclivity, where, if the sportsman can follow, it is easily shot. The klipspringer has a peculiar coat of hair, of nearly the appearance and texture of that of a young hedgehog; it is much prized by us for stuffing saddles. There is a singular legend connected with these hills. The story runs, that there is in one of them a cave or grotto, from which all animals, including man himself, first issued. They affirm, in support of the legend, that there are strong impressions of the footmarks of many animals, as well as of the human species, in the solid rock within the cave, which they call the abode of God (Morimo). Many of the Boers testify to the truth of this statement, but they are fond of the marvellous, and very credulous. It is, however, not improbable that the cave may formerly have contained a pond of water, to which game were in the habit of resorting, and that some local convulsion of nature may have displaced huge rocks, which, falling, partially covered it, and the clay, full of footmarks, might with the drying up of the water have become thus consolidated, and eventually petrified.

We were met on our road by a party of Balala, poor people of the desert, whose keen eyes detecting some vultures in the distance, they had run to the spot, and had robbed a tiger* of his prey (a jackal, which he had just killed), and lay watching its decomposition. They assured us it was a great delicacy, and were astonished to hear that we did not eat it.

The country here is covered with dense forests; the traveller sees for miles scarcely anything but the underwood and trees surrounding him, with the blue canopy of the skies above, and has no landmarks to direct him onward. On the 28th we were glad to leave the Notuani, and, steering northward, encamped at a small pool of water swarming with leeches, tadpoles, and other animalculæ. Our farther progress was still northward, through a vast forest, within which we were at no time able to see above eighty yards ahead. The forest generally consists of various kinds of mimosa, acacia, and other thorny trees, such as the "wagt-een-beetje" (wait-a-bit, which name Mr. Gordon Cumming has erroneously applied to a species of mimosa called "haak-doorn," or "mañana"), "kat-doorn," &c. There are also a variety of timber

* The panther, or "Cape tiger," generally does not devour his prey immediately, but, like the crocodile, watches over it until it attains a certain degree of decomposition. At first it only drinks the blood.

trees, such as the boekenhoud and the tambootie—a valuable and ornamental kind of wood, somewhat resembling mahogany—the seringa, and many others. Whole groves of sandal-wood, emitting delicious odours, are found at intervals.

The 1st of June, 1852, will ever be a memorable day with me, for then it was that I first saw, and had the good fortune to bring down, the elephant. On that morning we fell in with fresh spoors of the mighty game. On these we soon followed, accompanied by an emaciated skeleton of a Bushman we had picked up on the road, and who, on inquiry, informed us he had neither friends nor relatives, but dwelt alone. We had been also joined, to our great annoyance, by the posse of Boers. Having diligently traced the spoor for four hours through the forest, over the vast logs and branches which the animals had recently broken down, strewing our path with tokens of their strength, we emerged on a sandy elevation or "buet," overlooking an extensive undulation, or "leegte," covered with brakes of that hunter's pest, the ever-recurring mañana, with patches of open ground running parallel to the bush. Here we dispersed in search of the spoors; but one of our party, ascending a tree to look out, nearly fell from the topmost branch with the excitement of having at last caught sight of the herd. For myself, as I had never before seen a live elephant in his native wilds, my feelings may be imagined when, ascending the tree, I first beheld those gigantic beasts roving about in tranquil and happy unconsciousness of approaching danger. I descended quietly, but every pulse throbbing with violent action; indeed, so overpowering was the emotion, that for some time I could scarcely speak or even breathe. The countenances of all around me betrayed their like eager and anxious expectations; and, although the stupendous beasts were browsing peacefully at the distance of a mile or two, slowly flapping their prodigious ears to disperse the flies, we spoke in whispers. Our mouths were parched, but we thought not of it. Never before had I the presumption to think that I could kill an elephant, but now my desire to make the effort was great and overpowering, and I could scarcely be restrained from starting at once at full speed to the attack.

Vilgoen, however, having gathered all our wandering party, proposed that we should rest the horses for half an hour, while some of us watched the movements of the elephants. This was decided on, but ten minutes was the maximum our impatience allowed, and in that short space we were again in the saddle.

From our look-out we had only seen a few straggling elephants, but now a troop of at least 200 broke on our view. They were marching in single file, and had evidently been alarmed at something in their rear, or had scented us. Nearly all appeared to be cows and young bulls, but one very large male seemed three or four feet higher than the rest. Next to their immense size, their gait struck me as peculiar —ambling and shuffling along like a number of baboons on all fours. In an instant twelve mounted men dashed up to their sides, and, dismounting at the distance of sixty yards, poured the contents of their rifles into the dense phalanx. Several turned, with elevated trunks and distended ears, as if they felt inclined to retaliate, but as the firing ceased they shuffled and hobbled off at an increased pace, and entered a strip of mañana forest. Pouring a handful of gunpowder into our rifles we reloaded, and, following up the herd, I soon perceived that the large elephant at which Vilgoen and I had fired was turning away from his fellows, and pursuing a different course at a very slow pace. This gave me an idea that he must have been struck hard, and was at our mercy, and Vilgoen calling out to me to ride up and kill him, I was confirmed in that belief. We soon, however, discovered our mistake, and an accident happening to Vilgoen's gun, Wirsing and I followed the elephant. In vain I rode up to and round the beast to get a side shot at his head: all that I accomplished by hard riding was only to make him turn his stern towards me, and thus render my manœuvre abortive. I could not head him, although he was only moving at a slow walk. Ten ineffectual shots had I fired into his hind-quarters, and on dismounting for the eleventh time, at a distance of twenty paces—for I laboured under the impression that he could go no faster than a walk—I flattered myself that I was now going to get a fine broadside. As he turned towards me I fired point-blank, and before the smoke had evaporated from the muzzle of my gun, to my great consternation he raised aloft his huge trunk, and, uttering a shrill cry like the sound from some unearthly trumpet in the sky, rushed upon me, with distended ears, at a pace that I should not have given even an unwounded elephant credit for. Nothing could have reminded me more effectually of the last day and the last trumpet than this sudden rush and cry of the beast; and on reviewing the circumstance at this remote period, I think myself entitled to some credit for the agility I had the presence of mind to display. Although the elephant was only twenty paces distant when I fired, I managed to scramble on my saddle before he was down upon me, being then within three or four paces of my back. The terrific and appalling cry which he uttered while pursuing me gave additional impetus to the speed of my willing little steed, and I felt more misgiving about the unevenness of the ground, which had been cut up by the elephants, than the speed of my horse. Three or four times, when his cry ceased, I ventured to look round, believing he had given up the pursuit, but as often did he increase his trumpeting, until after running about 400 yards I had gained about forty paces in safety, though, to my great annoyance, my horse stumbled several times. At this juncture I turned suddenly my horse's head into a small thicket, when, becoming entangled in the bush, I deserted my steed and crept through the underwood, to make security doubly sure. I had the presence of mind to keep under the wind, and my enemy, failing to discover my retreat, returned sullenly on his march. Under any other circumstances I should perhaps have abandoned the chase, but I was urged on by a spirit of emulation and dogged enterprise to renew the attack, and I mentally vowed that I would not lose my prey. My friend Wirsing, who had also put four bullets into the beast, now joined me, and, gnashing my teeth, I followed the enemy for another quarter of a mile, when, coming up to a small tree, with a tuft of boughs sufficient to conceal his head, he came to a stand. This was my opportunity: to make sure of my victim I approached to within thirty paces of him, and, dismounting, took a most deliberate aim for his heart. I pulled the trigger, when, to my most unspeakable chagrin, my gun missed, and its report brought the elephant down upon me more furiously than before. My horse, startled at his cry, would not permit me to mount; but as it was a case of life and death, without throwing the reins over his neck I made a desperate leap for the saddle, which I only partially gained, the spur on my right foot catching in its seat. The elephant was on me when I was not yet firmly seated; I let loose the reins and held on like grim death; my horse flew over the broken ground, bending his back at every cry the elephant uttered, while the enraged beast coiled up and threw his trunk after me, vehemently sounding its trumpet. At length, with a strength which only the fear of instant death could give, I gained my saddle, and kept in the open fields till my huge pursuer was completely winded.

My first thought was for the safety of Wirsing, as the elephant, in charging me, brushed closely by him within a few paces, but without touching him. Though he was well mounted, and on the open plain, he had a most miraculous escape. He endeavoured to fly, but his horse was seized with a panic at the animal's thrilling cry as he swept after me—it became petrified with terror, and, putting its nose to the ground, trembled and perspired violently: and, as a natural consequence, the reader may readily imagine the rider was not in a much better plight. This little adventure convinced us of the danger of hunting such game on untrained horses; and we were subsequently informed that many very fine ones are subject to these panics at the sound of an elephant's cry. Otto Wirsing, though brave and daring, was but a novice on horseback, and a very indifferent shot, and depended more on the speed of his horse than the fatal effects of his rifle; and had it not been that the elephant had suffered the greatest annoyance from me, poor Otto would never have lived to tell the tale.

But to return to my own share in the adventure. The elephant entered a dense bush and concealed himself behind the first tree in his way, and as we unconsciously approached within a few yards, the tree bent down in our faces, and the beast again charged with an awful cry; but, having a clear field, we now easily escaped. Vilgoen, coming up, proposed our abandoning the chase. Just then a shot was heard in the bush, and the trumpeting of an elephant, and we feared that some of our companions had been taken by surprise. On hastening to the spot, we found that another of our party, firing at a different elephant, had nearly been crushed by the one we were attacking charging him, and he had lost the skirts of his coat in the thorn bushes. Our elephant's course being pointed out to us, we soon detected him, when, leaving my horse, I headed him on foot, and, concealing myself behind a tree, allowed him to approach within a few yards; I then sent a bullet close behind his shoulder, which brought him down with a long groan and a thundering crash, shaking the earth, and causing me involuntarily to leap aside. I took to my heels as fast as I could to regain my horse, and then rode up, and, having examined and admired the gigantic brute, I took notice of the most vulnerable parts as they were pointed out. We then turned in search of a mineral fountain, called *Lotla kaniñ*, where we were to meet our wagons.

While on our way, and we were all engrossed with the adventures of the day, I was mentally regretting that I had not had the satisfaction of slaying one of these animals without any sort of assistance. At that moment, looking up, I perceived a small troop of cows rushing past on my left in full flight. I only uttered "There!" and my horse, feeling the spur, bounded off before my companions knew what was the matter. Having headed the elephants, I dismounted, and drawing the sight of my gun past the shoulder of the largest of the troop, I fired, and before the smoke had dispersed from before the barrel, I was elated to hear my friends from behind shout, "Dáar-legt-zy!" and there truly she did lie, enveloped in a cloud of dust. The rest of the herd started for a moment, and, with their trunks and tails erect, charged off in another direction into the mañana forest, where we did not follow them, the lot having very poor tusks. The one I shot carried two tusks, weighing each 19 lb. Dutch, which, for a cow, is very good. The bull's tusks weighed 50 lb. each. By degrees we gathered all our lost companions, and found that "Piet Greyling" and "Old Booy" of our party had each shot a bull-elephant, carrying 40 and 100 lb. of ivory each. The party of Boers who followed us had shot some twelve or thirteen cows; their ivory, however, did not realise half the weight of our four.

At dusk we arrived at a reedy fountain, emitting a very strong sulphurous odour. We knee-haltered our horses and kindled a fire, as our wagons, much to our astonishment, had not arrived. As the darkness increased, the lions commenced roaring in every direction, and became at length so daring that we were obliged to keep sentry with firebrands round our horses, and we discharged blank shots and made bonfires, for fear of losing them. We could not think of tying them up, as they had worked hard and starved all day. Having fasted ourselves, we had reckoned on having a warm and comfortable dinner on our arrival; but we were grievously disappointed, and doomed, weary, tired, and hungry as we were, to remain " out in the cold," guarding our good horses from the lions and other ravenous beasts of the forest.

At about 11 o'clock, while still firing our guns to drive off the unwelcome intruders, we were delighted to hear the report of a gun in the distance, which we knew proceeded from the approaching wagons, and soon we heard the cracking of whips and the shouting of the men; but still an hour elapsed before their arrival. To add to our disappointment we now found the people in a glorious state of inebriation, my men having helped themselves to my store of brandy, and stood treat all round, in consequence of which our wagon was injured and had caused the delay. This affair satisfied me that the best of servants cannot be trusted with spirits, and I resolved, if ever I should make another expedition, not to take any intoxicating liquor with me. After our very severe day's work we had to prepare our own supper, or, I might call it, the three meals of the day combined in one.

Next morning we started early for the dead elephants, followed by a wagon, and a number of men carrying hatchets, axes, and spears. As we neared the first elephant we observed a large hyena within a few yards of it, his attention being entirely devoted to the carcase; and a number of jackals had assembled a few yards nearer. The trees all round were bending with the weight of some hundreds of vultures, patiently awaiting the decomposition of the beasts.

On approaching the carcase, we were startled by the angry growl of two male lions, who, bounding from it, were soon concealed in the neighbouring jungle. They had extracted the entrails of the elephant, but had not yet consumed much. We now assisted our men in extracting the tusks. The trunk, being first divided into four or five pieces, was carried away, and the hide and flesh all cleared off as far as the eyes. The spongy bone in which the tusks are embedded was then carefully chipped off, a labour occupying an hour or two in the male species. The tusks of the female being so much smaller, they are, of course, far less troublesome to extract; but when we remain stationary for a few days we prefer leaving the tusks in the head of the animal, and then they can be drawn easily and without blemish from the sockets. Coming to the next elephant, we met with our Bushman of the preceding day's spoor, who seemed suddenly to be possessed with some mysterious fear, and evinced a strong inclination to decamp. On inquiring of " Old Booy " why the man trembled so violently, he informed me that the wife and children of the Bushman were concealed in the bush, and that he was afraid we should discover them, for he had the previous day told us he had neither, and explained to " Old Booy " that his alarm proceeded from the fact that some Boers had, on a former occasion, taken the children of a neighbouring family, and that he feared we should do the same. I gave him some tobacco, and informed him that we only hunted for elephants. On a more intimate acquaintance he was recognised by Vilgoen, on being reminded that when that worthy field-cornet passed this way with Pieter Jacobs, the previous year, a lion had just killed and dragged away one of his wives, and, having shown the Boers the fresh trail, they followed and slew the lion with a single shot before he had devoured his prey.

Next morning we awoke at a late hour, and breakfasted

on part of the elephant's trunk, which was baked in a pit during the night, in the way before described, and was as soft as a jelly, resembling very much the flavour of ox tongue. The foot, a joint from which twenty men can dine, is also exceedingly fine, being a white, crisp, and grizzly kind of substance, strongly ingrained with fat, and, though rich as marrow, one may eat any quantity without ever feeling surfeited; but a certain portion of the head and cheek of a fat elephant cow is by far the most delicious morsel.

On Sunday, the men amused themselves by following the honey-birds (*Cuculus indicator*), during which they discovered several bee-hives, and brought us an abundance of honey of an exquisite quality; and on Monday the wagons steered north. Accompanied by Vilgoen and others, and taking in our pockets a crust of bread and a tin box of pepper and salt, we sallied forth, as usual riding in a line parallel with, though at some distance from, the road. Amongst a variety of game we marked a troop of elands, of which Vilgoen brought down one, and another fell to me. One of these was conveyed to the road, and, stripping off the hide preparatory to the arrival of the wagons, we amused ourselves by roasting tit-bits, stuck upon skewers cut from the bush, before our fire.

During the day we saw several snakes, one of which was very long and ash-coloured, and, after actually frightening our oxen, sped away at such a swift rate that we could scarcely follow it with our eyes. One of our party described a snake of a dark colour, with a yellow cross behind the head, which he had killed during the day. Our native servants described other large serpents as abounding in these parts, such as the Tloaric, a species of boa-constrictor, and an enormous black and shiny snake, called Picka, with a horny hook at the tail. They have an ingenious way of killing them by plunging a very sharp knife through the earth, and allowing it to protrude about half an inch inside the hole. The snake being in the habit of turning as soon as it gets its head out of the hole, and then turning on its back, the knife cuts open the whole length of its belly as it crawls out, and it soon expires. This snake, which is said to be accompanied by a numerous host of flies flitting about its head, is believed to be exceedingly dangerous. They are very rare. One has been known to inhabit for years a cave in the neighbourhood of Motieto, and it has already caused the death of two or three persons. Of a party of boys who were hunting some animal into the hole, one followed it, and was only heard to groan, and the next day, on the return of his companions, they discovered their comrade ejected from it, dead. By dint of much watching, they found the perpetrator to be one of these large snakes, and a man of their family shot it shortly afterwards.

In the afternoon, we arrived at a small pool of water which seemed barely sufficient for our cattle; but we took the precaution to fill our casks first, and then with whips and shambugs occupied the centre of the pond, and kept the cattle drinking on the margin only of the pool; otherwise, as oxen will do, they would have rushed in, and in the course of a few minutes converted it into a puddle of muddy water unfit for man or beast. All animals of the ox kind are very nice and fastidious about their water; horses and zebras, on the contrary, will swallow mud if they can obtain nothing better.

At this pond we found an outpost of Sekomi's scouts, who informed us his town was still distant more than a day's journey. We dined sumptuously from elephant, rhinoceros, and eland flesh, and, having got rid of our prejudices, enjoyed our new viands with great zest, though cooked in our rude and primitive manner. But I feel confident that if a rhinoceros hump was cooked by an artist the whole world of epicures would pronounce it a dainty dish.

Lounging on a mat with a pipe, after coffee, my attention was attracted by a slight movement in the ground, and shortly rose a small round lid with a hinge, like the lid of a snuff-box. This opening disclosed a large hairy spider emerging from a neat cell beneath, which was lined with a white, smooth, silky substance: having crawled away for a short distance, the spider returned, bearing a dead locust, which it dropped into its hole, and, then following, closed the lid with its foot. I must say I was never more delighted with any discovery than that of an insect endowed with the wonderful faculty of constructing this curious cell. We opened the lid several times, to the great annoyance of its owner, who as often returned to close it. I intended digging up the whole fabric the next morning; but before daylight we were on the march, and, though my companions assured me I would find many more, as they were common, I never saw another.*

Having travelled the whole of the next day, we reached at night the base of the Bamañwato hills,† at the pass and fountain of the principal town and residence of Sekomi, the chief of the Bamañwato nation. The town, which extends for more than a mile in length, though very narrow, is built at the base of a long and rugged range of rocky mountains running north-west and south-east, and is the stronghold and safe retreat of this tribe when threatened by danger. This

---

* This insect is, I find, well known by the name of the *Trap-door Spider*.

† The Bamañwato hills are part of the range called Bakad; they rise about 700 or 800 feet above the plains, and are composed of great masses of black basalt. At the eastern end these hills have curious fungoid, or cup-shaped hollows, of a size which suggest the idea of craters. This mass of basalt is about six miles long.—*Livingstone's Missionary Travels*, p. 149.—[ED.]

range is said to contain several extensive caves, with fountains of water; and in these caves Sekomi is reported to be in the habit of concealing considerable quantities of guns and ammunition, as well as all surplus grain. The town contains several thousand huts, with a population of, perhaps, between 12,000 and 15,000 souls, and is the largest native town I have seen. The houses are generally about twelve or fifteen feet in diameter, of a circular form, and constructed of "wattle-and-daub" walls about four feet high, covered with a pyramidical thatch of grass. This thatch forms eaves three or four feet wide all round the walls, presenting the appearance of a Malay hat. Supported by poles, the overhanging roof forms a shady verandah. The floor, inside as well as outside, is plastered with clay. A circular enclosure of wickerwork or thorn-branches surrounds the several huts belonging to each family, as well as the sheds or granaries.

Immediately after our arrival at Sekomi's Town some hundreds of men and women assembled round our wagons, inconveniently crowding us, and discolouring our clothes, tent, and wagons with their filthy and greasy skin-clothing, besmeared with red ochre. They became exceedingly noisy and boisterous, and, from the emphatic tones of their conversation, any one not understanding the language would fancy they were quarrelling. The principal clothing of the men consisted of a small softened goat-skin tied around the loins, one corner of which, being drawn between the legs, was tucked in at the back, and the skin of a hartebeest, divested of the hair, save at the tail, covered their shoulders. Otherwise they were entirely naked. They wore a few ornaments of beads and brass wire, chiefly armlets. The women were clothed by a sheep-skin round their loins, and wore a small skin-apron in front. They adorned themselves with coils of beads round their necks, waists, arms, and legs. The children ran about in a state of perfect nudity.

Most of our copper-coloured visitors recognised Vilgoen, and presently one of them gave him the salute of "Lumela jannie," while he was engaged in buying grain for our horses. Vilgoen uncourteously paid no attention to this salute. A death-like silence prevailed for a moment, when some one exclaimed, "Is it not the chief that speaks?" Vilgoen turning abruptly recognised Sekomi, and introduced me to the great chief of the Bamanwato in due form. I must confess there was nothing to announce his rank in his garb, which was, if anything, inferior to that of his followers. The distinguishing features in his personal appearance were a scowling, sarcastic face, and a wall-eye. His figure was, however, very perfect, and he stood about 5 ft. 10 in. in height. Having complimented us on our success in elephant-shooting, he bade us good night, with a promise to see us again in the morning.

Next day the natives brought early abundance of Kaffir corn, with mealies, beans, pumpkins, and water-melons, in exchange for beads; but they would have preferred the fat of the elephants, which we had, in our ignorance, deemed useless. In the midst of our traffic the chief, true to his word, visited and partook of breakfast with us. He seemed not at all at home in the use of knife and fork. Plunging the fork into his meat, he held it up in the air and cut slices from it, which went flying in all directions, falling on the heads of his admiring followers. I advised him to put the meat in his plate and cut it there, but he soon upset the plate, which lay in his lap, and, nearly plunging the fork into his thigh, spilt the gravy over his naked legs, to be licked off by his nearest follower. After breakfast Sekomi retired, fan * in hand, with some fifty or sixty followers, forming a very imposing single file. A few feathers, wild beans, and berries

* The fan was formed by the tail of a jackal, stretched over a short wooden handle.

were also brought for sale; the chief being supplied with a large dish of coffee, tea, and sugar, as well as a bucket of meal, besides which he extorted from us articles of clothing, and other things of small account.

The manner in which these people persevere in begging from their white visitors is very annoying. Beginning with your hat, they continue to ask for every article of your dress, down to your shoes and stockings; becoming so troublesome and importunate that you must either strip yourself naked to please them, or lose your temper. All my articles of dress were very much admired, and I had a hundred kind and considerate offers to rid me of them. The chief was as great a beggar as the rest. During his visits to us I observed that he was invariably attended by a remarkable old personage, of a most diabolical appearance. His neck was ornamented with amulets of lions', lizards' and other reptiles' claws, with snakes' heads, and roots, supposed to possess infallible remedies against injury which the evil-disposed may contemplate against the chief or his tribe. Four small pieces of ivory, figured over with black spots, are used as dice; and at any time when they feel disposed to look into the past or future these dice are consulted, the natives believing implicitly in its pretended prophecies, instead of obeying the dictates of reason and prudence when assailed by dangers. Many are the lives that have been sacrificed by their implicit faith in this. In later times, I have known a party of Bushmen, with whom I was hunting, inquire in the stillness of night, by throwing their dice, if there was any danger from a troop of elephants which were loudly breaking the trees round us, and having ascertained there was no danger, go to sleep fearlessly. In war, too, it often happens that, relying on the security indicated by the dice, they suffer themselves to be attacked by surprise. Even then the survivors of the fray, instead of discrediting the efficacy of the dice, will attribute their failure to the enemy's doctors possessing stronger medicines (charms) than their own.

The Bamañwato are generally of a middling stature, and of dark copper colour. Some, however, are much lighter. Many of them have a crisp beard. They are a much more inactive and effeminate people than the Zulus of Natal. All their occupation seems to be sewing karosses, laying under the shade of a tree, drinking beer (bogaloa), or taking snuff immoderately. The chase is the only manly pursuit they

NATIVE WOMEN CULTIVATING THEIR GARDENS.

engage in. To the women's lot falls the toil of tilling the ground, reaping, thrashing, and stowing the corn, building their houses, and basket and mat making, besides all their domestic duties, amongst which brewing is not the lightest.

The government of the Bamañwato, like most of the Bechuana tribes, is patriarchal and absolute. The kings or chiefs are called khosi. They generally have great influence over their people, more so than any other tribe, except perhaps the Zulus, or Matabele. A great deal, however, depends upon their talent, energy, and wisdom, which either makes them liberal or despotic rulers.

Like all the Bechuana tribes, the Bamañwato speak a dialect of the same language—the Sechuana. The language is called Sechuana, according to their idiom, which requires the name of every language to commence with the prefix "Se." Thus Makua is the name given to white men generally, and Sekua is the language spoken by white men. Baengelesi is the name they have adopted for Englishmen, and Seengelesi is the English language.

The Sechuana is but a dialect of the language spoken by the Kaffir family generally. The dialects spoken by the Bechuanas to the north, and the Kaffir tribes to the east, are so similar in their construction, that it has been said one grammar, with a few notes to explain differences, might be drawn up for both these dialects; and the vocables are so alike that probably out of every hundred words spoken by these two classes of tribes fifty are exactly alike, or differ only in certain mutations of letters, which take place according to fixed definite rules, the organs of pronunciation in each class having a peculiar facility for expressing certain letters. Thus among Kaffirs the sound R is never heard, its place being supplied by an L; whereas among the Bechuanas, though L is pronounced, it very frequently glides into R.

The great peculiarities of this language, which separate it from other dialects, and entitle it to be called a distinct language, are two. First, it is prefixal; that is, every noun commences with a prefix, and the distinctions of number are made in the changes of these prefixes, not in the roots themselves. Thus "mo-nona," a man, becomes "ba-nona," men; "um-fazi," a woman, "abaf-azi," women.

Secondly, these prefixes involve the principle of alliteration or euphonic concord. The noun is the principal word in the sentence: its prefixes are numerous, and most of them have their corresponding euphonic letters, which control all the prefixes in the sentence. Thus "Izinto zonke ezilungileyo zidalwa ku Tixo:" "All things that are good are made by God." Here because the noun "nto" (things) has "izi" for its prefix, the adjectives and verb take a similar prefix—zonke, ezilungileyo, zidalwa. Now, if we change the noun and retain the other words, we shall see the change effected in their prefixes. Take "abantu," men, plural form of "umtu," man, person, and we have "abantu bonke abalungileyo badalwe ku Tixo," meaning "people, all who are good," or "all good people are made by God."

The peculiarities of the language are more completely developed, and at the same time there is a greater harmony and softness of sound in the Kaffir than in the Sechuana; the only drawback being the occasional occurrence of the click, which, however, is gently glided over by the Kaffirs, and has probably been borrowed by them from their Hottentot neighbours.

Another characteristic is its copiousness. Its variety of expression is truly surprising, and this renders the thorough mastery of the language a somewhat difficult attainment. One who has acquired a tolerable acquaintance with it is astonished at the frequency with which he hears new words used. He can hardly listen to a conversation of any length without catching a word not formerly known to him; some of these words express fine distinctions, which we should hardly have expected from a people so low in the scale of civilization.

CHAPTER III.

Rain-makers and Doctors—Arrival at Sekomi's of Griqua Traders—Ceremonies of the " Boyali "—Bamañwato Superstitions—Proceed on the Journey—Fall in with Giraffes—Palmyra Trees—The Kalahari Desert—Bushmen—Salt-pans of Shogotsa—The Bakurutsie—Chapo's Village and Marsh.

LIKE all heathens, the Bamañwato are exceedingly superstitious, believing implicitly in their doctors and rain-makers. When one of the tribe is sick the doctor prescribes roots and herbs, and they consider the virtue of the medicine is not lost if another takes it on behalf of the patient. They use vapour baths, but believe the virtue to consist in some herbs infused by their doctors, without which they would not resort to them. Cupping and tattooing, the only mode of bleeding they understand, are extensively practised. In tattooing for rheumatic and other pains, they raise the skin and cut it through, describing sometimes ornamental figures on their bodies. They cup chiefly on the temples and neck, and suck the blood through a horn. When a wife is sick she is excluded from all intercourse, except with the old woman who nurses her. The husband dare not see her, nor tread within the enclosure surrounding her hut; any intrusion would destroy the effect of the doctor's charms. Sometimes the doctors impose upon the credulity of their patients by informing them that, unless they make sacrifice of a fat ox, and give them a certain portion to make medicine, there can be no cure. The doctor, of course, is not forgotten on such occasions, as they think it necessary to propitiate him with a share. From the poor people a sheep or goat is frequently extorted in the same way; and when at length, as not unfrequently happens, the doctor finds his skill and cunning unavailing, and that he cannot effect a cure, and the friends and relations of his patients become clamorous and importunate, he further imposes on their credulity by informing them that he must have some ferocious animal without blemish, which he well knows they cannot procure, such as a live tiger, baboon, or even a live crocodile—a plan the rain-makers also resort to when they fail to bring down rain from the clouds.

When a death occurs, the mother or the wife of the deceased is generally kept in ignorance of the fact for some days, as she is not allowed to visit the sick chamber; and the first intimation she receives is the food-bearer throwing down the wooden bowl, out of which they eat, before her door, upside down. Polygamy, as in every other tribe, exists here also. The chief himself has only seven wives, which for a man of his importance is a very moderate allowance. The females are generally betrothed while mere children, or even infants—sometimes even before they are born. A girl thus betrothed, as we should say, is reared by her parents until she is circumcised, or until the birth of her first child, when she is handed over to her husband, amidst feasts and rejoicings, and he pays the balance due on his bargain. I was not aware of this practice of early marriages until the wife of an old man I had engaged here to accompany us, a child of about eight years of age, was pointed out to me, and in my ignorance I laughed outright, until my interpreter explained the matrimonial usages of their people.*

During our stay at Sekomi's, about a hundred Griqua and Bechuana wagons from Kuruman, and from Griqua Town and Phillipolis, arrived here on their way towards Moselikatze's country. The wagons, which were very showy, and

* The author seems not to have been aware that such early marriages are common among the Hindus.—[ED.]

too good for this rough work, were ostentatiously drawn up in two long lines facing the town, to display them to the best advantage, so that the people remarked that the Griquas must be a powerful and rich nation. From the conversation of these Griquas we learnt that their object in bringing such a force was to deter Sekomi from opposing them in their intended visit to Moselikatze, he (Sekomi) having on a former occasion nearly cut off the whole of their party by a stratagem during a like expedition. He did not attempt to disguise his jealousy and dislike to their having intercourse with tribes beyond his territory, for fear they should also become possessed of fire-arms, which at present were his only defence against the powerful Matabele nation.

Before leaving Sekomi's Town, I visited, in company with Vilgoen, the khotla, or council-yard, a circular enclosure of strong stakes, about 100 yards in diameter. The chief not being present, we waited by the inevitable fire, smoking our pipes till our patience was exhausted. At last Vilgoen got up, and, strolling about in an absent mood, was about entering an adjoining kraal, over the gate of which was suspended a scarecrow made of the carcase or wings of a crow, or some other bird; but his progress was intercepted by a rush of old doctors, who, to his great amazement, stood jumping and dancing before him, and shouting and yelling in a most ridiculous manner. "Old Booy," our interpreter, soon explained matters to Vilgoen, informing him that the adjoining court-yard was the sacred chamber of circumcision, to which no one but the doctors and patients were allowed admission.

On making our parting visit to the chief, our attention was directed to the female candidates for circumcision, or the "boyali," consisting of some fifty girls, about the age of fourteen, dressed in a most fantastic manner, with thick coils of grass round their waists, and thorn branches in their hands. Armed with these implements of punishment, they were going to the outskirts of the town, where the lads were undergoing a similar ceremony, called "bogoèra," to inflict on them a castigation, and, as it was termed, "make men of them." These girls are, at this period, the dread of all the young men, who are subject to every torture they are capable of inflicting; and not till they can endure it are they considered men. While undergoing this important ceremony the young men are not allowed to go near the town. Both sexes are strictly guarded, and have no communication with each other until the term which they are to pass under the doctor's hands expires, this being for the females one, and for the males two, moons. In the meantime they are considered unclean. The first day of their release a general rejoicing takes place. They are decorated in all the beads, finery, and ornaments they can command, and, being thoroughly anointed with butter, oil, and ochre, and bespangled with *sebelo*, a kind of sparkling mineral which has a dazzling effect in the sun, they have a grand dance which lasts all the night. Being now initiated into the mysteries of manhood and womankind, their parents, relatives, and friends express their regard by lavishing on them presents of all kinds of clothing and trinkets, and, taking them by the hand, the old people honour them as their equals.

The wagons having started were now out of sight, and I followed on horseback in the afternoon with Vilgoen and "Booy," in the midst of a thunder-storm, a common occurrence at this season. A great number of women, employed in reaping the extensive corn-fields through which we passed, were raising their hoes and voices to Heaven, and, yelling furiously, cursed "Morimo" (God), as the terrific thunderclaps succeeded each vivid flash of lightning. On inquiry, I was informed by "Old Booy" that they were indignant at the interruption of their labours, and that they therefore cursed and menaced the cause. Such blasphemy was awful, even among heathens, and I fully expected to see the wrath of God fall upon them. Indeed "Old Booy" took the opportunity of telling me, that about twenty years before he had seen a woman struck dead in the midst of her anathemas upon the Great Giver of rain and thunder, and that he was always in expectation of witnessing a similar catastrophe. So inconsistent are these people, that at one time, when in need of rain, they spare no sacrifice in propitiating man as the means of procuring it, but when they are inconvenienced by it, they curse God as the cause. All Bechuanas believe in God (Morimo), whom they laud or execrate as good or bad luck attends them. Ask them directly whether they believe in a future state, and they will tell you they do not; but their common expressions on a death-bed tend to prove that they do, or that their forefathers did. For instance, when a chief dies, they say he has gone to the "Barimo," which literally means "Gods;" but, according to their own interpretation, a town in the invisible world, where all the great and good men assemble and live after death. When any one dies, they never say he is dead, but that he has "tied up his sandals," or "girded his loins and departed to the shades of his fathers." The same persuasion is manifested by many other modes of expression, and confirmed by a variety of circumstances.

The Bamanŵato have a tradition of a wise king among them, who acted the same part that Solomon did towards the two harlots who contested the right to a child, and others of their traditions exhibit equally curious coincidence with the events of Israelitish story. All the Bechuana tribes worship "Biena," some animal or thing, after which they are generally called, with a degree of reverence and fear equal to that paid to the Deity himself (Morimo). Thus the Bakwains hold crocodiles sacred; the Batlapien, fishes; the Barolongs, iron. But the crocodile is regarded by all the tribes with great dread and much aversion. They will not even look upon those animals if they can help it; for fear of some evil befalling them. Those who "biena" (dance to or worship) the buffalo will not slay, eat, or in any way make use of any portion of it; they will not even lend their spears or needles to cut or saw the flesh or hide. And it is the same with all the tribes, as regards those animals which they have learnt to look upon somewhat in the light of their special patrons.

On the other hand, all the tribes entertain a violent antipathy to the chameleon and lizard, blaming those creatures as the cause of all their misfortunes. According to a tradition of theirs, God sent the chameleon to inform the first Bechuanas that when they died they would live again; but something inducing the Almighty to alter his mind, he sent a lizard with the unwelcome tidings that after death they would never return to existence. Hence the lizard was killed for outrunning the chameleon and bringing such unwelcome intelligence, while the chameleon was condemned for allowing the lizard to outstrip him: and this vindictive feeling prevails against these harmless reptiles to this day.

The Bamañwato bury their dead. They put the body in a sitting posture before it becomes cold and stiff, and sew it in a blanket. In the case of a male, they bury with it the club and spear, as also the dish and spoon of the deceased. Herbs are deposited in the grave to propitiate the clouds, and a general wailing ensues. Strangers who may die at their town, however, are not buried, but are cast out to the vultures and hyenas; so also the bodies of their own tribe who are killed in battle, or by an animal, or drowned in a river, are left where they happened to perish. These people do not believe in natural deaths; everybody that dies is supposed to be killed by or through the instrumentality of somebody else. They do not like to be told that they must die, and if you tell them so, they think that you have some design upon their life.

There is much in the superstitious beliefs and usages of the

Bamañwato that exhibits affinity—in some cases amounting almost to absolute identity—with the practices of the Israelitish nation;* and much also to encourage hope in the practicability of their conversion to the tenets of a purer faith than that to which they now adhere with such remarkable tenacity. We find in their every-day life and expressions an involuntary belief in God, instead of their own cunning wizards and superstitions. For instance, I have frequently met with natives in a perfectly destitute state entering on a long journey without an article of food, or the prospect of any, or going into other dangers, and, asking them how they will manage to get through, they unhesitatingly reply: "*Gaki etsi Ra-Morimo itla tusa*," "I don't know; Father-God will help." *Ra*, or "father," is a respectful or affectionate term of address. A father often calls his own child *ra*, or *ma*, "mother," even though it be a boy. Amongst such a people, missionary labour ought, if properly employed, to effect a deal of good, and it is in combination with efforts for the extension of legitimate trade, "the golden girdle of the globe," that it might, in my opinion, be most advantageously employed. There is surely great scope for it among such a nation as the Bamañwato, who, including all the villages of refuge and subordinate tribes through their territory, probably amount to more than 100,000 souls.

But to return to our narrative. The first day's march from Sekomi's Town was about twelve miles north, by a winding road through a pass between high and rugged mountains. The following day, after a long journey, we arrived at Lutlochoè, a water in a small rocky valley, where another road

* The injunctions embodied in the 12th and 15th chapters of Leviticus, which are almost literally in force amongst the Bamañwato, may be cited as an example. Dr. Livingstone, however, seems to consider that such usages cannot even be traced to a Mahomedan source—"there being no continuous chain of tribes practising these rites between the Arabs and the Bechuanas or Kaffirs." [May they not rather be referred in all cases to some earlier and common traditions?—ED.]

to the town of Sechelli, chief of the Bakwains, joined ours. Here we stayed a few days, making preparations for the sport we expected, and recruiting our oxen for the journey across the great Kalahari desert, now before us, our intention being to cross it at this its eastern extremity, as the narrowest and most practicable part at which to encounter its dangers.

We shot a fine male giraffe, a few springboks, and some ducks and pheasants, and found a kind of wild fig-tree, with fruit very like that we find in the colony, and of which the mountainous country eastward produces abundance. There is also a sort of hemp (*Sanseviera angolensis*), which the natives convert into cords; and we met with one solitary euphorbia, but a variety of aloes, cacti, and bulbous plants grew freely within the limits of the low rocky range bordering this valley. Amongst our other preparations was making bullets: these we hardened with type-metal, which I tried instead of tin, and it answered so much better that I have adopted it ever since.

On the 21st of June we started early, and in about five hours reached a village of Balala (Bakalihari, or poor people of the desert), who live under a petty chief named Tlabala. Observing the women obtain water by making deep holes, and so filling their earthen vessels and calabashes, we enlarged the holes with our spades, and fenced them in with branches till the water accumulated for the next day. In the afternoon Vilgoen and I, with "Booy," rode out in search of giraffes, some of which we soon found. Our guides warned us to be careful of falling into the pitfalls ahead; but our excitement got the better of us, and we galloped on the trail at full speed, and soon approached the long fences made by the natives, with apertures here and there, leading to a covered pit, some of them eight feet long and fifteen feet deep, with sharp stakes planted in each.

We managed to shoot a brace of fine bulls, one of which was dragged to the wagon by a long team of oxen. It was curious to witness the mode in which the natives dissected the animal, their spear-shafts quivering over the crowd of heads as they plunged them into and cut up the immense carcase, their women standing by joyous spectators, dancing and singing, and clapping their hands in the anticipation of a hearty feast.

Here we bartered two or three cows for about a ton or two of the corn (*Holchus sorghus*) for our horses, but when we had stowed away one half of the corn bargained for, it was found that they could not supply the rest. However, they promised faithfully to make up the quantity on our return, when they will have reaped all their corn.

We started at 1 o'clock the next day, and laboured along through the heavy sand and dense bush, the oxen coughing and nearly choked; while our Hottentot drivers were laying on the whips unmercifully, and shouting themselves hoarse. Before dark we changed oxen, had tea, and then proceeded again, travelling all night. The next morning, about 8 o'clock, we reached a vley called Matulani, having gradually entered a more open country. Although everything in this so-called "desert" grows luxuriantly, and grass and herbs most exuberantly, we found no water. Vilgoen was aware of the existence of some wells five hours ahead, called Inkaoani; but as we found that at these wells the water for the supply of more than 100 head of cattle must be drawn up in buckets, we started for a vley about the same distance westward on another road, hearing that it still contained some rain-water. In crossing the plain five of us shot each a fat eland. On reaching the vley we found barely sufficient water left there for one drink, another travelling party having just left it. Our oxen, soon returning for another drink, as they generally do when they have thirsted long, and finding none, lowed and bellowed for more, and we gathered all the bitter water-melons we could find for them.

After such a long drag of nearly 50 miles without water,

our cattle needed a rest, which we allowed them, still without water, till the next morning. At 10 o'clock we started, accompanied by some wretched-looking Bakalihari, who promised to conduct us to elephants still farther westward. At 4 o'clock we found a place where there seemed a likelihood of obtaining water, and dug to the depth of twelve feet, but finding that the water percolated very slowly, we started again, and, travelling for three hours without any road, we were compelled by darkness to unyoke for the night.

Next morning, seeing some flocks of goats and sheep, we concluded that there must be water in the neighbourhood; but the natives in charge informed us that their flocks never got any water, and that they themselves eat bitter watermelons to assuage their thirst. On tasting the melons, such was their bitterness that I could scarcely believe that even these wretched people could swallow them.* Here we were joined by about sixty Bushmen and Bakalihari, some of them stout fellows, carrying on their shoulders spears with melons stuck upon them. At length, having travelled about 12 miles, we were delighted to find a fine vley of water, with numerous elephant spoors about it. We pitched our tents about half a mile from the vley, before the wind, for fear of alarming the elephants in case of their coming to drink.

The next day a Bushman brought a child for sale, informing us that its parents were long dead, and he was tired of feeding it. A Dutchman of our party had agreed to buy it for a bunch of beads, but one of the men whom we engaged at Sekomi's threatening in consequence to leave us and return to Sekomi's, the bargain was cancelled. I shall never forget the avaricious twinkling of the old Bushman's eyes

* Some of these melons, however, have a tart and pleasant taste. The Bushmen affirm that they become sweet or bitter according as their seeds may happen to be deposited in the dung of the white or black rhinoceros, or in elephants' dung.

when the beads were shown him, contrasting dreadfully with his look of disappointment when the bargain was declined.

On the 28th we had another and a fruitless hunt after elephants, but returning from it we were rewarded with one of the most magnificent sights that it ever fell to the lot of a traveller, even in these regions, to behold. A troop of between 200 and 300 giraffes, evidently migrating, came in view, a Bushboy with bow and arrows stalking among them. The dense mass, half a mile in length, formed a complete and impervious wall, and all the Bushmen and Bechuanas declared they had never seen the like, though they were born in the country. They attributed the migration of this herd to the want of pasture driving them out of the desert. Eight of these gigantic and beautiful animals we shot for our famished followers. On the whole, the Bushmen were more willing to work for us than the Bechuanas, and quite proud of being permitted to carry our guns and march ahead of our horses.

The Bushman who had been stalking giraffes was caught by some of our party while attempting to make his escape. He was desperately frightened, and prepared to defend himself with his poisoned arrows, when the other Bushmen arriving, informed him there was no danger; but never having seen white men before, he seemed in doubt whether they were to be trusted, and sullenly returned his arrow to his quiver, still trembling so violently that it was some time before he could speak. We gave him some tobacco, and asked him to show us water, as we were thirsty; but he told us he never drank water, living upon the lerush, a juicy root, and bitter melons. Although we were aware that this is common in the desert, we did not feel inclined to believe him, having heard there were "sucking-places," which it is their policy to conceal.

These "sucking-waters" are generally damp spots in the earth, where, nevertheless, no water can be found by digging. But by inserting a tube of reed, with a sponge at one end, the depth of a couple of feet, abundance of clear and cool water may be obtained by sucking at these tubes. I have often since surprised Bushmen at these places, and got them to suck by this ingenious contrivance sufficient water for three or four horses, rewarding them with a little tobacco.

From this Bushman I endeavoured to purchase the whole of his worldly possessions, which, like all other Bushmen's, consisted of a bow with quiver, containing about twenty arrows, the blades of which are mostly made of bone, some few being of iron, in the shape of a harpoon, and well plastered with poison. These darts are so arranged in the shafts that they disconnect themselves from the latter on striking an animal, without injury to either. Besides these fatal instruments they carry generally a worn-out spear, a sharpened stick to dig up roots, two pieces of wood with which they make fire by friction, an awl, and a needle, with some medicinal roots, chiefly antidotes for snake-bites, and some surplus poison. The poison is obtained from roots, also from a kind of grub, and sometimes from a species of bulb (*Amaryllis toxicaria*). One sort is extracted from a species of euphorbia (*Euphorbia arborescens*). Besides these, they have several kinds which they keep a profound secret. No antidote is known for these poisons, but they sometimes cure a wound by sucking it out. All these articles are packed together in a sort of knapsack, with a small tortoise-shell for a spoon. Nothing would tempt the Bushboy to dispose of them.

On the 22nd we steered for the Great Lake road, which we reached in six hours, and after travelling again in the dark for four hours more, we halted and slept. Here two of our party were bitten by scorpions, and I found one nearly six inches long crawling on my neck, which I happily seized and committed to the flames.

The country over which we had been travelling since

leaving the Bamañwato is called "the desert," and travellers going to "the lake" from the colony are obliged to go round it by the course we were now steering, as the most practicable. Could they make direct for the lake through the country of the Bakwain or Bawañketze tribes they might reach their destination in half the time. Though called a desert, the reader must not picture to himself another Sahara; for although a sandy country and devoid of water, excepting a few scanty wells at intervals of 40 or 50 miles, it is nevertheless fertile, the grass growing most luxuriantly, and large forests of trees of many kinds abound. The principal trees are the mokala or camel-thorn (*Acacia giraffe*), the mañanathorn, or hook-thorn (*Acacia detinens*), the gum mimosa (*Grewia flava*), and several other mimosas; the vaal bush, or mogonono, the kroutzer bush and moretloa, with sandalwood and several berry-bearing bushes. These comprise the favourite food of elephants and giraffes, as well as of the eland and gemsbok, with all which animals this vast wilderness abounds. Great quantities of bitter melons are scattered over the surface, and at certain seasons several smaller kinds of gourds, affording a pleasant substitute for water to both man and beast. Besides the generally known mimosas there is one in particular, more troublesome than the rest, which I call "sickle-thorn," a kind of briar spreading from the root in semicircular bows of long tough shoots, thickly set with thorns, like the larger thorns on rose-briars, and studded together like the teeth of a saw. These inflict severe punishment on horses and cattle, as well as on man, cutting to the bone when they come in contact with the body, and sometimes stripping the whole skirt of a coat. When, as frequently happens, one of these thorns catches a horse by the nose, he stops suddenly, throwing the rider over his head. Cattle and horses are much frightened at these bushes, and generally avoid them.

The next morning we reached a vley called Mañana, in a valley leading to a spring called Kokoñyani. This fortunately contained water. Large flocks of the Namaqua partridges (a plump and fleshy bird) were drinking at the vley; nine of these we killed at a shot.

From Mañana our track led down the valley to Kokoñyani, but, the road being very winding and sandy, we steered during the afternoon and night across the veld to the westward, cutting off a great bend. We now travelled through a new kind of bush, called mopani (*Bahunia*), or fly-bush. The timber was evidently a kind of mahogany of a bright colour, and susceptible of a very high polish. Where the poisonous flies are found, they generally infest this kind of bush in preference to any other; on the leaves an insect deposits its larva, a kind of glutinous substance partaking of the nature and properties of shellac. This scabby-looking substance is gathered in large quantities, and greedily devoured by the natives as an article of food. In the night we slept again in the valley leading to Kokoñyani, and next morning reached the fountain, an excavation made by the natives in the earth.

Here we sighted for the first time the tall palmyra trees of the tropics. They generally attain the height of about sixty feet, forming a striking feature in the scenery by their contrast with the character of the rest of the surrounding vegetation. At the summit of its straight and lofty trunk the palmyra is crowned with a tuft of enormous radiating fan-like leaves, of several feet in length. The trunk of the tree is exceedingly hard and fibrous, attaining the circumference of from three to five feet. It bears clusters of a kind of brown fruit, the size of an apple, and partaking internally the nature and appearance of cocoa-nuts. The fibrous covering of the seed is eaten by the natives, and tastes like gingerbread, and the milk inside resembles and tastes like that of the cocoa-nut. From the trunks of the young trees, as well as those of the date palm, which we only find farther north, the core is sometimes taken when they are only a foot out of the ground. This is a juicy, soft, and white substance, very much resembling the inside of a cabbage stalk. By the advice of our servants, our cattle were allowed to have a rest here; and in the afternoon we sallied forth, accompanied by some Bushmen, to a neighbouring reedy fountain called Lotlakane. From thence we followed the trail of some rhinoceroses that had recently drunk there, and in half an hour we started a fine large male in a dense patch of mañana, which my horse Dreadnaught alone would face. Being the best leaper, I had the advantage, and after a run of about two miles secured him with five heavy shots. The Bushmen soon joined me, and, according to a superstitious custom, threw a handful of dust (not salt) on the animal's tail, spitting in its eyes before they touched it, muttering meanwhile some gibberish to "Torra"—a charm to insure them future luck. Some of my friends had administered a potation of cognac to one of these Bushmen enough to prostrate any European; but, unlike the Hottentots, who are so easily inebriated, this fellow was not in the least affected. He walked all the firmer, as if his feet spurned the earth; and it was settled amongst them that cognac is the medicine Europeans drink for good luck. I have never passed this place since without being importuned to barter some of this charm for ostrich feathers. Observing one of these people to be a cripple, probably from a paralytic stroke, I inquired the cause; they replied, "God had struck him."

On the 3rd of July, being aware we should find no water for another 40 miles, we remained till almost 10 o'clock, to let our oxen drink before they started. In the evening we overtook a Bushman planted beside an enormous dead giraffe, which he had wounded with an arrow some 50 miles back, and had followed thus far. The active fellow, on being convinced that we were not going to rob him of his booty, as the Bamañwato would have done, started back to report his success, and bring up his family. The reader can imagine the state in which the carcase must be on his return; but it must be understood, once for all, that meat, no matter in what state of decomposition, is never discarded by Bushmen.

On the 4th, late in the afternoon, we reached the salt-pans, or lakes and fountains, of Shogotsa, incorrectly called Inchokotsa. These are filled for miles with salt water, and thousands of red-winged flamingoes and other aquatic birds are scattered over the broad expanse.

Having previously to commencing this journey ascertained from Mr. Inglis the result of Mr. Oswell and Dr. Livingstone's discoveries in Sebetoane's country, inhabited by one of the most powerful, if not, indeed, the most powerful, of the tribes in South Africa, and living on the Chobé, a tributary of the Zambesi, I had all along resolved to penetrate into, and, if possible, beyond, the territory of this chief, in search of elephants, which I was told abounded in that region. I now, therefore, determined to obtain information respecting the country directly ahead, and to endeavour to procure guides to take us direct through the desert, instead of going round by the lake, a distance of more than 300 miles to the north-west. So, after pitching our tent, I despatched a party to a tribe of Bakurutsie living in the centre of a very extensive marsh or bog, on what is commonly called the termination of the Botletlie river; for, although the stream still flows far to the east, yet, owing to the strong saline impregnation of the earth, the water becomes perfectly salt at about 30 miles north-west.

Owing to feuds among themselves, the Bakurutsie of the

Mariqua, formerly the most powerful tribe there, were divided into several branches. Two or three of these tribes remaining in the Mariqua were conquered or destroyed, or driven out by Moselikatze, and one portion of this latter fled to the east, where they now live on the Shashe river. These, adopting the name of Bakurutsie, again divided, and the father of Chapo, their present chief, led this portion of the band to the spot just mentioned, where, surrounded by the dangerous bogs and ever-burning turfs, they have lived in security for forty or fifty years, acknowledging, however, the supremacy of the Bamañwato nation, to whom they pay a yearly tribute of karosses, &c. Since their residence here they have become amalgamated with the Botletlie, the aborigines of the country after which the river is named. They speak a "click" language, something resembling that of the Bushmen. Many villages of Makalaka, Bakalihari, and Bushmen refugees, have also placed themselves under Chapo's protection; and every day is adding fresh numbers, so that they bid fair yet to outstrip many of the neighbouring tribes, and become an independent people. The islands within the marsh possess a most fruitful soil, producing abundant crops of grain (*H. Sorghum*), maize, beans, pumpkins, water-melons, calabashes, and sugar-canes (*Holchus Sacharatus*). They seldom venture out of the marsh, which is about 60 or 70 miles in circumference—indeed they have little occasion so to do; and for strangers unacquainted with the intricacies of the only practicable entrances to penetrate into it would be a perilous undertaking. Abundance of buffaloes, a few sea-cows, and thousands of waterbucks (*Luchee*) inhabit the marshes, grazing peacefully in sight of the native dwellings. Some of these are daily taken in pitfalls by these refugees, and they possess numerous flocks of sheep and goats.

We remained at Shogotsa till the 8th of July, suffering

severely, as well as the cattle and horses, from diarrhœa, which may be attributed to the strong mineral properties of the water of the fountain. Meanwhile we had ascertained that we might safely travel through the desert, as heavy and late rains had recently fallen. This, however, the natives expected would bring on severe attacks of fever. We shot here a few zebras for some poor famishing Bushmen we fell in with. One of these proved to have a fine ear for music, this being a faculty for which Bushmen and Hottentots throughout South Africa are noted. This Bushman could imitate any tune which my friend Wirsing's cornopean was made to sound.

We steered about three hours north-east without a road, but guided by a Bushman. I and my friend had an unsuccessful chase after a fine troop of giraffes, as on dismounting to fire our horses ran away. We afterwards fell in with a troop of about fifty gemsboks, which we could not approach on foot.

CHAPTER IV.

A Bushman Village—Cross the Botletlie River—The Ntwetwe Salt-pan—A magnificent Baobab Tree—Reach Thumtha, or Goroge's Post—A superior Tribe of Bushmen—The Makalakas—Trek Northwards—A Bushman Smoking-bout—Elephant-hunting in the Madenisana Desert—A dangerous Encounter.

DURING the day we lighted upon a Bushman village of seven or eight huts, consisting of a few forked sticks, about three or four feet long, set up against each other, covered over with a few handsful of long grass. A small broken earthen pot stood boiling some roots over a fire; a calabash of water, a few tortoise-shells for dishes, and the hulls of innumerable bitter melons, lay scattered about, together with a few bones; but the inhabitants, saving one blind old man, had fled, leaving their bows and arrows still hanging on the trees. We also discovered some Bushmen graves, being excavations made in the large ant-heaps, raised by the white ant (*Lintoa Termes*). In these the body is placed, and the hill enclosed with strong thorn branches to protect it from the vultures and hyenas.

Having crossed another salt lake about two miles in length, we outspanned a little mineral fountain in a rock, called Makhamma. The pass over which we crossed, being dry, was covered with a white nitrous or saline incrustation, the dust rising from which was very painful to the eyes, and nearly blinded men, dogs, and cattle. A few hours to the northward of this fountain we crossed the bed of the Botletlie river, containing only pools of very salt water. Fish, however, must abound when the river is on flood; for here were many stone kraals made to cut off the retreat of the fish at the ebbing of the river.

On ascending the opposite bank, we perceived a troop of gemsboks ahead; some of my party, being mounted, gave chase, while I preferred remaining to witness the chase from the wagons—a stirring spectacle, which I enjoyed more than if I had been engaged in the adventure myself. Shortly after this chase, which turned out unsuccessfully, a black rhinoceros (*Ketloa*) was shot by Piet Greyling, and "Old Booy" shot the fattest cow-giraffe that any of the party had ever seen, it having three inches of fat between the shoulders. We encamped for the night in the neighbourhood, despatching one wagon to bring up the best of the flesh. The rest was given to our Bushman guide, who buried it; and, in order to conceal it from the hyenas, covered it first with a layer of earth about a foot deep, and then with the dung from the animal's paunch, which again was overspread with earth. He informed me that this is a stratagem frequently resorted to, and that the wolf, scratching to the depth of a foot, and finding only dung, generally abandons the search, and turns away in disgust.

On the morning of the 9th of July we steered rather more eastward, to a permanent pool or spring of water, called *Kobé*, when we took breakfast, and, proceeding on our journey N. and N. by W., fell in with a troop of eight giraffes, to which we all gave chase, and eventually I succeeded in bringing down a fine old male. During the chase, this giraffe threw out his heels several times before me as I rode too close alongside, and when I brought him to a stand, turned upon and endeavoured to paw me, convincing me that these animals, when brought to bay, are not quite harmless. We travelled till sundown over a limestone country, dotted over and densely sprinkled with what the natives call mochueri trees and mokala (the camel-thorn), and killed two puff-adders as we outspanned for the night.

On the 10th, before daybreak, we unyoked, and as the dawn appeared entered the Ntwetwe salt-pan, about 18 miles broad, and extending for a considerable distance east. When in the middle of this pan, we appeared to be in the broad expanse of a calm and white ocean. The air was bitter cold, and, although we wrapped ourselves in all our blankets, we could not feel comfortable. The keen wind lifted the salt dust arising from the crackling incrustations, and drove it in the eyes of man and beast, who all showed symptoms of the pain they were suffering. However, after a six hours' journey, we found ourselves, to our great delight, huddled together round a blazing fire on the opposite shore of the pan, enjoying our warm coffee and some delicious giraffe steaks. Having feasted and warmed ourselves, a second trek brought us to a spring-pond in the limestone, called *Gootsa*, having stopped midway to admire the first moana tree (*Baobab*) we had ever seen. We were lost in amazement, truly, at the stupendous grandeur of this mighty monarch of the forest, in comparison with which the largest of hundreds of surrounding palm trees sunk into apparent insignificance. The dimensions, which we took with a measuring-tape, proved its circumference at the base to be twenty-nine yards. It had shed all its leaves, but bore fruit from five to nine inches long, containing inside a brittle shell, seeds, and fibres like the tamarind, enclosed in a white acetic powder or pulp, which, mixed with sugar and water, makes a very pleasant drink. The natives eat the fruit when thirsty, and at other times boil the pulp into a sour porridge. The flower, as I have since observed, is, when in season, lily-white, consisting of four leaves, and about eight inches in diameter. The leaves are of a glossy green, three always joined to one stem. The wood is soft and useless; the bark is more than six inches thick. The roots spread in every direction, and where the ground is hard project two feet above the surface, running for more than 100 yards. These trees are the resort of innumerable squirrels, mice, lizards, snakes, and also a small winged insect, called palele. When these insects alight on the human frame, they generally deposit a fluid, which soon induces a painful irritation and small white blisters. It seems to be a kind of vine-bug. Very fine hives, with honey, are often found in the baobab, and many of the larger birds delight to build their nests in them. Amongst these are the koro (*hornbill*), grosbeak, green and black shrews, crows, and the loveliest of birds, with plumage of celestial blue, the rajaike, which often cheers one with its sweet song at the dead hour of midnight.

The spring-pond (*Gootsa*) at which we rested was also surrounded by several moana trees of rather smaller dimensions, with several mokolani (palmyra trees).

Recovering from their panic, several Bushmen came to see the white men and their wagons, starting at our every movement. The children seemed very fond of their mothers, clinging to their naked and dusty bodies while squatting on the rough stones, actively engaged with hands and feet, amidst clouds of dust, in grubbing amongst the loose limestone for bulbs; these they seemed to find without the aid of their eyes, which were never withdrawn from us, and conveyed to their mouths with great rapidity, free of pebbles. Covered, as I have said, with dust from their baboon-like employment, they had a truly hideous and diabolical appearance. These wretched creatures informed us very civilly that our cattle were in danger of getting killed by some poisoned stakes they had planted by the water to kill game. Having promised to remove them, we shot for them several gnus and zebras, of which animals we saw many thousands, and they undertook to conduct us to the next water.

On the margin of the pond we found a tree, bearing very pleasant-looking and luscious-smelling flowers and fruit. It is called "moralla," and they said it was poisonous. The water in the pond was still plentiful, but a troop of buffaloes had so wallowed in it that the cattle would not drink freely. We presented the Bushmen with a jackal and a hare, which our dogs had caught on the margin of the pan. These they readily converted into a stew, and devoured on the spot. They admired very much a little dog of ours, inquiring why we did not eat it, it was so fat. This dog, which had followed us from the Notuani river, was so like the silver jackals that many even of the natives mistook it for one. I conclude it must have been a hybrid, having heard of such instances, as well as of hybrids between wild and domestic dogs.

Two hours and a half more to the north-east brought us to another spring-pond, called Thumtha, or Gorogé's Post, where we shot many wild ducks and geese. We were surprised at observing the remarkable difference between these Bushmen, called Madenisana, and those we had formerly met, both in personal appearance and in stature. We could not help admiring their independent and elastic gait, dexterity, and skill. They wore their hair in long twisted cords, trimmed to one uniform length, and shining with grease. Their average stature was 5 ft. 10 in., but it often exceeded 6 ft., and their features more nearly resembled the European in point of delicacy than those of any other black tribe I had seen. These people seemed to be the pride of their masters, the Bamañwato, for whom they had swept the country of all the elephants with their spears alone. They scorned the idea of serving as guides for the remuneration of game, brandishing their spears to show they were expert in the use of them and could kill game for themselves. Thus we were doomed to submit to their own terms, which were the acquisition of a dog. In the presence of their masters, the Bamañwato, however, they aspire to no higher position than filling the offices of dogs, pack-oxen, and horses; and, to add to their degradation, they are never designated otherwise by the Bamañwato than in terms applicable to beasts, such as bulls, cows, and calves, never being called man or woman—dogs, inasmuch as they are required to kill game for their masters; pack-oxen, as they are required to carry the ivory, flesh, and hides which they kill to Sekomi's residence, some hundreds of miles; and horses, because they are obliged to convey the most trifling news they gather with the speed of a horse to their great chief, who by this means becomes acquainted with the most trivial occurrences transpiring within or beyond his territories. With all their boastings, however, some of the most wretched among them would gladly have followed our fortunes home, for the sake of eating plenty of fat, with which they had heard our country abounded.

On the following Monday, as game was still abundant, and there were plenty of mouths to eat the flesh, our party shot several head of gnus, zebras, and springboks, and on nearing the next water, Thageara, some five hours distant N. by W., I shot an ostrich. Thageara, or Thouñara, was, like most of the other spring-ponds, surrounded by a number of moana trees, in one of which a room had been excavated —the habitation of some Bushmen in times when lions became troublesome. Whole families of Bushmen are said sometimes to build their habitations in branches of these trees, for fear of those savage beasts. In most of the trees I observed two rows of wooden pegs, by which they ascend to rob bee-hives and pluck the fruit.

On the 15th of July we travelled for six hours N.NW., passing several spring-ponds, or rather small lakes, swarming with hundreds of waterfowl of every description. On the following day, five hours in a more westerly direction

VOL. I. F

brought us to a village of Makalakas, under their chief Khaetsa, at a spring-pond called Zoutharra.

These Makalakas are a tall, athletic race of black men, cleaner and more industrious than the Bechuanas. They are a portion of a powerful tribe, formerly dwelling to the east, and had been conquered by Moselikatze, the terror of the South African nations; and they were constantly flying in small parties to escape his iron sway, seeking shelter among the more independent neighbouring tribes, where they prefer paying a yearly tribute to giving up their sons to fight the wars of a tyrant. This village mustered about 300 souls. Their weapons were battle-axes of their own making, hideously barbed spears, and knobkerries. Several of them wore, besides their skin-clothing, a kind of knitted rug, made by themselves, of an indigenous cotton, and coloured with a plant which I presume must be a species of indigo. They had a small herd of goats, and seemed to depend for their sustenance upon their gardens, containing maize, beans, earth-nuts, and a species of millet called *lebèlèbèlè*, a very prolific grain, growing in the most barren soil, some of which I have introduced at the Cape, where it has yielded from one seed several thousand fold.

This tribe, as well as all the neighbouring tribes, are subject to a fixed yearly, half yearly, or sometimes indefinite tribute to their masters, the Bamañwato, who are in the habit of making constant visits for the purpose of extorting from them whatever they can lay hands on. The principal articles of tribute consist of ivory, dried flesh, fat, hides, and skins, feathers, mopani gum,* caterpillars, and even the skulls of all the jackals they kill, from which a little fat is obtained by boiling it out. Still they consider these exactions more tolerable than Moselikatze's sway.

As the whole country now before us for 200 miles was a

* An insect on the Bahunia.

perfect desert, without any perennial waters, I endeavoured to obtain guides from Khaetsa, but he assured us that the waters in the desert (Motlabba) were already dried up, and that his Bushmen had a few days before picked up some of the Makololo (Sebetoane's people) on their way home in a dying state from thirst. Had I at that time possessed a sufficient knowledge of native character, I should not have been so credulous as to have listened to this report, for the idea of Bushmen carrying human beings whom they found half dead out of a desert implies an act of charity quite inconsistent with their natural disposition and habits. They are at all times rejoiced to abandon such unfortunate beings to the vultures and hyenas. However, our spirits drooped at this unwelcome news, and we submitted to be led away by these fellows in a north-easterly direction, where they informed us that we should meet elephants in numbers that would frighten us. This gladsome intelligence again raised our spirits, and we struck a bargain with the old chief for being conducted, under the guidance of his son, to a tribe of Mashubea, of which Chapatani was chief, and from these to a tribe of Mashuna, or Banabea, under Juankie (or Wankie), on a large river, which I concluded must be the Zambesi. We went to bed in a cheerful mood, giving up all idea of visiting Sebetoane. I was roused next morning from dreams, in which I had been transported home, by the crowing of fowls—an unusual sound here. These fowls had been brought by the Makalakas from the east.

On the 17th we broke up our encampment by the spring in the desert, after a week's sojourn, and under the guidance of Khaetsa's son, accompanied by about thirty of his followers, who led the way, and trekked to the eastward. In a few hours we entered a forest of richly-clad golden-leaved mopani trees, at times fancying ourselves driving through fine parks and avenues. Previously to our starting, the chief had impressed on our minds that we must not expect sport with the elephants until the fifth day's journey, and it was therefore matter of great surprise to find in the first few hours innumerable spoors of these animals traversing the country in every direction. Our horses were quickly saddled, and, while the wagons were being conducted to the nearest vley of rain-water, we followed the greatest spoor, and soon came up with a troop of eleven ponderous old bulls, of which we killed six of the finest. This was grand sport, and coming so unexpectedly it had a very salutary effect upon our spirits, cheering us, as well as our followers, with the prospect of a glorious hunt, during which everybody would find some opportunity of distinguishing himself. From this time I resolved to devote myself almost exclusively to elephant-shooting. I had brought down one of the finest bulls so quickly, that I was able to assist afterwards at the death of two others; and discovering that I really could kill elephants (a thing I never dreamt of when I left the Mooi river), and that it was not more difficult, though it might be more dangerous, than killing other large game, I concluded that to find elephants enough to kill in these parts was now the chief difficulty, and I believe I am not singular in that opinion.

The distance at which we were generally in the habit of firing at an elephant was from sixty to eighty paces, though the first shots were generally fired at the distance of about forty or fifty paces. I have heard and read of people riding up to elephants repeatedly, and firing at the distance of sixteen and twenty paces; but this I am not very ready to believe. It has been done with impunity to my knowledge, but very rarely. An elephant, when so closely pressed, is sure to turn upon you, and if you fire at too long a distance your shot is uncertain, and the sinewy and elastic qualities of the hide much deaden the force of the ball. You may fire a hundred shots and not kill him. My plan therefore is, as I have already stated, not to fire the first shot until I get within fifty or sixty paces, and have nearly the elephant's broadside towards me. By this means I can make sure of putting the bullet, if not through, at least near, the heart, and into some of the vitals. This breaks the creature's spirit, and soon brings him to a stand, and you may then fire at him almost with impunity. Some people fire from horseback; but this is a plan I do not approve of, as few can make so correct a shot on the back of an unsteady, panting, and frightened horse, as they can standing on the firm ground, and thus it must be that so many people throw away an unaccountable number of shots. Even Gordon Cumming himself, the great Nimrod, backed by a pack of curs, has fired more than forty shots at a single elephant; whereas the average number of shots of any fair marksman would not exceed six to each elephant he kills. But hardened bullets and a strong double charge of powder are indispensable. With a 2 oz. ball I generally allow from 6 to 8 drs. of good gunpowder. I have used nearly an ounce of bad, but it all depends upon your gun; some will do with less, while others must have more.

In the evening we felt so elated at our success, and so excited by the recapitulation of the adventures of the day, that we could scarcely eat our dinner; and on the following morning we mounted fresh horses, and followed the trail of a troop, supposed to consist of about 500 elephants, chiefly cows and calves. The trail extended a quarter of a mile in breadth, and hundreds of fine trees, recently broken down by the herd, lay across our path; but we found at length that they had taken fright, probably at the smell of the dead elephants, and we only overtook one straggling old bull at night, at which we fired, but lost him.

From this unsuccessful chase we returned through dense forests, recently trodden by hundreds of elephants; and, overwhelmed at last with the fearful darkness in which we had been travelling, with danger of breaking our necks over the fallen trees, or encountering a troop of elephants, or losing our way (for all had different opinions with regard to the position of the wagons), we lay down to sleep in the bitter frost, without any covering, having never eaten nor tasted a drop of water during the heat of the day, which had been intense. The following morning, as the day dawned, we continued our march in search of the wagons; but there being different opinions as to the course we should take, and each stubbornly maintaining his own, we parted company, till after a time the whole party, out of regard, as they averred, for my safety, followed my track, although they felt confident I was going wrong. However, in the course of a few hours we had the pleasure of hearing the barking of our dogs, and of perceiving the white tilts of the wagons peering through the trees. A cup of coffee, our first drink since the night before last, soon revived us.

On the 20th of July we rode to the field of blood, on which lay the scattered carcases of the eleven elephants we had killed. We found most of the natives in a perfectly helpless condition, having gorged themselves to surfeit on the flesh. They had extracted the tusks, and, cutting all the flesh into long strips, had hung it up to dry on racks, rudely constructed of branches of trees. Thus there were acres of raw flesh, under the shade of which lay many of the Makalakas, full to repletion. Others were still employed in disgusting operations, and still more odious feeding, on the foul and mangled fragments and remains of the huge animals, the details of which would be revolting.

On the 21st we had another tedious and unsuccessful hunt, following the trails of a hundred elephants or more; but we could perceive by their processional sort of movement that the animals must have taken fright, and were on the move into another district. Indeed, many other troops of elephants that seemed to have been browsing fearlessly, discovering their trail, seem also to have taken the alarm,* and followed it at once.

Before daybreak on the 22nd we were roused by the crowing of a hundred pheasants, and, as soon as we could, followed the track of another troop of elephants; but we were abroad too early, as the frost at our fingers' ends soon testified. It was so sharp that we could scarcely hold the reins or handle our guns. We travelled all day without falling in with the elephants; and at night, having kindled a fire, we lay down around it—hungry, thirsty, tired, and cold. We generally took the precaution to make a Bushman carry water for us, but he too often lagged behind, and at last returned to the wagon for fear of being benighted, so that we got none.

On the 23rd, the Makalakas being disengaged, and the flesh all carried away by their wives and children, we expected to make a more extensive hunt. We had all along been afraid of venturing too far without their guidance, they having persuaded us that the country ahead swarmed with the tsetse (*Glossina morsitans*), or poison-fly, and that it would be very unsafe to travel, without a thorough knowledge of the country, into regions where we should run the risk of losing our cattle, horses, and dogs, by the fatal bite of this insect. We had very patiently waited the convenience of our guides, giving them time to boil all the fat from the elephants' bones, though we little thought what wealth we were bestowing upon them. We afterwards learnt that with the fat alone we had put them in a condition of purchasing a number of wives, blankets, and supplying other wants. We waited in vain, for the whole gang had now disappeared in a most mysterious

---

* These sagacious animals always know by the appearance of the track of their own species whether it has been made in flight, though the track be several days old, and they are sure to follow.

and unaccountable manner, taking with them their wives and children, not forgetting every particle of fat, flesh, and hide of the six huge elephants.

Here a complete stop was put to our progress. We knew not which way to steer for water, nor in what direction to avoid the fly; while, had I been as well-informed then as I am now, we might have travelled safely in any direction, as the country infested by the insect was still 200 miles distant, and as for water, it abounded in the track we wished to take. In this dilemma, and while we were still speculating on the probable motives for the disappearance of our Makalakas, but, being yet novices in African travelling, and unskilled in native craft, could arrive at no satisfactory conclusion, a fortuitous incident revealed what had escaped our attention. As it exhibits a faithful picture of the deceits and cunning almost universally prevalent among these tribes, and others in the interior, of whom we have accounts—a most common object being to delay the travellers either by hopes and promises of assistance, or by practising on his ignorant fears—I may be allowed perhaps to detail the circumstances.

While we were still in a state of suspense a Mañawato, evidently of some importance, with a train of followers, carrying their spears, calabashes of water, and other travelling accompaniments, which showed they had come a long distance, drew near our wagons. They squatted, according to custom, some little distance from us, the leading man, however, nearer than the rest, and evidently expecting, as is usual, somebody from our side to meet and welcome him, it not being their custom to present themselves uninvited. It was more than half an hour before our Bamañwato guides, though of inferior condition, condescended to notice the strangers. They then went over to greet the leader, asking him the news. This, as usual, was detailed in a style similar to that of a diary or regular journal, every occurrence since we left Bamañwato being related in succession; but as on these occasions the relation of every trivial circumstance might exhaust the listener's patience, the informant, during this first narrative, merely sketches the outlines, and afterwards, and during the night, while the pot is cooking, dilates upon the particulars. This often seems to be done in a very graphic and humorous style, to judge from their merriment.

But to return. By what we heard from these people, by the aid of our servants, we discovered that Sekomi, fearing we were going to tribes beyond his country, and jealous of our intercourse with them, had raised obstacles to our progress by these stories of the tsetse and the scarcity of water. The very men who now sat in assumed unconsciousness of our position, and were secretly glorying in it, were the workers of the iniquity. We found that they had dogged our footsteps from the chief's homestead for the purpose of frustrating our plans. We accused their leader of being guilty of this treachery, offering large rewards for a guide to some new and unexplored region, but nothing would induce him to aid us in our project. However, shortly afterwards some others of the tribe passed, carrying brass wire and beads to some distant tribe north-east to exchange for ivory; and, having discovered this fact, we determined to follow their footmarks. This we did at first on horseback, for a considerable distance north-east, and then brought the wagons on to some vleys of rain-water. We had abandoned the idea of remaining any longer in this neighbourhood to hunt elephants, finding it useless, as they seemed to have left the country, owing to the smell of those that had been slaughtered, and to the waters being tainted with the stench of entrails and garbage which the natives had washed in the pools. Fortunately we now caught a Bushman, who, won over by a present of tobacco and beads, became reconciled to us, and promised, as well as we could understand from his gibberish, to conduct us to game.

On the 29th of July we travelled by a well-trodden elephant path several hours north-east. Such paths radiate from every vley or watering-place to all points of the compass, each leading to some other vley, and a careful pursuit of the most beaten track is certain eventually to lead to some permanent water or river. The whole country is intersected with these paths. Our guide took the lead, carrying, according to custom, a fire-stick in his hands to warm them and his naked breast against the bitter frost. Every now and then, when the wagons delayed in order that the forest trees might be felled which stood in the way, he deliberately squatted to light a fire and smoke his pipe, or broil some beef. However, as he was the only friend we had in this wilderness, we were obliged to put up with his ways, and treat him kindly, for fear of his running away and leaving us in the lurch. I could not help admiring this fellow for his fine manly figure, his graceful, composed, and dignified movements, his fine forehead, dangling with shiny corded hair, delicate features, and bright black eyes, and teeth that any European lady might have envied, besides an exquisite little moustache, which gave him quite a military air. I could almost imagine that he was some half-breed, with European blood in his veins, so different in all points was he from my preconceived ideas and observations of Bushmen.

Led on by our guide without knowing whither, we could only follow in mute silence, having abandoned in despair all hopes of understanding his gibberish. We merely comprehended by the signs he made that he would bring us to some place, which we should reach when the sun had sunk about 45° towards the western horizon. At noon, having, with much labour, cleared a path through a dense forest in a heavy sand bult,* he made signs for us to stop. He turned into a by-path, and, following him, I observed that he occasionally examined something which I discovered to be the spoor of Bushmen; having tracked this for half an hour on tiptoe, we both came suddenly to a halt. The Bushman, bobbing down, motioned me to do the same. It appeared he had discovered the Bushmen ahead, and, crawling backwards on hands and knees, we returned to the wagons, and, having saddled our horses, some of our party returned, in company with the guide, to speak to them. On our approach they were dreadfully alarmed; the women, taking their children on their backs, fled; while the men, seeing no chance of retreat, adjusting their arrows in their bows, and assuming an attitude of defence, took aim, and would have discharged them at our breasts, but for our guide calling out that we were friends, and had come to kill elephants, not men. In an instant their attitude was changed. They drew their arrows, and stood trembling before us. However, we knew of something possessing a greater charm in dispelling fear than any speeches and protestations we could have made, and, handing them some tobacco, under its magic influence they speedily became reconciled to us. The water here being very muddy, we enlisted some of the band to take our horses to a better water. Shouldering their spears, they marched off, calling and whistling for the horses and cattle to follow; but these animals exhibiting no symptoms of the intelligence and sagacity they seemed disposed to attribute to them, the Bushmen supposed that the fault lay with them, as not possessing any of the white man's medicine or charms. They gazed with wonder on ourselves, our wagons, and our horses, "things that had never been dreamt of in their philosophy," and had

* We find, at intervals throughout this desert, these long sandy ranges running parallel to each other, and reminding one of the heavy swell of the ocean. They are generally covered with larger trees or denser vegetation.

only heard related as marvellous legends unworthy of belief. Lighting on an old fellow who had held some intercourse with the Bamañwato in former times, and still retained a smattering of their language, we ascertained that they had at first mistaken man and horse for one being, a cannibal mounted on a god (Torra)!

Having recovered from their panic, a young girl approached our guide, and anointed and sprinkled him with a powder made from a red root, repeating some unintelligible words; this, we were informed, was a usual ceremony, which would act as a charm against Porrah, the evil spirit, doing him injury for having brought so great a surprise on his friends. This ceremony being ended, another girl brought a dish of pounded sweet berries for our guide to eat, and several for ourselves, and, this done, he had to relate the news, which he did, as is usual, in a sort of rhythm consisting of measured sentences, each containing a certain number of syllables, to which the listeners made one and the same antiphonal response. The news related was addressed to the father of the family only, and then the respective parties greeted each other by clapping hands all round. Bushmen do not exchange this greeting until the news has been told, so that it may be understood from the intelligence given whether the errand is peaceful and friendly. No one dare give any information in the absence of the chief or father of the clan, and Bushmen and other natives never expect it, knowing their customs. In my early travels I have frequently met with young Bushmen who, when asked questions, made me no reply than "I don't know." Being better acquainted with their customs, I have never, of later days, asked them for news, but have inquired for their father, to whom I first tell my own news as well as I can, and then get his story. Travellers unaccustomed to their ways are apt to become impatient and uncivil. The first version of what a Bushman or any native has to say can never be relied on; whatever you ask him about, he invariably says first, "I don't know," and then promises to tell you all he does know. Ask him for news, and he says, "No; we have got no news," and shortly afterwards he will tell you news of perhaps great interest. Ask him for a pinch of snuff, he always says he has got none, but presently produces it.

The preliminary ceremonies being over, the Bushmen indulge in a bout of smoking from a rude clay pipe, which being passed round, each inhales one mouthful. A fit of violent intoxication ensues, the stomach distends, the breast heaves, the eyes turn their whites to view, a quivering motion seizes the whole frame, and they fall back in terrible convulsions; kicking and writhing, their faces assume the most hideous contortions, and the foam issues from their mouths, while the more hardened of the party try to restore the senses of their fellows by squirting water from their mouths on their faces, and pulling at a tuft of hair in the crown of their heads. This is one of the most disgusting spectacles that can be witnessed. It occasionally happens that some of them die in these convulsions; others, on recovery, say they have been in an ecstacy of delight, and desire a repetition; and it is every young Bushman's greatest boast to have been drunk from tobacco. When smoking alone, they frequently fall into the fire, and are sometimes burnt to death. In the course of my rambles, I have seen hundreds who have been injured by fire, into which they have fallen during this state of delirium; and they are too lazy or thoughtless to take any precautions before they commence these dangerous orgies. The Bushmen generally obtain tobacco by sending a few jackal skins to Chapo, a distance of 150 miles, in barter for it. The first time I observed one of these people in this state, not knowing the cause, I turned to inquire from the others, but found they were all in the same state of stupor, one excepted, who looked particularly foolish, and smiled at my dismay, though his head was fast bobbing. Presently he rolled over amongst the rest. Appalled at the symptoms, I seized one of their tortoise-shells, ran for water, which I dashed unsparingly over them, and on their being restored found that this immoderate use of tobacco had caused them.

Having refreshed ourselves with a hearty breakfast, we started, accompanied by all the Bushmen, and soon falling in with a troop of fine elands, slew five, and also a fine male gemsbok or oryx, the flesh of which, hastily cut into strips and stored in the wagons, confirmed the attachment of our new friends, who stuck faithfully to us. As we bivouacked in the afternoon on a wide plain, dotted over with single trees and bushes, a description of country called "Bonk-Veld," or "Eiland-Veld," some bees were detected darting past to the eastward, and we lay down with our faces near the ground and looking westward, to mark the approach of the swarm above the horizon and discover the course of its flight. This we soon found directed towards a fallen log, about a quarter of a mile to the eastward. Thither we followed them, lying down at intervals of every fifty yards to watch other bees collecting, and be certain of our game; we were rewarded with two gallons of delicious honey, having well feasted on which we lay down happy again and well contented.

The Bushmen, also, having gorged to excess on the luscious prize, squatted round a large fire, drawing their knees close up to the chin, and cowering under their only garment, a piece of skin barely sufficient to cover their backs, and hanging on them like a sheet of paper; they did not sleep, however, for more than an hour at a time, waking at intervals to empty and replenish their pots, which they kept boiling all night. With these fellows, as well as other natives, no portion of an animal is lost. To secure the blood is their first care, and if they have not a vessel to catch up that which is spilt, they scrape it together in a hole in the sand and gravel.

The pith of the horns and hoofs are likewise eaten, and the hide is set aside for the last eating, or a portion of it converted into sandals.

On the 26th we travelled all day under the guidance of these Bushmen, and, having cleared a road with our axes, arrived safely at a number of rather large pools of water. We found numerous spoor of elephants, buffaloes, elands, and lions, and during the night a fine troop of bull-elephants drank at one of the pools; but as we were in the habit of resting on the Sabbath, we did not follow the spoor, and on the Monday none came. In consequence we pushed our way on horseback farther into the interior, our wagons following; but, owing to the hard work our men had to cut a road, they did not reach the spot where we were waiting for them, so we were obliged to return in the night to get a supper and a blanket. Some of our party had, during the day, discovered that the country was inhabited, and having "caught" two women and children, the men taking to flight regardless of the fate of their families, we dispatched them with some tobacco to their husbands.

Early on the 30th we sallied forth on the spoor of elephants, accompanied by a number of Bushmen, carrying sacks made out of pieces of elands' paunches, and full of water. We had determined, as a general rule, not to ride on horseback after cow-elephants, not considering them worth the risk of life, or the ruination of our horses; but as no sport had fallen to our lot for some time, we agreed to pursue this herd, and, if possible, to drive them about according to a plan we learnt from the natives, that, being winded and knocked-up, we might be able to *bag* easily the whole lot. About 3 o'clock we came up with a troop, fleeing about a mile before us, to which we gave chase, shouting and yelling at the utmost of our powers. Their speed soon slackened, and they commenced dashing the water with their trunks from their stomachs on to their backs. After half an hour's further chase their water became exhausted, and, coming to a stand, they sprinkled each other with dust instead, throwing it in clouds over each other's backs. Commencing the attack, the first shot brought the largest cow upon us, with trunk extended, and terrific trumpeting. "Old Booy," not being able to avoid the charge, owing to the density of the mañana bushes, would have fallen a victim to the elephant's wrath, but for a timely and well-directed shot from Piet Greyling, which laid her low. The rest of the herd now marched off in single file; and, riding up, I gave the next best of the troop a shot in the ribs, which sent her, shaking herself, in another direction for a hundred yards, when she fell dead. The rest now dispersed, and my companions after them, and selecting another young bull-elephant, about eight feet high, my first shot brought him down with a velocity which one would expect only from a much smaller animal; in an instant, however, he was up again, and two more mortal shots from my large rifle entered his body. These brought him to a stand under the shade of a tree; but, with an activity and perseverance scarcely to be expected, he chased me three times consecutively, but without trumpeting, as I was making up to him for another shot. At the fourth shot, shuffling back a pace or two as if to obtain an impetus, and charging with renewed vigour, the elephant fell dead at my mare's heels, with a crash that urged her to a sharp spurt, although, being accustomed to the habits of the animal, she generally needed no urging.

While seated on the carcase of my victim, waiting the return of my companions, one of them rode up in pursuit of a fine cow, and, dismounting in full view, one well-directed shot struck her in the region of her heart; still she ran for about a hundred yards, with trunk and tail erect, sweeping down in her career of death a tree of three or four feet in circumference, with apparently as little effort as I should use to pull up a twig of grass, and fell at last with a loud crash, uttering a piercing cry.

The Bushmen, who had been silent though excited observers, from the boughs of some tall trees, of our mode of warfare with these huge beasts, now, according to custom, threw handfuls of sand in the eyes of the dead elephants, muttering something that might be construed into thanks to the Good Spirit, and a propitiation for future favours. They then strove with emulation for the first plunge of their spears into the carcase, and, having opened a small part to see if the beast was fat, their snake-like eyes sparkling joyfully at the result, they proceeded to cut open and dismember the carcase in the manner already referred to—a process too unsavoury and disgusting to be further dwelt upon.

I think I have already taken occasion to remark that the elephant's head is considered the perquisite of the father of the family, as a token of his superior skill, and a mark of the great respect with which he is always treated. This being the choicest part, is devoured only by the elders, who persuade the young men into the belief that they will die, or otherwise come to grief, if they eat the forbidden parts, which are called *miela*. Traditions of this kind, and in which they place implicit belief, respecting certain parts of other animals besides the elephant, are also prevalent amongst all the tribes; not the young men of the tribe only, but their women, being also included in the prohibition, which, it may be observed, extends generally to the most delicate parts of the different animals. There are some on which chiefs and chieftains only are privileged to feed, and are forbidden the others under the assumed penalty, and with the denomination of *miela*.

Having counted the results of our day's work, and found them to be eight elephants, we returned at sunset in search of the wagons, not reaching them till after midnight, as is usual in all our wanderings through the dense forest under the guidance of Bushmen. Never having been very strong, the day's exhaustion and fatigue proved too much for me, and on arriving at the wagons I fell powerless from my horse. A cup of strong coffee, however, with a touch of cognac in it, proved a powerful restorative, and before going to bed I was again speculating what sport the morrow would bring.

The following morning, on our return from an unsuccessful search for spoors, we stumbled upon two Bushmen, who took to their heels. I gave chase, in order to catch and win them to the dreaded intercourse with white men, so as, if possible, to obtain information respecting the country ahead; but the trees broken down by elephants impeded my progress, and my horse, who had cleared two logs, fell with me over the third, so that we lost sight of them. In the afternoon, however, our friendly Bushmen detected some of their wild countrymen crawling up to our wagons and eagerly spying us. Being caught, they were with difficulty prevailed upon to come towards the wagons, when tobacco, beans, and flesh, very potent "medicines," being duly administered, they took courage and ventured to approach within fifty yards of our tent, starting ever and anon at our slightest movement. Never did I witness such a picture of wonder and amazement, of fear and surprise, as was exhibited in the strained eyes and open mouths, suspended breath and tiptoe attitude, of these wretched beings. They were clapping their hands in testimony of their wonder and admiration, when one of them, having ventured too near, with his eyes bent on the wagons, accidentally trod upon the *trek-touw*.* Upon dis-

* The trek-touw is a long rope which, fastened by a sort of splinter-bar to the wagon, is extended between the teams of oxen, the yokes of the several "spans" being attached to the rope. Thus the vehicle is "trekked" or "touwed," to borrow the word from its usual adaptation.—ED. [B.]

covering this, the poor fellow bounded off in a panic, and was seen no more. From the rest we ascertained that, having found the spoor of our wagons, they were at a loss to make out what animals they could be; and others, seeing our cattle, thought they must be game belonging to God (*Torra*), and were doubtful whether it would be safe to discharge their arrows at them. Having discussed amongst themselves all they saw, the wagons proved their chief puzzle; and we were asked through our old interpreter, with the most consummate gravity, whether the big wagons with the broad tires were not the mothers of the little ones having slender proportions.

On the 1st of August, having now an addition of several fine Bushmen to our party, we followed for four hours the trail of a large troop of elephants, chiefly cows, and had devised a scheme for destroying the whole, by surrounding them with fires, which we were told they would be afraid to face in making their escape. But our state of excitement and eagerness overruled this plan, and we sent a party of armed natives on foot into the bush to drive them out. After the shots had been fired, seeing or hearing nothing of the elephants, we fancied they had taken another course, and on dashing into the forest were met by the whole troop. They came on thundering and crashing down the trees in our very faces, and we were compelled to make a precipitate retreat, which was effected with difficulty, owing to the rankness of the underwood and briars thickly matted together. Emerging from these thickets into the plain, and racing on the flying spoors of a troop, Vilgoen espied one solitary bull making away at a great pace about a mile ahead, to which we gave chase. My companion, however, hearing numerous shots behind, left me to continue the pursuit, returning himself to the main body. After following it for three miles farther, the wood became so dense that

I could not penetrate it, except by the path the elephant made, and as in so doing I should have to run the risk of the sagacious brute turning on his footsteps, as is their habit, and perhaps crushing me before I had time to beat a retreat, I was forced to give up the chase.

Following now my own, or rather my horse's spoor, as is our custom in any emergency, to find my way back, I was rejoiced to see a troop of about ten elephant-cows making straight towards me at a distance, and at a pace that plainly showed they were heated and blown. While I halted a moment to rest myself and my horse, the animals continued approaching at a slower pace, discharging copious showers of water from their trunks over their backs and shoulders. Allowing them to come within 100 paces, I shouted, thinking to turn them from me. Owing, probably, to my stationary attitude, they did not appear to discover me, but on springing to my saddle the motion caught their attention, and they altered their course and increased their speed, flapping their huge ears.

Again I dismounted, and with a well-directed aim laid the foremost cow low by a single shot in the temple, slightly above the line of the eye with the ear. The rest, raising their trunks and tails aloft, charged at random in every direction, and then returning to their prostrate companion gently touched her up with their tusks and trunk, as if to stir her up and urge her to flight with them; and I freely confess that this kindly touch of nature in the sagacious animals caused me a momentary pang of remorse.

The troop, finding their prostrate fellow deaf to their entreaties, moved on, and I soon followed, but was repeatedly driven back by a worthless old hag of a cow, who, with the air of a vixen, would not allow me to come near the troop; and I found it was necessary to kill her first, if I wished to get at the rest. So, dashing towards them, she turned upon me just as I had dismounted, uttering a fearful cry. This was a desperate move; it sealed her fate, though mine seemed in greater jeopardy, as my horse, terrified at the elephants' thrilling cry, pulled the reins out of my hand and left me on the open plain, staring death in the face without any cover to flee to. At this critical moment I despaired of life, but presence of mind, together with an unusual firmness, were now vouchsafed to me. I felt that I had but one chance for life, and that I held in my hand. Now for courage and a steady shot. It was the courage of despair, and it was Providence that directed my aim. I awaited the furious animal's approach with my gun at my shoulder; but my hand shook so violently that I could take no sure aim, and I felt reluctant to pull the trigger. Still the enemy approached, with outstretched trunk; her loud trumpeting had ceased, but she uttered a series of short-fetched grunts, which sounded in my ears like exclamations of triumph at having her enemy in her power—a victim she would grasp in another moment with her powerful trunk, and crush to atoms with her ponderous feet. At this juncture she happened to lower her trunk from before her, and the slight movement leaving her forehead exposed, I instantly took advantage of it, and a bullet from my rifle crashed right into the centre of her skull, and she came down with overwhelming violence at the distance of seven paces from the spot where I was planted. But if before I had been sustained by Providence, and indeed I felt that something beyond my physical power had saved me, I now lost all my fortitude, and stood for a moment perfectly aghast, trembling, and most horribly bewildered. But now, again recovering myself, and inspired by the first law of nature to lose no time in retreating from a possible death-struggle with my prostrate foe, I ran to my horse, standing at the distance of 200 yards, and cocking his ears in amazement at the fray: I then reloaded, and began to speculate on the doubtful prudence of following the rest of the troop.

Ambition, however, to achieve a glorious success in this field of sport, and the spirit of emulation, which urged me on to rival my companions, who were mostly "old hands" at this work, would not allow me to suffer the rest of the elephants, exhausted as I perceived they nearly were, to escape unscathed without making a fresh attack on them, particularly after seeing them come to a stand; and firing this time at longer distances from my horse, after many shots, which brought my companions to the spot, I killed two and wounded two others, all cows. The rest, taking a fresh start, were soon shot down in a row; but on approaching the last, a young heifer, we were all ingloriously dispersed for the moment by a little calf only three feet high. The heifer to all appearance was dead, but I observed her eye wink as I stood before her, and a simultaneous exclamation from Vilgoen to "take care!" caused me to start back from the grasp of her trunk, as she raised herself and trumpeted. Then, passing my gun under her lifted trunk, I discharged its contents into her breast, and she fell instantaneously with a groan. Having shot eleven elephants, six of which by the laws of hunting I was entitled to claim, we made for the wagons; but I felt so exhausted with over-exertion, fatigue, and excitement, that my companions were obliged to leave me to recover myself under the shade of some trees.

CHAPTER V.

Turn back, Water failing—Travel North-east—Goroge's Post—A Bushman Dance—Attack of Fever—Return to Sekomi's—Native Politics—A fearful Tragedy averted—Proceed to Sechelli's Town—The Chief Sechelli: his People massacred by the Boers—Arrival at Kolobeng—The sack of Dr. Livingstone's House—Leave Kolobeng—Arrival in the Trans-Vaal—The Boers: their Position and Prospects.

In the early days of August, remaining at our last encampment, we continued our pursuit of game, with many fatiguing rides and but little success; but not being able to get any further intelligence regarding the country ahead, and the waters in our rear, which had supplied our wants thus far, now fast drying up, we were compelled to make our retreat while we could. Leaving, therefore, our hunting-field, and abandoning for the present the prosecution of our proposed journey to the northward, we returned on our path to the Makalaka village from which we had started on the expedition.

Although we had travelled over a considerable extent of unknown country towards Wankie's, we met with no fountains, streams, or rivers of any description. The whole country is a perfect waste of forest, interrupted occasionally with open plains, having little of interest but the game which abounded on them. All the waters we found were rain-waters, dispersed in pools all over the country, but often at long distances; and these have such barbarous Bushmen names, that I found it impossible to put them to paper.

Driving fast, and travelling all one day without any water, we reached Zontharra, the Makalaka village, on the 11th of

August, having shot a few buffaloes on the way. Here we brought the Makalaka guides to account for their scandalous conduct; and a scene of crimination and recrimination ensuing between them and some Bamanwato, it came to light that Goroge, a nephew of Sekomi, who had come up after us, was, by his underhand dealings, the chief cause of our misadventures.

Having received private information of the existence of a river called Shua, about ten days' journey east, which forms the boundary of Sekomi's and Moselikatze's country, I determined upon returning on our way as far as Goroge's Post, and making that our starting-point for exploring these regions far eastward in search for game, or, if that should be found impracticable, to follow up the course of the Botletlie river to Lake Ngami.

On the 12th, while the wagons took our old route back, I made a circuit on horseback, and discovered a trail of the preceding day of four large bull-elephants. This I followed for several miles, discovering where they drank, and during my ride shot three ostriches out of a large flock, which had just come from the water. I might have killed many more had I felt inclined; but, being young hens, their feathers were of no value. Ostriches, although the fleetest of all animals, may be easily run down with a good horse, when coming from the water on a warm day.

On the 14th the four elephants had again come to drink, as we expected. In some places and in some seasons these animals drink only every other night, and then go far to feed. In parts where they are much hunted, they remain two nights away from the water, and return every fourth. We fell in with them in the course of three hours; and Vilgoen, Piet Greyling, and I succeeded in killing each a patriarchal old bull, besides assisting five of our party in killing the fourth, on which they had already expended more than eighty bullets, with but little effect. On arriving at the spot, my first shot fortunately broke the hind leg of the beast, who at once came to a stand; a few more shots, and he fell apparently dead. Leaving my horse, in a few minutes I was mounting the elephant's back; but, feeling a convulsive motion within the beast, I slid from him, and as my feet touched the ground a blow from his powerful trunk had nearly crushed my head against his side. I discharged my gun at random in my hasty flight, stumbling over the twigs and briars, and displaying a wonderful agility. However, when away from his side, there was no danger, for, though rocking his body on three legs and trumpeting loudly, he could not move a single step forward. A volley of eleven more shots brought him down, this time quite dead. I resolved to be more careful in future; and it is worth mentioning, by way of caution to South African hunters, that most elephants recover their powers of doing mischief after appearing to be mortally wounded, unless they have uttered their death-groan.* I have known them to fall four or five times as if dying, and then rise again.

On the 14th we had great sport among the countless multitudes of springboks. In the course of half an hour I shot five, and all our people were in great spirits with their good day's sport, which at length became tame, and as for myself, but for the sake of the Bushmen I would have prohibited shooting these animals. During the night we were plagued with a lion prowling and roaring round our camp, while seeking to find an entrance into the kraal; and next morning we discovered that he had followed immediately on the spoor of one of our men, for a long way, the night before.

On the 17th we arrived at Goroge's Post, where we found two of his wives in charge of a number of Bushmen, over

* It is dangerous also to approach a buffalo until he has uttered his death-groan.

whom they were domineering in a scandalous manner during the chief's absence. I found much difficulty in procuring guides, to carry out my object of proceeding eastward; and as, besides, my Dutch companions had no particular desire to go any nearer to Moselikatze's territory, I proposed making direct for the lake. But as the fever season was expected soon to set in, that also proved a disagreeable proposal; and I felt rather perplexed as to my future course. The idea of returning with the tusks of only thirty-two elephants, after the preparations I had made, was not satisfactory, and I wished at least to see more of the country, if not to secure more game. The heat was every day becoming more intense, and our beards, now of several months' growth, on being combed, emitted, probably owing to some peculiar state of the atmosphere, electric sparks and sounds similar to those produced by stroking the back of a cat.

The next day I persuaded my companions to travel for some distance up the Botletlie; but our intention becoming known to Goroge's wives, they prevailed upon us to wait till their husband, who was out to the north, employed in gathering tribute, returned to them. They informed us that they had heard him express a desire to show us a splendid field for elephants. Such agreeable intelligence induced us to remain here some days longer; little thinking, in our ignorance of the native character, that this was but a ruse designed to give an opportunity of extorting from us what they could during our stay, and to get us to shoot game for their own use. But these flattering hopes were never to be realised; though day after day we were out following the game, and, besides numbers of the smaller kinds, we shot a brace of fine eland bulls and several giraffes. I was under the necessity of speaking rather sharply to my Dutch friends, in censure of a cruel practice common with the Boers of leaving animals they had wounded to die in agonies, and be the prey of the vultures, perhaps before their time, because they were too lazy to kill the poor animals and cover their carcases with branches.

I found it advisable to keep my people actively employed in sport, fearing that if not so occupied they might become dissatisfied, and, their minds dwelling upon the tales they heard about savages and fever besetting our path, might combine to force me to abandon my plans and turn the faces of the party homewards.

In one of these hunting expeditions we fell in with a large pack of wild dogs, as many as forty or fifty, making terrible havoc among the springboks. These dogs did not betray the least symptom of fear for us; on the contrary, when we rode between the pack and the bucks, they turned to bay, and prepared to make an attack upon ourselves and our horses: a shot or two, however, sent them barking in all directions. During some seasons these animals are said to be dangerous when thus congregated. We saw also a vulture, or eagle, which the Dutch call lamb-fanger (lamb-catcher), seize a rather large kid of a springbok with its talons, and soar away with it into the air. The same day, on our arrival at the wagons, we were invited to see a Bushman dance.

About sixty women, decorated with all the beads and trinkets at their command, and formed into a semicircle, were clapping their hands and singing to the time and tune of a quick waltz; while about thirty young men, adorned with plumes of black ostrich and eagle feathers on their heads, a few ancient beads on their bodies, a fan (the tail of a gnu) in their hands, and the fruit of the moana tied to the calves of their legs, bore them company. The shells of the moana fruit, being perforated with holes, and the pulp sifted out, made a rattling noise as the men kept time to the monotonous music performed by the women, in front of whom the men were tramping and marching to and fro, grunting and singing, and describing all sorts of uncouth antics, gestures, and gymnastic evolutions and contortions, with occasionally a graceful movement, the result of accident rather than design. Now and then a wild peal burst from the chief dancers, and the females, bowing low in acknowledgment of the compliment, responded with thrilling yells. The principal dancer, who seemed to command general admiration, commenced like a reasonable being, but ended like a man in a frenzy, gesticulating with hands and feet while he lay on the ground with his face to the sun. At these contortions, made by severe muscular exertion, he laboured for a long time, enveloped in a cloud of dust which, continually settling on his body, had, when the exhibition ended, left a pretty considerable plaster of clay on his skin. These savage gestures and capers over, the company turned their attention to some dishes of bulbs (*lunches*), and the insect deposits of the mopani trees, which they greedily devoured.

Some days having passed in this manner, while we were still waiting the return of the chief Goroge we happened to request his wives and followers to procure us some salt at a neighbouring pan. This was soon forthcoming, but the demand in payment was so exorbitant that we declined the bargain. In truth, we could not afford to give an equal measure of beads for the same quantity of salt. These people were most disagreeably importunate in begging for flesh (although we had killed more than a hundred animals for them), and for all manner of things, beads, tobacco, &c.; and finding us not disposed to yield to their demands, they poured volleys of abuse on our heads. We were under the necessity of driving them off with our dogs, of which, owing to their superiority in size and bark over their own dogs, they have a great dread, as much almost as if they were lions. At the same time, these wretches, as well as the Bushmen, are so ravenous, that if they saw a chance of doing it with impunity they would rob our dogs even of a bone. The dogs kept by the natives are all a very diminutive mongrel species; very weak in giving tongue, and with little spirit. This is not to be wondered at, considering that they get no food but what they steal, and are always in wretched condition. Like their masters, these brutes always lay near the fire, or in the ashes, and are generally disfigured by large burnt patches on their bodies.

On the 21st we shot about twenty head of game, which we gave to the poor Bushmen, as more deserving of our favours than the Bamañwato. But they were not left long in the possession of their abundance—their mistresses, the chief's wives, soon confiscating the whole, besides two young lions which the Bushmen had killed during the day. To judge by the clothing of these Bushmen, which consists chiefly of lions' skins, they must kill a great number of these animals in the course of a year. In some parts, however, the natives do not kill either lions, wolves, or wild dogs, regarding these animals as hunters of game, which they turn to their own account. Indeed, instead of killing, I have known them to release a wild dog which had fallen into one of their pitfalls. For these destructive animals, as well as for vultures, whose whirling flight often helps them to find the carcase of a dead animal, they have a great respect.

At length, on the 27th of August, Goroge, otherwise Mogorooi, who has the superintendence of the district of the Madenisana, as other subordinate chieftains, or khosis, have over other districts, arrived, and the object of our visit was made known to him; but in answer to all our importunities, accompanied by promises of a handsome reward if he would help us on our way, we got no answer but that he was "a child," a native way of expressing incompetency, and for reply to our further questionings, "his mouth had no word." He averred that there were no elephants nor any water eastward, but plenty of the tsetse fly. We had good reason to believe that in this refusal Goroge was influenced by Sekomi's desire to prevent all communication with Moselikatze by the white men, and he feared we should soon discover his town after reaching the Shua river.

On the 29th, having spent three days in these rather humiliating negotiations without any satisfactory result, we determined on finding our way eastward with one wagon only, in spite of the fly, and the absence of guides, Goroge having carefully ordered all the Bushmen out of our way. Seeing we had inyoked and were prepared to start, he came over to say that he would join us if we waited till the Monday; and although he gave us no reason for this delay, and would not promise that we should meet with elephants, we consented, and having shot a number of quaggas to feast the natives, they danced and smoked and sang without intermission.

A hundred or more Bushmen now arrived carrying large bundles of elephant and other flesh, which Goroge had confiscated, with some sixty fine tusks of ivory, also tiger (panther), jackal, and other skins, properly prepared and softened. In preparing skins for clothing, if it be a thick hide it is first scraped with a sharp little adze, till it becomes very thin and pliable; then, having wetted it, they scrape it, and rasp it with a rough stone, or a bone, and afterwards rub it dexterously between their fists till it is dry and soft. The skins are coloured with some kind of moistened bark.

On the 31st of August some of our companions, the hangers-on from Matebe, fearful of approaching Moselikatze's territory, parted company and proceeded towards the lake; while Vilgoen, Greyling, De Waal, and myself, steered eastwards with one wagon and our own servants, leaving my friend Wirsing in charge of the remaining servants and the other wagons. Travelling in a north-easterly direction, parallel with the Ntwetwe salt-pan, which we left a few miles south, we passed three or four fine spring-ponds swarming with waterfowl. We shot several quaggas, gnus, and springboks, which, in the course of the day, crossed our path in countless droves, or stood watching our strange cavalcade trespassing on their hitherto unexplored domains.

Thus we continued travelling for four days, passing every day one or two spring-ponds, in limestone beds, and always bivouacking near one at night.* We generally laid wait in turns by the sides of these ponds to kill rhinoceroses, covered by a small circular wall of stones about a foot high, constructed on the bank. Here we lay with slight shelter from the night frosts, while the lazy Bamañwato were stretched on beds of soft grass round their fires, with their feet towards it, and stones or wooden logs for their pillows. On our way hither we shot some seven or eight giraffes, and a number of gnus, quaggas, springboks, &c.

News now came from all quarters that a commando of Moselikatze had destroyed a whole town only six miles beyond our present post, and that, having heard we were hunting in these parts, they were in pursuit of us, and would fall upon us the next day. On receiving this intelligence one of my Dutch companions, saddling two of the best horses, fled by himself back to Thumtha, where he alarmed my friend Wirsing with the report that we had been, or must be, murdered.

Our Bamañwato pretended that it was incumbent on us to send out spies in all directions, and that it would not be safe for us to hunt, as the report of our guns might draw upon us a body of men whose numbers they magnified to thousands. But the equivocations and conflicting statements

* This chain of ponds, lying parallel with the Ntwetwe salt-pan, seems to be formed by springs fed by some subterranean stream. They are called by the natives Matlomaganyani.

of these people had given me some insight into their character, and being convinced that these manœuvres were all intended to get us back, I accused them of the deceit, and threatened to go on without them. My companions, however, who had lost many of their friends by Kaffir treachery, felt not so confident as myself, and as they seemed inclined to believe that danger existed, I offered no opposition to their desire to return to the Trans-Vaal.

While this was pending, a party of Makalakas, who, coming from the eastward, met us here one evening, reported that they were flying from the supposed commando which they had seen the day before. Finding, however, that these men returned home during the night, laden with flesh, and having observed, during their conversation with the Bamañwato round their fires, gleams of triumph, as I fancied, light up the countenances of our pretended friends, as if they exulted in the success of some project, I felt satisfied that all these accounts of threatened disaster were mere fabrications invented for the purpose of thwarting our plans. It can scarcely be credited that so much trouble would be taken to carry out their schemes, but it is well it should be understood that craft and falsehood are characteristics of most of these tribes, and I learnt subsequently in the course of my travels, from these same Makalakas, that my suspicions were well founded.

But now something occurred which, independently of any influence from these proceedings, compelled us to return without any delay. We, as well as the natives, were all, with one exception, suddenly prostrated by violent attacks of rheumatic fever; and I therefore hastened to rejoin my friend Wirsing at Thumtha, or Goroge's Post, as it is also called. This station we reached in a few days, when we found that, although we had been so far separated, the party there had also been attacked by the same disorder, and on the same day. We learnt afterwards that our friends from Matebe, who had got as far as Chapo's, were likewise attacked at the same time. To add to our troubles, on arriving at Thumtha we found that our oxen and horses were suffering from sore eyes to such a degree that some of them were already blind. This disorder was caused by a small moth in the weeds near the spring. We found salt water very effective in restoring the sight of the rest, and after staying at Goroge's Post for a few days we bent our way homeward.

From Koobye we took a different route from that followed on the outward journey; thereby cutting off an angle, so that in five hours we crossed the Botletlie river, farther to the eastward than before, and in a direct line for Lotlokani. In six hours more, travelling for a considerable distance on the margin of a wide lake covered with a saline efflorescence, we reached Kabatie's Post, so named from a native who was professedly the mate or friend (kala)* of Vilgoen. Here we found the natives unaccountably uncivil, robbing us of the water which we had spent the whole night in procuring by digging. In the morning we despatched a party on horseback to procure salt from some neighbouring pans, on which the incrustation was for several miles nearly two inches thick, and very fine it proved. Meanwhile we set to work again digging for water, of which we stood very much in need, but were again robbed of it by the natives, who exhibited a spirit of hostility, united with insolence, of the most provoking description.

* This is a very curious connection. Every native has a dear friend, his "kala," in every tribe with which he may have sojourned when travelling, and this friend is allowed to make free with anything belonging to the other, such as killing a sheep or ox for food; he is even welcome to the use of the house and one of the wives of the other. The bond between such friends is generally reciprocated. In Vilgoen's case, however, the privilege was not exercised further than the Dutchman's appropriating a tusk of ivory or a kaross, and the other something of about equal value, as a fine dog or cow.

Our men, returning with the salt, brought with them, packed in their trousers, which they had stripped off and converted into saddle-bags, a quantity of fresh ostrich eggs, of which they had robbed a nest. We left this fountain in disgust, and steered for Lotlokani. Shortly after starting several giraffes were shot in sight of the wagons, and, travelling all night across the country, we found ourselves next morning in the main road still used by the natives. Here our cattle drank for the first time since the 11th of September.

On the 17th we were delighted to be again on the march, and, travelling all day without water, in the evening we passed Mañana, where also the water had long dried up. It was, therefore, necessary to push on, and before daybreak we passed a wagon with a number of natives crowding round it, but we went on without stopping. Soon after sunrise we fell in with a Hottentot carrying a small keg, and trudging over his ankles in the deep white sand on his way in search of water, about 30 miles ahead. This man informed us that the wagon we had passed belonged to his master, a Mr. Simpson, who was returning from Sebetoane's country, where his cattle had been bitten by the fly, and had since all died, leaving his wagons a wreck in the wilderness, 30 miles from water on either side. I went back to offer any assistance in my power, when I found that, to add to his distress, one of his servants had run off with a flock of thirty sheep and goats.

Pressing forward as fast as we could through the heavy sands, we passed another man of Simpson's on the road, where he had been digging in vain three days for water. After travelling all day under a scorching sun, and exerting myself at night with "Booy" to shoot some fine eland bulls, however, I found myself again knocked up. The next day we were able to reach Inkaoani, and, the oxen having drunk, I sent old Joachim, with a span of the best, to fetch Simpson's wagon. Three days afterwards he returned with the oxen in a miserable plight to get fresh ones, having succeeded in bringing the wagon within a few miles; this being accomplished, I despatched some men to shoot elands for Simpson and his servants, as he informed me he was a very bad sportsman, and had no horse.

On the 21st of September we left Inkaoani, accompanied by Simpson, to whom I had promised the further loan of oxen to bring him on as far as Tlabala, a distance of 50 miles. This place we reached in three days, not having found any water during the journey. As Simpson had not a single ox that could pull, I now made a span of twelve or fourteen for him, by taking three or four of the largest out of each of my teams, and letting him have them at a moderate rate of payment, in order to get him on; though I could scarcely hope they would accomplish much after the severe work they had lately done.

After resting our oxen we proceeded to Lutlochoé, a distance of five hours. I shot two giraffes on the way. At Lutlochoé some milk was brought to us in a lekuka—a leathern bag, with a small rude tap at the bottom to strain off the whey. The milk was of the consistency of new cheese, and, though rich, was sour and rancid.

On the 28th we made for Sekomi's, and "Old Booy" fell in with a woman, who informed him that the Boers had attacked and routed Sechelli, who had sent an express to Sekomi to say he must kill all the Boers that pass him, especially Vilgoen and all that are with him. This intelligence startled us, but we did not know whether to believe it or not. Travelling till 12 o'clock by a fine moonlight, we were under the necessity of keeping guard all night round our horses, firing at the hyenas, which were so daring as to come within a few yards of us, keeping company with us, and striving every now and then to attack the snorting horses in spite of our firing.* After a short rest we again proceeded through the rugged passes of the Bamañwato hills, and outspanned at daylight; but shortly after sunrise we were obliged to press on to get to water, and having met two of Sekomi's nearest relations mounted on horses, we inquired about the news we had heard, and were informed that there was no truth in it.

On reaching Sekomi's, Vilgoen, "Booy," and I visited the chief, who received us very kindly, and, after complimenting us upon the success of our expedition, proceeded frankly to relate all the circumstances tending to confirm the news we had heard from the woman on the road, strongly expressing his indignation at the proceedings of the Boers, who attacked Sechelli for no other reason, that he could see, than to rob the tribe of their wives, children, cattle, and guns, and scatter and destroy them in such a manner as to compel them to seek a subsistence by offering their services as labourers amongst the Boers.

During our conversation some large earthen pots, varnished and painted on the outside, were set down before us, full of beer (*bujaaloa*) made of fermented Kaffir corn, and furnished with ladles, for which a species of calabash formed a convenient shape. The parties who made the beer came with them, according to custom, and drank a ladle full of each, to show there was no poison in it.

Sekomi then went on with the details of the war, informing Vilgoen that he himself was now also at war with the Boers—that Sechelli was his relative, and that they would jointly resent the aggression. Then turning to Vilgoen, he said, "You, Vilgoen, are a dead man at this moment, and if I kill you it will not be I that killed you, but your own country-

---

* Wolves or hyenas in the neighbourhood of towns are exceedingly daring; they prowl about and feast almost nightly upon the corpses that are carted away, and frequently carry off old men or women, and many children, and get so accustomed to human flesh that they are considered dangerous at all times.

men, who knew when they were making the attack upon our relatives that you and many others were amongst us and in our power."

Vilgoen, having great confidence in the friendship of Sekomi and his principal men, replied firmly that he could not exonerate Sechelli and blame his own countrymen until he heard their version of the story, and that, to judge from what he knew of Sechelli's insulting conduct towards the Boers at various times, he fancied his countrymen must have had serious cause for making reprisals; adding, however, that he felt his life was now in Sekomi's hands.

The chief, after further dilating on Boer aggressions, turned to Vilgoen, and addressed him to the following effect, sentence for sentence being repeated to him by "Old Booy":—"You have ever been kind to me and my people; your life is spared; and although, if I mistake not, had you been at home you would have joined your countrymen in this unjust war, and after you get home you will, in all probability, come back and kill me, that is nothing. Go, and carry my defiance to your countrymen. I know I have but one year to live, and will prepare myself to die—but to die the death of a warrior. Go tell those who left you to be killed, that he who should have done the deed has been your preserver. Sleep well this night, and as the day dawns I shall supply you with a faithful guide. Make for the Limpopo; from thence cross the Mariqua, and proceed cautiously along the southern banks homewards. Sechelli's men are waiting outside to see you killed, and expect to take back the tidings. They have come here to urge me to do it, but I will not stain my hands with the blood of a friend."*

On returning to our wagons, hearing that an Englishman had outspanned a few hundred yards off, I hastened, in com-

---

* Vilgoen was the only Boer that Sekomi tolerated, and the friendship he entertained for him was a warm one.

pany with Wirsing, to pay him a visit. From him we received intelligence that, had we been half an hour earlier on our way, we should have been massacred to a man that same morning, by a party who had been lurking several days in the Poort for that purpose. We further learnt that meetings had been held daily to decide upon the propriety of the measure; and that, although Sekomi was always opposed to it, in consequence of his friend Vilgoen being one of the party, as well as two Englishmen, Wirsing and myself, he, Sekomi, had been greatly blamed by his warriors and councillors for his leaning towards white men. It was argued that, as a Boer, Vilgoen would naturally sympathise with his countrymen; and as any one living with Boers and coming through their country must be friends of theirs, and as they had been so cruel in their war upon Sechelli, the least they, the Bamañwato, could do was to take the revenge now in their power, and kill us all. Thus pressed, a reluctant consent had been wrung from Sekomi; but two or three days passing without our arrival, he happily, on further reflection, altered his mind, and, disregarding the invectives of his councillors, despatched a countermand of the fatal order, which reached the party lying in wait for us just about half an hour before our appearance, and much to their disappointment.

Next morning Vilgoen, having given over his wagon and oxen to "Booy's" charge, took leave of us, in company with the other Dutch companions of our journey, whose lives were spared through Sekomi's friendship for Vilgoen, on his intercession. Vilgoen also strongly interceded for Steyn and Potgieter's party, who were still behind, being spared; but Sekomi remained firm to his purpose, declaring that the Boers could not be suffered to escape.

Having gathered from various sources a favourable view of the character of the chief Sechèlli, I determined, on the advice of Reeder, the Englishman we met here, to steer direct for his town, rather than risk my life and property by exhibiting any fear of him, or giving him occasion to think we were his enemies. This display of confidence may be considered unwarrantable, considering that he had sent instructions to his friend and relative, Sekomi, to put us to death. We might, indeed, instead of taking this bold step, have mounted our horses and fled with Vilgoen, but in so doing I should have been under the necessity of abandoning my wagons, oxen, and ivory, a sacrifice too great for me to make. Rather than that I resolved to run all risks, and go boldly to Sechelli's Town as soon as Sekomi sanctioned our departure, although his advice was against us taking this step, and he said that Sechelli would certainly kill me for a Boer, as I had lived among that people.

Meanwhile the Bamañwato were in daily expectation of the arrival of the other Boers from their expedition to the lake, and delighted at the prospect of having them in their power. Thus excited, they continually practised firing at targets, expending daily no small amount of powder and shot. They drank beer to intoxication at their daily meetings in the khotla, cutting all manner of heathenish capers, said to be the rehearsal of the slaughter of the Boers. A great concourse of the natives assembled at these meetings, with their faces and limbs painted most hideously in fantastic streaks of red, black, and white. Their heads and bodies were decorated with black ostrich plumes, and strips of the skin of that bird, some with eagles' feathers, their legs and arms encircled with the bushy tails of oxen; others were dressed in karosses of cats' tails, about five feet long, hung round their necks, through which at times their naked arms and legs appeared, the whole imparting an almost diabolical aspect to their swarthy forms. Occasionally some renowned warrior rushed out from among the dense masses, and bounded and capered about the circus, to the great delight of the females, who stood in the background, cheering him with deafening yells. These subsiding, he would proceed to speak. At such meetings the natives are generally suffered to say anything they like in approval or disapproval of the chief's conduct, no matter how derogatory; all his follies and neglects are descanted on with most unsparing severity, the speaker stopping occasionally to take breath and receive the applause and approbation of his admiring audience. His eloquence is further employed in lauding himself and his warlike deeds, calling upon the listeners to observe the notches on his knob-stick, denoting the ten or fifteen men he has killed in battle, and exciting them to endeavour to achieve the like glory. Then bounding about and describing the most wild and unaccountable figures, he retires, amidst the yelling of the women, into the crowd, and another takes his place.

Those who have distinguished themselves in war, having something to talk about, are generally the most eager to be heard, and all the abusive epithets they can command are heaped upon their less distinguished fellows, chiefly the young men. The latter, on their part, being cut to the core by these studied speeches, the strife often results in their immediate application for permission to attack some tribe or other. Sometimes they are driven by these taunts to seek a quarrel with any unfortunate wretch they meet in the veld, and kill him in cold blood, merely to escape the disgrace of being pointed at or *bienaed*, immortalized in some song of derision, instead of glorified, as those who can boast of having killed a man generally are. The young men, therefore, are always thirsting and eager for war, and the old men, having had enough of fighting, and aware of the danger, are generally slow in encouraging deeds of violence. But they seldom inquire into the justice of the case, and mercy is a thing unknown to the native mind, as the more they slay the more honour they obtain. Had it not been for the caution, prudence, or cowardice of the old men of the tribe, who probably feared terrible retribution either from the Boers or the British Government, we should all have been massacred—to have killed white men enhancing the honour tenfold.

Several meetings were held, and we were extremely anxious to learn their decision as to the fate of the expected Boers, which at every council was pronounced in the words—they must die! I had pleaded for them to the utmost of my ability without effect; but about 11 o'clock one night Sekomi sent for me, much to my surprise, and, taking me by the hand, led me, accompanied by my interpreter, through a number of winding passages, and, having dismissed his guide, we entered a small barn alone, where he squatted, and desired me to do the same.

Having given me to understand the decision of his councillors, he said, to my no small embarrassment, that he had sent for me to ask my advice respecting the measure, adding that his people were unanimous in their resolution to kill the Boers, as no such chance of revenge would offer again. I was startled at this communication, not knowing whether I might attribute his seeking my advice to his being really disposed to be guided by it, or, what seemed more probable, that, with native cunning, it was merely a stratagem to discover how far we sympathised with the Boers. I told him I was incompetent to advise a great chief like himself, the more so as I had never in my own country interfered or engaged in public affairs of such importance; but that I would take the opportunity of informing him that it would be revolting to the ideas of Englishmen to cut off a party of unsuspecting people, who had not taken part in the hostile commando, but were at present hunting confidently and peacefully within his own dominions. In our country such an act would gain neither approbation nor glory, for we should consider it a womanly act, unworthy of a really brave man; and that as we should greatly despise a chief for consenting to what we should consider cowardly conduct, such things are never heard of among us. Knowing that Sekomi entertained a very high opinion of British valour, and that he was an admirer of our ways and customs, I deemed it the most advisable course to endeavour to rouse him to the display of a magnanimity which, though seldom exercised, is still greatly admired by these heathens. Having also impressed upon him, as a matter of policy, that there was great risk of his incurring the forfeiture of the sympathy and good-feeling which his acknowledged friends the English had for him, he thanked me, and bade me good evening.

Next morning, hearing cries of "Maboers! Maboers!" (Boers! Boers!), and learning, from the excitement prevalent, that the Boers were near, I again despatched old Saul, a civilized Bechuana from Kuruman, to endeavour to dissuade Sekomi from his murderous undertaking, for I began to be apprehensive that I might have yet to witness a bloody massacre on their arrival. But a message came to me from the chief, saying: "Fear not; I have heard your mouth, and, although I have been advised by many to kill them, as they are the worst of the Boers belonging to Enslin's party, who have done great injuries to the black tribes, and deserve death by our law, and although our kindred have been murdered by their friends at home, still I will take your advice, and not be the first aggressor. I shall, nevertheless, let the Boers know of my displeasure, and, being determined to have no friendly intercourse with them, I shall warn them to keep beyond the limits of my boundary on pain of death."

I had argued with Sekomi that by killing the Boers, and saving us, he might become unconsciously the author of our destruction, for that the friends of the Boers at home would conclude that we had in some way or other connived at their destruction, or have suspicions of that kind, and our lives and property would not be very safe amongst them again; and I even went so far as to observe that, if he intended killing the Boers, he might as well begin with killing us, as we could not think of entering the Boer country under such a cloud. This, combined with my former conversation, operated well, and Sekomi particularly expressed a wish to see me again and form a lasting friendship with me—the wish, perhaps, originating in the representations of "Old Booy" and others that I was very rich.

Another plan had suggested itself to my mind in the event of everything else failing, which, however, I was determined to keep a profound secret, until all other efforts to save the lives of these poor fellows should prove unavailing. My plan, which I trusted would tend to the advantage of both parties, depending only on the conduct of the Boers at home for its happy result, was that Sekomi should hold the party on the expedition as prisoners, and, treating them well, should demand, as a ransom for their lives, the restoration to their mothers of the 200 or more children captured in the attack upon Sechelli. But I was glad to have succeeded so well without resorting to this step, as, even if it proved successful, the affair would have entailed on me various anxieties which I would rather have avoided.

In the afternoon the Boers arrived in company with Mr. Simpson, all alike ignorant of the fate that had awaited them. The trader, Mr. Reeder, lost no time in informing Mr. Simpson of the fact that the Boers were to be murdered, as yet knowing no better, and he advised Simpson to separate himself from their company without delay. This fearful intelligence was, of course, instantly communicated to the Boers, who hurried to me, trembling with alarm. I received them cordially, and after offering them dinner and coffee, of which they would not or could not, in their present state of mind, partake, I informed them that they need be under no apprehensions, and communicated to them the result of my conferences with Sekomi. I told them again and again that they need fear nothing; but their panic was not to be allayed by any such assurances, and they started up to look for their horses and fly at once. The horses, however, had been sent to the water, a distance of more than a mile, and could not be expected for an hour or two. They sat down again, trembling violently, and their terror exceeding all bounds, it was observed by the natives, who, with most malicious looks, were sitting on their haunches in a circle round us, peeping out of the corners of their eyes, and gloating at the spectacle; insomuch that I felt under the necessity of exhorting the Boers to a more manly behaviour in the presence of the natives. But nothing could persuade them they were not to be killed, and I thought that one of the party, Hendrik Prinsloe, would have died on the spot.

On their return to their wagons the natives crowded round in great numbers, the Boers very freely distributing gunpowder and bullets among them. But the natives, who would have admired a display of courage and firmness, began now to entertain a very contemptuous feeling towards them, and bandied among themselves most humiliating remarks upon the cowardice they betrayed—the dread which they had hitherto felt of the Boers being converted into ridicule and scorn. And now the natives, taking advantage of the utter prostration of the party, began to steal without disguise, and soon appropriated all the bullets, lead, and clothes they could lay their hands on. Hastening to the spot, I caught one of the ruffians in the act of stealing a bag of bullets, and took it from him, throwing the bag back into the wagon, and thereupon sent a message to Sekomi to beg him to order his people off. One of the Boers (De Beer), with a black patriarchal beard, ought to be mentioned with exceptional honour for his firmness; but both his good example and the encouragement he gave to his fellows were of no avail.

Fancying that the Boers entertained ungrounded suspicions against us, I requested them, if they feared any danger, to remain in my tent, assuring them that in case they were attacked we would fight and die together; and as they seemed to place no confidence in the promises of Sekomi, I proposed a plan to them by which we were to start in company with our wagons, and, making a long circuit eastward, go home together. De Beer acquiesced gratefully in my proposal; but Prinsloe was determined to fly at once, and, like King Richard, would have given all he possessed for "a horse." A few shots being now fired in the direction of the water, they made sure that the member of their party left in charge of the horses had been killed. The natives meanwhile having withdrawn from their wagons now congregated round mine, till the tumult became somewhat alarming, and had they not known my influence with their chief I certainly should have met with serious annoyances.

The Boers now freely parted with any article of dress that was begged of them; while the natives, evidently vexed and disappointed at Sekomi's altered views, and frustrated in their hopes of being allowed to massacre the Boers, vied with each other in giving them offence, in order to create some occasion for a quarrel, which might lead to something more serious than words. But just then a startling announcement was made to the Boers which shook their inmost souls. Sekomi had sent to invite them to pay him a visit, and exchange greetings with him. This reminding them of the treacherous murder of Retief by the Zulus, they thought all was over, and insisted upon bidding us adieu for ever. I endeavoured to persuade them that their fears were groundless; but all to no purpose. Shortly after we followed them to the khotla to watch the tide of events, through the high palisades, where also hundreds of natives were assembled in expectation of a scene. The Boers, seated near the chief, and surrounded by about 300 of his councillors and warriors, cast quick and restless glances in every direction, in order to ascertain whether any hidden spears or battle-axes could be detected under the karosses of the people. As to the Boers, the chief, with that wonderful command of temper and passion which only savages possess, received them so courteously as only to increase their fears, considering that this, as indeed had often been the case, was only a mask or a prelude to the barbarities which were to follow. According to their own account, they had made up their minds that they were to die. Sekomi inquired, in a very friendly manner, after their luck in sport, their health, and so on; and expatiated in such a merry and agreeable manner on elephant-shooting, that I did not know what to make of it myself, and how it would all end, and returned to the wagons, anxiously awaiting the result of the interview.

Shortly afterwards, the Boers returned in safety from the khotla, but their horses not having yet made their appearance, their alarm increased, though I did my best to reassure them. My friend Wirsing, very justly perhaps, blamed me for the great confidence I placed in the natives, and advised me not to persuade the Boers to remain against their will, as something might yet occur to make them change their minds, and I might have to repent of it. Acting on this piece of advice, I no longer attempted to prevail on them to stay, and, at their earnest request to sell them three horses, I even offered to lend them three to carry them home, at whatever risk of Sekomi's displeasure, should he become acquainted with the transaction. They were very grateful, and, overwhelming me with thanks, offered me the whole of their ivory as a reward for the part I had taken. This, of course,

I declined; but as they were in my debt, I offered to accept of a certain quantity in payment, but they were too frightened to bring it over, and so the matter ended. A few Bibles and hymn books and a broken gun were left in my charge. At dusk they came to fetch the horses I had promised them, and, while my boys were gone to bring them in, they returned to saddle their own. My horses soon arriving, I sent to inform the Boers of it; but what was my surprise at finding they were gone, decamping without bidding us adieu, and leaving their wagons, oxen, and property on the plain, at the mercy of the natives. For this most strange and unaccountable conduct I have been unable to find any sufficient cause.

Next morning, the flight of the Boers becoming known, the chief and a great concourse of his people assembled round the wagons, squatting over large pots of beer, while the ivory and other articles of the Boers were being handed out; and having emptied the beer, they returned, carrying home all the ivory, and drew the wagons up to the chief's house. In the meantime Sekomi's brother and commander-in-chief, Chukuru (rhinoceros), came over to my wagon, and inquired whether I knew that the Boers were going to fly; and on replying in the affirmative, he wished to know why I did not kill them, or let the chief know of their intentions. To this I replied,* "Did I come here to kill people? Didn't I come here to kill elephants?" after which the great warrior pretended the utmost cordiality and friendship, and shook hands with me; but the blood ran cold through my veins when it occurred to me that this same fellow had used all his influence with Sekomi to put both Wirsing and myself to death along with the Boers of our own party.

The natives, disappointed by the flight of their expected victims, became hourly more uncivil towards us, crowding round our fires, till old Joachim, losing his temper, would

* The natives always answer a question by asking another.

have struck down one of the great men if I had not fortunately arrested his arm in the act. A serious quarrel ensued, which however ended upon the approach of the chief, who, having begged the blankets off my bed, and various other necessaries, returned to his town.

The following day we got permission to leave, and being visited by the eldest son of Sekomi, a lad of about sixteen, of very prepossessing appearance, I dressed him out in a suit of my own, and presented him with a nice little pony, to the delight of the natives generally. I thus established myself, quite unintentionally, in the position of *mate*, or bosom friend, to Sekomi's heir, and a change came over the whole black race, whose demeanour was instantly converted from their late insolence to the most cringing obsequiousness; and I was known only as "Chapiman Kala," or "Khama," Chapman the friend of Khama (Hartebeest), which name I have retained ever since in these regions. My "mate's" father, finding me so good-natured, asked whether he might not some day expect a wagon and a few horses from me, and he at once set to work, though without success, to discover the thief who had taken a bar of 68 lbs. of lead and some bags of powder from my wagons the night before; and a man, who was found breaking the lock of one of my guns, was at once stripped in the khotla of his clothing, being all his worldly possessions, the most disgraceful punishment that can be inflicted.

On the 15th of October we were delighted to be under way, steering for Sechelli's Town, which, after several days' march through heavy sands and dense forests, in parts well stocked with game, we reached on the 28th. Wirsing and I proceeded to Sechelli's residence on horseback, riding forward the last stage through rugged glens and among rocky hills, never venturing to move faster than a walk. We found the chief at his residence, perched on a hillock composed of blocks of sandstone, loosely piled upon each other, a fit abode for baboons only.

Sechelli, chief of the Bakwains, a tribe mustering about 500 men, stands about 5 ft. 10 in. high, has a pleasing countenance, and is rather stout. He was dressed in moleskin trousers, a duffel jacket, a wide-awake hat, and military boots. In address and behaviour Sechelli is a perfect gentleman. He can read and write, having learnt within the last few years, and is an accepted member of the Kuruman church. He was instructed by Dr. Livingstone, who lived with him for four or five years. Sechelli is said to be very quick at learning, and anxious to substitute more civilized customs among his tribe in the place of their own heathenish practices. He is also said to be good-natured and generous. He presented us with a fat ox for slaughter, a custom prevailing among all the tribes that can afford it.

Sechelli at once pronounced us to be Englishmen; and having corroborated the intelligence we had already heard from Sekomi respecting his disasters, he apologised for not being able to receive us as he would like; but he entertained us with roast beef, sweet and sour milk, served in clean dishes, and with silver spoons, also with sweet earth-nuts; and while we were doing justice to his hospitality, a man stood fanning away the flies with a bunch of white ostrich feathers. His loss, he informed us, was sixty-eight men killed of his own tribe, besides a number of women, and between 200 and 300 children carried away captives. He lost, also, about 1500 head of cattle, and several thousand sheep and goats. For his cattle he seemed not to care so much, although his people were starving. He hoped to be able to replace them by the profits of huntings for ivory; but his people felt sorely the loss of their children. Ninety wagon-loads of corn had been carried off by the Boers, and the rest they had burnt in his town. Besides his own property, they had carried off several wagons, oxen, and other property belonging to English gentlemen at that time travelling to the lake.

From Sechelli we learnt that the war originated with Masellcelie, chief of the Batkatla tribe at Mabotoa, who had often been promised by the Boers that if he supplied them with a number of servants he would be exempted from further demands; but on giving one supply after another, still more was demanded, in spite of the promises made him. At length he refused, and became surly, thinking probably, with many others of the natives, that the late fever had so diminished the numbers of the Boers that he could successfully resist their authority. The Batkatla chief having ascertained, however, that the Boers intended to punish him, and being an arrant coward,* fled to Sechelli for protection, it being a custom amongst those races that when one tribe flies to another and solicits protection it must be given; so that on the Boers demanding that Masellcelie should be delivered up, Sechelli refused, saying he "could not do it unless he was to cut open his own bowels and let them fall out."

Most of the people of Sechelli's tribe were out during the day grubbing for roots, their only food at present. Famine, "the meagre fiend," that "blows mildew from between the

* The Rev. Mr. Edwards, who has been labouring for some years among this tribe, has not much to say in favour of this chief. An anecdote which he told me of him may be interesting:—One day, in the midst of a dreadful thunder-storm, the chief sent a messenger to Mr. Edwards, begging that he would not be offended because a cow of Mr. E.'s, as well as four or five of his own, had been killed by lightning of his manufacture. But he assured Mr. E. that he did not mean to injure him; he was only making rain for the gardens. Mr. E. smiled at this; but shortly afterwards another messenger arrived in great haste to say that Mr. E.'s cow had risen again, on which Mr. E. observed, "Because I did not believe in your charms my cow has been spared; but because you have believed in them, and attributed to yourself a power that belongs to God, you are punished."

shrivelled lips," had already made great havoc among them. Several mothers had followed the Boers home, and, hiding themselves during the day, endeavoured at night to steal away their children; a few only had succeeded and returned.

On the 1st of November we obtained a guide from Sechelli to conduct us to the main road, our wagons having been brought since our own arrival up to his town. We accordingly departed, and at night overtook some emaciated Bakwains, roasting the roots they had gathered during the day. I ate one of these roots, but I thought I should have died from the effects it produced, creating a lather like soap, and blistering the inside of my mouth in a few minutes. I drank water to cure it, but that only aggravated the symptoms. The pain I suffered was at last allayed by putting some fat into my mouth.

Next day we travelled still south, and reached Kolobeng in the forenoon. This is the site of the town where Dr. Livingstone lived with the tribe.* His house had been pillaged, and presented a melancholy picture of wanton destruction. The Boers had taken away everything that was valuable to them in the shape of furniture, utensils, and implements, and destroyed some hundreds of volumes of Sechuana Testaments, and other religious works and tracts, the leaves of which still lay scattered for nearly a mile in every direction. Even the window and door frames had been taken out, and the floor was strewed with bottles of valuable medicines, the use of which the Boers did not understand. The town where Sechelli was attacked, and which was burnt to the ground, a few miles from Kolobeng, presented a melancholy scene of desolation, bestrewn with the unburied carcases and bleaching bones of the natives who fell.

In the afternoon we started from Kolobeng, and, traversing

* See an interesting account of Sechelli and his people in Livingstone's "Missionary Travels," pp. 14 *et seq.*—ED.

the side of a very high mountain, the next day pushed on again. Having given away the greater part of Sechelli's presents to his own famishing subjects, we had suffered all the preceding day from want of food, and this day could kill nothing. I rode in advance, firing at and wounding tsèsèbies and quaggas, but with no better success. Arriving before the wagons at a fountain near Aasvoegel Krantz, I knee-haltered, or tethered, my horse, and lay down to wait their arrival. By the fountain, and about some hundred yards from where I lay, stood a solitary but dense bush of the waght-en-beetje (wait-a-bit) kind, and about ten minutes after, to my great surprise, about thirty savages, armed with spears, shields, and knobkerries, emerged from the bush. They came on, eyeing me with great suspicion, and inquiring whence I came; but on learning that I came from Sechelli's, they seemed satisfied; and from what I could understand they had heard of our being on the way from "Old Booy," whose wagon had preceded us a couple of days. Otherwise these fellows would probably have put an end to my travels, having been placed here by the chief for the purpose of cutting off spies or straggling parties of Boers.

The next day Wirsing and I left the wagons to follow, and proceeded on horseback to Mr. Edwards's station at Mabotoa, where, arriving in the afternoon, we were hospitably entertained with the first food we had eaten for three days, except a few gwarrie berries, which had made us sick. Having heard the news, and been shown all the damage done to Mr. Edwards's property by the Boer commando, we passed on to Mr. Inglis's station at Matebe, where we were also kindly received, and reached Vilgoen's place on the 5th of November, greatly to the astonishment of the Boers, who had long given us up as dead, and who overwhelmed us with inquiries about the movements of the natives, of whose attacks they were under great apprehension—wondering by what miracle we, having ventured to go to Sechelli's, had come away unhurt.

My stay among the Boers was limited to the time necessary to the winding up of the commercial part of my transactions with them, on the completion of which, after a final interview with Pretorius (then in chief authority amongst them)—my wagons being already across the Vaal river—I saddled my horses and bade a final adieu to the Trans-Vaal country.

Having conveyed incidentally the results of some of my observations on the habits and manners of the Boers, I leave the reader to review in his own mind, as well as he is able to do from these inadequate notes, their past and present condition, and the degenerating tendency their daily course is taking. This arises partly from their nomadic life, their natural taste for that wild and roving state of existence in preference to any other, and their frequent collision and hostilities with the natives, for whom their hatred and contempt seems to have become quite inherent, and has been and is exaggerated in proportion as the tribes meet with sympathy from the Christian world.

It must I think be admitted, if such a man as Judge Watermeyer is an authority, that the first trekking of the Boers was not an act forced upon them by the British government, but the spontaneous inclination of a people fond of a roving life, and wishing to escape the control of law—a feeling which was already in operation during the time of their own Dutch government, and from the very beginning of the colony, and that in those days already inroads were made upon the territory of Hottentots and Kaffirs.

According to Judge Cloete, it will further be seen that the great impetus given to the trekking of the Dutch farmers commenced seriously after the merited condemnation to the gallows, by British authority, of some of their countrymen for acts of a criminal nature. These and other acts admit of no palliation. However much we may regret the necessity of such extreme measures, nobody will dare to dispute the legality of the punishments inflicted, which were only similar to many that have at various times occurred in England and Ireland, and which would very likely have been dealt more stringently upon rebellious Englishmen. The act of hanging the Boers on "Slagter's Nek" was a very painful necessity, but at the same time will never reflect dishonour on the British name, although it may excite pity for the sufferers, and their unfortunate relations who saw the sad end of their foolish and misguided countrymen.

It will probably not be denied that, on the part of the many Boers who have left the colony since the British administration, there were some *real* causes of complaint amongst many fancied grievances. Through a change of ministry in England, or a change of governors in the colony, it has happened that faith has not always been kept with some of the colonists, British as well as Boers, but it has been the peculiar failing of succeeding governors, as well as secretaries of state for the colonies, to upset, upon first coming into office, all the good work their predecessors have done.

Cases of unwarranted aggression on the part of Kaffirs towards the colonists have also transpired, such as may well wring the heart of many an honest farmer—English as well as Dutch. But even in these matters, when we come to look back into the first causes of things, and to review the many injustices that have been done to the native tribes of South Africa in the beginning, and take into consideration the unforgiving and unforgetting disposition (the great characteristic) of a savage, we cannot wonder at his persevering hatred towards the white man, however much we may pity the innocent, and, as it were, modern colonist, made the scapegoat for the sins of a former generation, and suffering grievously in the loss of their property and their friends at the hands of a savage and ruthless but, it has been asserted, in many respects still a noble enemy. The native races, it must in justice be allowed, have had cause for their hatred to the white man, and, as I have before remarked, a Kaffir may forget an injury for a time, but he never forgives. It is a national characteristic, and until their nationality is utterly destroyed this prominent feature will never be eradicated.

Vacillating as the government of this colony has ever been, there was one remarkable feature which had hitherto always pervaded British rule, and that was, the earnest solicitude it evinced for the welfare of the native tribes, as well as the desire it had shown to regain their confidence, and the inclination to redress their wrongs; but in one short year a blow was struck that has proved, as it will continue to prove, an injury to British subjects, a disgrace to a civilized and Christian government, and the bane to the British name in South Africa. A high commissioner, inexperienced in the affairs of the colony, in the person of Sir George R. Clerk, was entrusted to break up Her Majesty's government in the Sovereignty (now Free State)—a task which he is said to have performed with a tact and precision that did him great credit. This was in truth the greatest act of cruelty and injustice, not alone towards natives, but towards Europeans, that any government could have been guilty of, and was virtually equivalent to the discarding a great number of Her Majesty's faithful and loyal subjects! These men strongly remonstrated against the act, but too late, and their voices were too feeble. They indeed sent to England two delegates to represent their grievances, in the persons of Dr. Frazer and the Rev. A. Murray, jun., but it has never transpired satisfactorily in what the labours of these gentlemen resulted. It may be asserted that the orders Sir George R. Clerk received were imperative; but if he had represented the feelings of the people to the home government, and remonstrated with them, he would have done humanity a service, saved a jewel in the British crown, and gained honour for himself by advocating the retention of the Sovereignty. By leading people to believe that he was, if anything, in favour of retention, he misled them, so that they had no chance of remonstrating until the act was done; and this piece of injustice, unprecedented in the history of civilized nations, has not even been equalled by the Trans-Vaal Boers. From such humiliating injustice and pusillanimous behaviour on the part of a government to its own people —an act perpetrated by a government which, above all others, makes pretension to be the most humane and just towards its subjects, the protector of the oppressed, and the defender of the whole world against slavery—we must recoil in horror and hide our faces with shame. It is indeed a pity that the fair fame of our beloved sovereign should be compromised by a transaction which, to those who regard it from a distance, may seem of so trivial a nature, and should thereby receive in this colony an enduring blemish, brought about, we believe, by the misrule or misrepresentations of the individual or individuals employed to carry out this cruel and unwise policy.

Having alluded to the great change of feeling on the part of all the native tribes in South Africa—unfavourable to the English—I may mention that it is principally resulting from the injustice of the treaties made with the Boers, by which the natives are debarred from getting gunpowder and firearms (to them the *necessaries* of life as well as to the Boers), while every provision is made for the Boer getting as much as he requires. Native tribes within and out the colony, who are known only by their fidelity to Britain in her time of need, and were taught by us to fight against those very Boers who had rebelled against their queen and given so much trouble, are now spurned for their loyalty, and not alone spurned and discarded most ungraciously, but oppressed in the most arbitrary and ungrateful manner that ever a people were.

As soon as the tenor of the Sand River Treaty became known to the natives, and the subsequent abandonment of the Sovereignty, the news spread like wildfire from tribe to tribe, not alone along the whole border of British and Boer dominions, but even into the interior to the Bamañwato, Lake Ngami, and even to the Zambesi river, and meetings were held between the Griqua, Basutos, Barolong, Batlapies, Bakwain, Bahurutsi, Bamañwato, Baselika, and many other tribes: treaties were made between them for mutual protection and defence against the white man—Englishmen in particular—who seemed now to prove themselves the true oppressors of the black. By these treaties they are still bound to each other, nearly from one side of the continent to the other, and if any Kaffir war had unfortunately broken out after this crisis, woeful would have been the consequences to the northern parts of this colony, and the Trans-Vaal and Free State in particular. As it was, many travellers were handled roughly, and, being questioned about the strange and dishonest policy of their government, were overwhelmed with shame. Much is due to the missionaries of Kuruman and other parts for the part they took, not in palliating or exculpating the policy of government (which they could not in justice do), but in seeking to alleviate the pain it inflicted, by expressing a conviction, prompted by their own hopes, that these proceedings would be disapproved of at home: they were, however, grievously disappointed. Time has, indeed, softened the feeling of the natives to a considerable extent, and much praise is due to the more enlightened and liberal rule of the present government, which has in many cases, by its acts, given the natives reason to hope for better days: and but for this some sad tragedy would have darkened the pages of modern South African history.

To return, however, to the Boers of the Trans-Vaal, on whose probable future, their prospects of an improved state of civilization, and the chances of a prosperous and useful career, I may perhaps be expected to say something. This I shall briefly enter upon, but with great diffidence, as it is a subject that ought much rather to come from one skilled in politics and prophecy than myself.

In the present state of society in the Trans-Vaal, it seems to me that as there are many ways (although sometimes uncertain) of easily attaining a livelihood and a competency, emigrants will find their way thither from the colony; but the treatment hitherto has been such as to exclude enlightened foreign immigration, and rather to encourage the introduction of disaffected Dutch farmers from the colony, many of whom have recently sold their farms there, and, with their flocks and servants, trekked into what they consider a free country (because exempt from British rule, which gives equal rights to white and black—an equalization which they cannot conceive to be just), where they can obtain a title to a 6000 acre farm for 2s. 6d. Beyond this, they give only encouragement to British deserters from Bloemfontein, or from Natal, because these are often useful as bricklayers, carpenters, or schoolmasters, and are compelled by their peculiar position to work for almost any terms offered to them, which generally consist of little more than their wretched food, a hut to sleep in, and concealment from their pursuers, if any should be in search of them. Hollanders were at one time in great favour with them; but since the defalcations of one or two, who imposed seriously upon the credulity of the Boers by representing themselves emissaries from the King of Holland, &c., they are utterly at a discount.

I have often thought that as European immigrants, and especially missionaries to the native tribes, have met with so much opposition from the Boers in these quarters, it is a pity that missionary societies, instead of quarrelling with the Boers in their endeavours to instruct the natives, have not rather turned their attention and their means to the advancement and instruction of the Boers themselves, who at the time I speak of had not one permanent clergyman amongst them, and generally no better schoolmaster than a British deserter fresh from the ranks to teach their progeny the *Dutch* language. In other countries missionaries have become willing martyrs to the Christian cause; and if the missionary societies had but taken half the pains to instruct the Boers first, and spent as much money in edifying and enlightening them during the last twenty years as they have done for the natives, they would now have reaped by this time a glorious harvest, not alone from the white populations they had reclaimed from a retrograding state, but in the good fruit that that glorious harvest would again have yielded. The Boers of the Trans-Vaal then would not only have by this time been perfectly civilized themselves, but many of them might have gone forth pioneers of civilization, carrying the Gospel into the dark regions, and have proved a blessing to South Africa, instead of carrying destruction and desolation through the land. Real Christianity infused into the bosom of the Boers would have done away with that feeling of superiority with which the white man is prone to look down upon the black races, their contempt for whom would thus be eradicated, and, as a necessary result, slavery in every form would be abolished. Until this is done, it appears to me that no real substantial improvement will take place. They must be taught to believe first. At present ninety out of every hundred will not believe that the soul of a black man is esteemed by the Creator equal to a white man's, and a feeling of this kind was manifested only a few days ago, even within the borders of this colony, when some white men declined sitting on a jury with men of colour; but the judge, though a descendant of a Dutch colonist himself, to his honour be it said, reproved them in a very laudable and becoming manner.

In the present unsettled state of the Trans-Vaal country, missionary influence will do great good if exerted first to promote the refinement of the Boers themselves, and soften down their peculiar prejudices. In this manner they will gain the confidence of the natives, and break that chain of alliance which, since the giving up of the Sovereignty, has been formed amongst the native tribes on their immediate borders. Their *real* position needs but, I am sure, to be made known to their fellow-countrymen and relations in the Cape (amongst whom we have so many noble examples of the great and powerful advantages of education), and they will, I feel confident, use their utmost exertions and influence amongst the many members of the Dutch Reformed Church, to whom they are bound by the ties of faith and relationship, to aid in promoting the education of the Trans-Vaal youths, by sending thither the necessary means, ministers, and teachers. In so fruitful a region, every description of farmers and mechanics can do well, and such people would find their way into a peaceable country, but that in the present state of things they are never certain of what they possess, for various reasons, which will have been gleaned throughout the perusal of this journal. Missionaries to preach to the Boers, and to instruct their children, might make them in a few years all useful and intelligent members of society, and thus we might, at no distant period, hope to see these very men, now so narrow and selfish in their ideas, confer benefits upon the natives by teaching and instructing them, and sending missionaries from among themselves (as some of the native tribes have long done) into the dark regions beyond. In the hope that this may yet be their destiny, we may look forward trustfully to happy days for our expatriated fellow-colonists.

CHAPTER VI.

Another Expedition—Outfit—Start from the Trans-Vaal—Rhinoster-Kop—Junction of the Valsch and Vaal Rivers—Koranna Villages—Mahura's Town—Alarms of an Invasion of the Boers—Motieto—Mr. Moffat's Missionary Station—Arrival at Kuruman—Threatening Aspect of Affairs.

On the 6th of March, 1853, I left the Trans-Vaal, with the intention of proceeding to Bloemfontein, as the nearest British town,* to lay in a supply for another and a longer trip into the interior to hunt for elephants. I had purchased a stud of six good salted (acclimatised) horses, with which I flattered myself I might hunt at any season, and had made arrangements with a Mr. Thompson, who had left on a trading expedition down the Vaal river, to meet him at Kuruman on my return from Bloemfontein, and engaged the services of his wagon and servants for the whole term of my expedition.

Crossing the circuitous and rapid ford of the Vaal river, which here runs 500 yards wide at Coqui's drift, we next day travelled among hills covered with rank golden-coloured grass, richly studded with fine mimosa trees, giving the country round a grand and pleasing appearance, especially when the early rays of the rising sun shed a glorious lustre on the distant landscape. We saw many ostriches, several pauws (the great bustard), and a few hartebeests.

Next day we proceeded on our journey towards Bloem-

* The Orange River territory, within which Bloemfontein is situated, was, it will be remembered, at the time under the rule of the colonial government.—Ed.

fontein, and in four hours reached the drift of the Rhinoceros river, which had been long pointed out by a streak of willow and mimosa trees, running nearly parallel with the road. The banks abound in members of the feathered tribe, aquatic and others, the most remarkable being muscovies, geese, ducks, guinea-fowl, and pheasants. I started a number of hares and steinboks, and we saw large herds of quaggas ascending the sides of the distant hills, and innumerable gnus, blesboks, and springboks, frisking and prancing on the plains, or describing strange figures and circles as they move with wonderful agility, creating clouds of dust wherever they went. The Rhinoceros river, which is periodical, flows into the Vaal, and abounds in good edible fish of different kinds.

Having crossed the river, we followed the wagon road about south across the vast plains of the gnus (*Wildebeest vlakte*). The grass of these plains had been devoured by the multitudes of game, so that the country presented to view a dreary wilderness. In the evening we encamped in full view of a rugged and bushy conical hill called Rhinoster-Kop, it having been the favourite haunt of that animal in the days when they abounded here, some twelve years before.

A few hours on the morning of the 9th brought us beyond Rhinoster-Kop, where we found water in a valley or leegte. Here we met with a sad confirmation of the intelligence we had received respecting the loss of cattle, famished for want of pasture in the prevailing drought. On the outspan lay the carcases of several oxen, on which innumerable vultures were regaling, and the skeletons of several others that had recently died. In this state of affairs I determined to give up the prosecution of my journey to Bloemfontein, and, striking to the westward, make the best of my way to the Kuruman, where I hoped to procure some of the necessaries I meant to have laid in at Bloemfontein. The country we now passed is well adapted for all kinds of cattle. Sheep, in particular, must thrive well, to judge by the large flocks I have at different times seen on all the Valsch river and Rhinoster river farms.

Crossing the Valsch river, we fell in with a Koranna village, composed of portable tents or huts made of mats spread over a framework of canes. Thus a whole village can be dismantled in a few minutes, and in the course of an hour reconstructed in another spot distant some three or four miles, as is frequently done. The inhabitants seem to have a mixture of Hottentot, Bushman, and Kaffir blood; their features being more akin to the Hottentot—flat noses, wide mouths and thick lips, high and protruding cheek-bones, and small narrow eyes. The attire of these people, whether the native garb, or consisting of all kinds of cast-off European clothes, was as mixed as their blood, and ridiculously fantastic. I purchased of them a couple of goats for a table-knife each, and fell in with a Boer who was trading largely for cows, for each of which he gave a roll of about 6 lbs. of Boer tobacco, an article they prize above all the luxuries this world can afford, and which they literally imbibe with their mother's milk, for it is not unfrequent to see a child of three or four years old still suckling, and at the same time already an accomplished smoker.

On the 12th we encamped at the junction of the Valsch and Vaal rivers, and next day travelled down the Vaal, which is here very pretty, the banks being lined with drooping willows. We slept on the 14th at a cattle station in charge of a tribe called Baqueen or Bakwain, which is subject to the Korannas. These Bakwains resemble the Bechuanas, and speak the Sechuana language—some of them the Zulu also. Immense herds of cattle, numbering perhaps 1000 in each, were under their care.

We reached next day the Koranna village of Oolson Lynx,

the brother of Geert Lynx, the chief of the Koranna nation. The inhabitants begged several articles of me, but I found great difficulty in understanding them, not being aware of their inability to pronounce a word ending in a vowel without affixing a consonant; thus, when they asked for tea, they said *tip*, and for coffee, *coffip*.

Having once more crossed the Vaal river, at a ford called Bamboe's spruit, west of the Makwassiis-bergen, we travelled down its northern banks, and reached the farm of P. Bizindenhout, where we spent two or three days, and I made the valuable acquisition of a horse and some oxen.

On the 21st I started for Mahura's Town, and on my way fell in with my acquaintance, Mr. Thompson, of Natal, who had been at Mahura's and Gasibone's, trading for sheep, karosses, and feathers, a lucrative business when understood, in which many are engaged. During the last three or four days' journey, the country appeared exceedingly well adapted for the growth of wool; we passed large flocks of sheep every day. Next day, after travelling for two hours, we passed a native village under one Mantlanyani, where a native teacher had established himself. A little farther (two hours and a half) we came to the Berlin Missionary Station of the Rev. Mr. Schmidt, who preached in Dutch to some Korannas through an interpreter, and is of opinion that, to introduce the Gospel effectively amongst them, their own language must be totally eradicated, and the Dutch, or some other, substituted. Proceeding, in the afternoon, towards Mahura's Town on horseback, leaving my wagon to follow, I reached the place late that night, and met with a very inhospitable reception. I found that there was quite as much ill-feeling amongst Mahura's people towards the English as ever existed towards the Boers. It seems to have originated in the Sand River Treaty between the British government and the latter, the natives generally considering that the effect of that treaty

KORANNA PACK-OXEN.

was to hand them over to the tender mercies of their oppressors, the Boers. Until now, the English government and people have been held in the highest honour, and respected, I may say almost worshipped, by all these tribes, as the only protectors of the black man; and I cannot but regret that a people so docile as these Mahuras, who had already so wonderfully improved as to adopt European attire, and to abandon most of their barbarous and superstitious practices, should experience such a check in the midst of their course of advancement. Should they now be driven by the aggressions of their powerful and dreaded neighbours, the Boers, to become freebooters, it can only be attributed to this mistaken policy.

In features, Mahura resembles more the Koranna than Bechuana. He is good-natured, and by using gentle and civil means is found tractable enough, and after some explanations his behaviour to me became sufficiently courteous. Although his brother, Gasibone, is paramount chief, Mahura's authority seems to predominate, and his followers are the more numerous. He is not a Christian by profession, but his brother is. His town contains about 2000 or more huts of the usual style, with a population of some 10,000 or 12,000 souls. The town stands on the banks of the Hart river, on a slope, the summit of which is surrounded by a natural bulwark of rocks. His people possess abundance of milk, many cattle, and large gardens. They have lived for many years in peace with their white neighbours; and up to this time never gave offence in any way to the Boers of the east or south, nor to any Englishman, as far as I know. Their principal employments seem to be the chase and trade; many of them have bullock wagons worth £100 each, and great numbers are well clothed, and nearly all have fire-arms. The wealthier people frequently make long journeys to hunt elephants; the ivory thus obtained being a great source of profit to them, but at the same time of great jealousy to the Boers engaged in the same business. This tribe of Bechuanas is called Bathlapie (fishes). Most of them can read and write, and thousands of them, I believe, profess Christianity. This great change has been effected since the commencement of Mr. Moffat's labours among them. They are on very friendly terms with their neighbours, the Korannas and the Griquas, particularly the latter, with whom the foundation of an amalgamation seems already to have been laid by some important intermarriages.

The natives were labouring under an impression that the Boers were coming to attack them with a large commando, and were making daily preparations for such an event, holding meetings and councils of war in the Khotla, at which there was generally a gathering of several hundreds in their war dress and paintings, and armed with guns, spears, battle-axes, bows and poisoned arrows, clubs or kerries, and shields. Every now and then one of their renowned warriors jumped into the centre to address the mob, and was vehemently cheered or hissed by the women, who stood outside under parasols of black ostrich feathers, as he either flattered and encouraged the mob, or abused the Boers. The speaker dwelt much upon a declaration they had received from the Boers, that the British government had given them (the Boers) all authority over the land and tribes north of the Vaal and Orange rivers. Another grievance was that, while the Boers could procure as much powder as they chose, the natives were not allowed to purchase a grain in the colony, though they depended upon the use of fire-arms in the chase for subsistence as well as the Boers. They could not see why they were thus unfairly treated, no provocation having ever been given on their part. The oration gave rise to various discussions as to what was to be done. They complained indignantly of the Boers' assumption of authority over them. "The Boers," they said, "were refugees who had fled into their country and that of neighbouring tribes from the British government, after they were driven from beyond Natal by Sir Harry Smith, and were still making daily encroachment, and robbing them of all available farms containing fountains." They sympathised much with their neighbours, the Bakwains, under the forcible abduction of their children, seeming to expect the same treatment for themselves; and they remarked that the Boers had set them an example of killing women and children, an atrocity of which the natives were never guilty in their wars.

I was aware that the Boers intended going round to all the native tribes for the purpose of laying down the law to them, but I explained to Mahura that it was not intended to make war on them. The chief, however, came to the conclusion that it was best to send a notice to the Boers that, if their intentions were friendly, they had better not come with a party of 500 men, as a bloody war would be the consequence.

Leaving Mahura's on the 3rd of April, I made for Kuruman, by way of Motieto. The mission station there is charmingly situated on the slope of a hill, facing the south. A clear fountain, oozing out between the church and dwelling-house, flows into a fine fruit and vegetable garden; and the buildings are nearly buried in the dense foliage of a grove of gigantic Babylonian willows, planted by the former missionaries, Lemoen and others, whose graves are to be seen a few yards off. Some hundreds of native huts are scattered here and there, and the inhabitants, who were neatly clad, have large gardens in the valley beneath. The missionaries, who were expecting the invasion of the Boers, had removed all their most valuable property to the south.

Next day I rode over to Kuruman, where I found my friend, Mr. Thompson, who afterwards travelled in company with us. Here I was introduced to the worthy missionaries

Messrs. Moffat and Ashton, and their families, the memory of whose uniform kindness I shall ever cherish. Milk, new bread, and fresh butter, we were never in want of while near these good people, and of grapes, apples, peaches, and all other products of the garden, there was never any lack at our wagons. Everyone is struck with the beauty of Kuruman, although the site cannot boast of any natural charms. All we see is the result of well-directed labour. A street of about a quarter of a mile in length is lined on one side by the missionary gardens, enclosed with substantial walls, and teeming with fruit and vegetables of every description. A row of spreading willows are nourished by a fine watercourse, pouring a copious stream at their roots for nearly a mile, and beyond the gardens flows to the eastward the river Kuruman, between tall reeds, with flights of waterfowl splashing on its surface. The river issues a few miles south from a grotto said to be 100 yards long, and very spacious, the habitation of innumerable bats, owls, and serpents of a large size. Stalactites of various shapes and figures are to be found in this grotto. I have seen some beautiful specimens adorning mantelpieces. One party discovered in the roof of this grotto portions of a human skeleton perfectly petrified, and a part of which was broken off.

On the opposite side of the street, and facing the row of gardens, the willows, and the stream, is a spacious chapel, calculated to hold more than 500 people. It is built of stone, with a missionary dwelling-house on either side of it, and a trader's dwelling-house and store at the western end. All these, as well as the smaller but neat dwellings of the Bechuanas, built in the European style, and in good taste, have shady seringa trees planted in the front. At the back of the missionary premises there are store and school rooms, workshops, &c., with a smithy in front. Behind the chapel is a printing office, in which native compositors were setting type for the new editions of Mr. Moffat's bible. Thousands of Sechuana books have been as well printed and as neatly bound in this establishment, under the superintendence of Mr. Ashton, as they could be in England. The natives here are the most enlightened and civilized I have seen, the greater portion wearing clothes, and being able to read and write. It was pleasant on Sunday to see them neatly and cleanly clad going to church three times a day. In their tillage they are also making rapid progress, and, having adopted European practices, instead of the hoe they use the plough.

We were detained at Kuruman, or Latakoo, as it is still sometimes called, for about three weeks, owing to the delay of the wagons of the trader Hume, who had a store in the village from which we expected to obtain supplies in the shape of provisions, such as coffee, sugar, tea, &c.

Even here, in spite of the influence exerted by the worthy missionaries in our behalf, an opposition, instigated by the elephant-hunters, in league with the Griquas, was raised to the prosecution of our journey. A public meeting was held, at which it was resolved that, inasmuch as the British government and the English people had proved themselves oppressors of the black man, we should not be allowed to pass through their country to the elephant veld, but should be permitted to exchange our ammunition, &c., at Kuruman or Griqua Town for ivory, which article they alone would maintain the privilege of collecting for the future. This resolution was authoritatively announced to us by the field-cornet of Kuruman; but, disregarding it, I inspanned my wagons next day, informing the field-cornet that as the chief Mahura had given me permission to pass through his territory, though I was not to take any of his men with me, I should act upon it, and defied all interference. So, discharging those of Mahura's people I had in my employ, I proceeded with my two wagons, accompanied by Mr. Thompson. Mr. Campbell joined us, and we were followed by two discharged soldiers who had been plundered by the Griquas. The natives looked on in silent wonder at our temerity, but made no attempt to obstruct our progress. Our cattle and horses had enjoyed long rest, and, being quite frisky at starting, ran rampant in every direction, and gave our people great trouble to keep them together.

CHAPTER VII.

Leave Kuruman—Threatened Attack—The Bawankitze—Trial by Ordeal—At Sechelli's and Sekomi's again—Massacre of Boer-hunters—Proceed towards Lake Ngami—Chapo's Marsh—Perils of the Desert—Water fails—The Passage effected—The Chenamba Hills—Hunting Adventures—Native Cure for Madness.

On the evening of the 3rd of May we reached Motieto, a few miles beyond which we were overtaken by a party of twelve mounted Bathlapians armed to the teeth, who, emerging from the bush, and forming a circle around our wagons, ordered us to stop and turn back; but finding we were prepared, and determined to resist any attempt of the kind, they, after a parley, thought it prudent to draw off, and we trekked on without further interruption.

Starting for Chooi, a salt-pan which we reached in about six hours, having overtaken and passed the wagon of a trader from Kuruman, and losing our road, we had to steer by the stars, and did not arrive at Clayton's vley till 8 o'clock next morning. In the afternoon we got on to the Baralong pits, where we expected to be interrupted by the people of that name, but the rapidity of our movements deprived them of the opportunity of hostile encounter. Next day, after travelling for about ten hours, we reached Lohageñ, a cave on the Maritsani river, in a most romantic and lovely spot, with a pretty variety of rocky and wooded scenery. The grass grew green and rank, the water was excellent, and the cattle revelled in the change. Here we shot a springbok, wounded a pallah, and found a large bee-hive in a cave; but the bees defied with their stings all our attempts to rob them of their honey. Shortly after our arrival we discovered a party of about twelve Bakalihari, who, armed with broad-bladed spears, were creeping towards our wagons over the stream under cover of the bush. Observing that their movements were watched with guns in hand, they turned and fled. With some difficulty we prevailed upon them to stop, shouting that we were English and friends. At length some others came bringing berries and roots for barter, and we found that at first we were mistaken for Boer-hunters, to whom they intended to show no mercy.

This being the district in which it was generally supposed the Boers would try to intercept hunters, it was necessary to be always on the alert to escape molestation. Crossing on the way the Molopo river, we pushed on for several days, outspanning at one small pan or vley after another, experiencing in general rather a scarcity of water, and seldom hunting. Shortly before reaching Kolobeng, we passed the country occupied by Santhoohie, chief of the Bawankitze—the remnant of the great Makaloa spoken of by Moffat. It is said that among the Bawankitze tribe there is a curious way of testing the guilt or innocence of an accused person, by their doctor (or ngaka) preparing a pot of boiling fat, in which all those charged with the crime have to put their hands, and it is asserted that only the real culprit will be scalded.*

On the 20th of May, after a long day's travel, we reached

* This singular mode of trial strongly corresponds to similar forms of the *ordeal* imposed among European nations in early times as tests of innocence. Ordeal, under some form or other, appears to be an institution widely established in South Africa. Dr. Livingstone found it in Angola on the west, as well as through the whole of the country north of the Zambesi, applied to persons accused of witchcraft. In these cases, an infusion made from some poisonous plant is used, the drink being generally fatal, from the virulence of the poison. The *tanghin* of Madagascar constitutes another well-known form of ordeal.—ED.

the Bakwain town or village, Kolobeng, and stayed there over the following day. Sechelli, the Bakwain chief, was absent. His brother, Khosa-linsi, who acted for him, told me that, after having made peace with the Boers, he sent two men with pack-oxen to buy corn from the natives living amongst them, and that a party of Boers, having shot three of the Bakwains, and taken their oxen, he had sent some of his people to avenge the outrage. News now arrived that the Bakwains had fallen in with a party of Boer-hunters, had killed three of them, and were bringing home their wagons, oxen, and property. We were much distressed by this intelligence, well knowing that none of the Boers who had taken part in the war against Sechelli would so soon after have placed themselves in the power of the natives; and felt that the victims must have been either peaceable hunters, or traders innocent of any hostile designs on Sechelli's people. This supposition was afterwards confirmed. Disgusted at hearing the natives relate with brutal glee the cruel manner in which they had murdered their captives, I at once ordered my men to harness the oxen, much to the vexation of the chief, who seemed surprised that we would not stay to join them in gloating over the bloody trophies they were expecting. While we were hurrying away dreadful and deafening yells issued from the town, as the warriors returned decked out in strips of gory blankets and clothing of their victims on their heads, followed by hundreds of women, giving vent in the most barbarous manner to their feelings, and doing honour to the triumphs of their husbands in shrieks of savage joy which made our blood curdle. Khosa-linsi, on whom the responsibility lay, was the only one who appeared sad, and was evidently absorbed in thought. He begged me to write for him a letter to the Boers to the effect that he had killed the three Boers in revenge for the death of his Bakwains, caused by themselves, and that he was now satisfied. I declined compliance with this request.*

Fairly upon our way on the great high-road to Lake Ngami, our course lying nearly north on the meridian of 26°, and between the parallels of 22° and 24°, we slept the first night at the outspan of Kopong, where we stayed a day to hunt. Thence we made for Sekomi's Town, skirting the great Kalahari desert, and outspanning at the usual stations, where water was to be found. Two days afterwards we reached Boatlanami, and then Lupèpè. On the 29th of May we crossed the tropic, making two long tracks to Mashooi, Gordon Cumming's favourite elephant fountain. Remaining there next day, we shot a gnu and a springbok, and saw several gemsboks and elands. Pursuing our way, at Lobotani we met a party of horsemen despatched by Sekomi to stop our progress, and to conduct us to his town. They used rather threatening language; but I took it quietly, and next morning ordered my people to inspan, which done, I sent a civil excuse to the chief, and trekked on without hindrance.

I had reason to fear our being plundered if I had gone to Sekomi's; but finding at Lutlochoé, where we slept the next night, that another party had been sent to intercept us, I felt that it was imprudent to set the chief's authority altogether at defiance, and explained to his messenger that it would be inconvenient for us to make so wide a circuit, as the season was far advanced, and the water drying up, and sent the chief and his son (my mate) each a present, trusting

* I had subsequently the satisfaction of hearing that this affair was amicably settled by mutual explanations between the Boers and Sechelli, from whom they sought no amends, acknowledging that the three Bakwains had been shot by a party of uncontrollable young men without the sanction of their authorities. They only lamented the loss of three peaceable burghers, who, they said, had always entertained friendly sentiments towards the natives, asking Sechelli, as a favour, to return their wagons and oxen to their widows.

to their friendliness of feeling on my behalf to be spared from further annoyance on the part of his people. On the 2nd of June we left Kalaba's village, containing about 500 souls (Bakalihari), and entered the desert along the track towards the lake. Next day, travelling eight hours, we reached Matalani vley, which still contained some rain-water, and found that elephants had been drinking there a few days before. We fell in with a troop of about 100 elands, one of which I shot from the wagons, and drove in another to our place of bivouac.

Pursuing a general north-westerly course for some days, and outspanning at the well-known vleys, some of which, however, were now dry, we found ourselves by the evening of the 14th of June upon the margin of the vast reedy swamp referred to in a preceding chapter. This swamp, which is probably not less than 70 miles in circumference, includes several islands, upon one of which dwells in security the chief named Chapo. We had captured on the way an old man, whose wife fled on our approach, in the belief that we were Boers, of whom all the people in these regions, as far even as Sebetoane and the shores of Lake Ngami, entertain, since their attack on Sechelli, a well-grounded dread. He gave us to understand that the Botletlie river was dry, and that Chapo had sent a cow, accompanied by a prayer for rain, to the chief Lebele, who lives 500 miles north-west of the lake. This strange request was prompted by the belief, entertained by all the natives in these parts, that Lebele holds the keys of the river-sources, and has under his control the thunder-storms and showers, which occur here during the season of summer, and come from the direction of that chief's dominions.

We were not aware of any approach to the chief's town other than along the western side of the marsh, and had made a long day's travel, with the hourly expectation of coming to a well-known watering-place; but our hopes were disappointed, and at 9 o'clock at night we lay down, hungry, and almost dying with thirst. The wind was blowing extremely cold; and, being many miles away from the woods, we had to make a fire of such reeds as we could find, but, the quantity being insufficient to keep up the fire, we passed a most comfortless night, and next morning discovered that our cattle had strayed away to get water.

All the next morning we sought in vain for our cattle, as well as for water. Once I thought I had found them, but came upon a troop of buffaloes instead. Again I saw a troop of animals approaching in the distance, leaving a cloud of dust in their course, and I flattered myself they were the oxen being driven in haste towards the wagons; but in a few minutes after I had to clear the road for a troop of buffaloes, who had taken fright at my companion. Shortly afterwards, guided by information derived from some natives with whom we fell in, I found at a few miles' distance eight of our oxen bogged in the swamp, with just their nostrils above the pools of mire into which they had fallen. A little farther lay two oxen dead in a game-pit, and some were still missing. We were obliged to leave them in this position, not daring to venture near them. Having taken the remainder of our oxen (which we had brought with the purpose of leaving here during our absence at Sebetoane's), we returned for the wagon, which was taken to the outspan by the Botletlie river. With the aid of a fresh team, we toiled all night in extricating the oxen from the cold and miserable plight in which they had been lying since the previous night, and, this being accomplished before daylight, I was glad enough to make one of a wretched circle of Bushwomen and their children, who, with outspread hands and crowded knees, sat cowering over the sparks of a miserable reed fire, which required continual feeding to keep it burning, and employed many hands in collecting the fuel.

We outspanned on the bank of the Botletlie river, at the north-west end of the marsh. The river is here about sixty or seventy yards broad, and when full is navigable in boats all the way up to the lake. The river and the banks are frequented by the rare species of waterbucks, called by the natives luchee* (*Adenota Lechee*), before noticed in this neighbourhood. These animals are gregarious, rather larger than the rietbuck, with horns resembling those of the latter, but twice as large. The females have no horns. The hide is covered with a long warm hair, much prized by the natives, of a greyish colour, with a dark reddish tinge, white breasts and belly, and knees black. They are seen in troops of several hundreds, and have a peculiar action, owing to their high hind quarters. Their slower movements are rather ungraceful, especially when on the trot, carrying their heads near the ground. When startled they have a very lively and graceful action. Unlike most other animals, they generate all the year round. I was fortunate enough to secure one of them, and bring it alive to Cape Town.†

At the outspan we were visited by my old acquaintance Chapo, a good sort of fellow, with the reputation of being honest: hence we thought it advisable to intrust him with the care of our surplus oxen during our absence at Sebetoane's country, so that, in case of our meeting with the poison-fly and losing our teams, we should have this reserve to fall back on. So with a trifling remuneration, paid in advance, they were handed over to his charge. Here we purchased corn, maize, beans, water-melons, pumpkins, with some calabashes to be used for carrying water. For a few pounds of common beads we obtained a whole wagon-load of grain

---

* Dr. Livingstone writes it *leché* or *lechwi*. See his description, "Missionary Travels," p. 71.—ED.

† I believe it was afterwards presented by His Excellency Sir George Grey to the Zoological Gardens, London.

and beans. We also purchased some of their tobacco, which they manufacture by pounding the leaves into a paste in a wooden mortar, then mould it into the shape of a ball, and dry it. This tobacco is easily converted into snuff by grinding it with a round stone on a flat one, and mixing some wood ashes with it to sharpen it. This is a staple article of trade with the neighbouring tribes, who are all snuff-takers, and with this article they buy their weapons and implements from tribes distant 200 miles eastward. Sheep and goats were also brought for sale, as well as karosses, tiger-skins, feathers, and a little ivory. For these articles they chiefly demanded powder, which, owing to the stringency of the colonial law, we could not spare.

On the morning of the 21st, on leaving this place we met with two Griquas, the remnant of a party who had gone in search of, and discovered, Lebebe, when they met with Portuguese traders from the north-west. Ten of their number had died of fever; their oxen had all been killed by the fly, and they were compelled to abandon their wagons 700 miles from this place. The two survivors had been cared for by the chief of the lake during their recovery. We were glad to have it in our power to render these poor fellows some assistance on their journey home, still 700 or 800 miles distant. They must have been very enterprising and energetic people to have accomplished what they had done. But neither Griquas nor Hottentots are at all adapted for travelling in this country, where fever rages; their manner of living being too gross, as they eat scarcely anything but flesh, consuming immense quantities of fat from one year's end to the other.

We now pursued nearly the same track we had followed in 1852, by Kobe and the Ntwetwe salt-pans for Zoutharra, the village of the Makalaka chief, Khaetsa. This village stands near the southern extremity of the vast Madenisana desert, extending north from the Great Salt-pans to the Chobé and Zambesi rivers. The passage through this wilderness is, from the want of water, except after the rains, perhaps more impracticable than that of the great Kalahari desert, which approaches its confines, at least since the road on the skirts of the latter has been so well explored. The more northern desert has extensive tracts of dense forest, in which, it may be remembered, we hunted elephants, found there in vast numbers the preceding year. We were thus compelled to abandon the idea of farther penetrating it by the drying up of the springs, and the impossibility of obtaining guides, or gaining trustworthy information respecting the country ahead to the north or the east.

I had now resolved on renewing the attempt to cross this desert in a direct course for Sebetoane's country, lying on the Chobé river, and, as will presently appear, I was more fortunate than the year before, having now succeeded in accomplishing my plans, though with extreme difficulty.

Arriving at Zoutharra on the 30th of June, my first care was to send for the chief, Khaetsa; but he declined my invitation, and I could gather no information about the country ahead. I felt much disappointed, as the year before Khaetsa had, though with some reluctance, rendered us assistance, sending his own son, with a party of his people, to be our guides. However, having rested the next day, I shot half a dozen gnus, intending to fill their skins with water,* so that if the worst came to the worst I might be able to give my oxen each two buckets of water on the third day. Thus prepared, we started on the 2nd of July to pursue our adventurous journey of 2000 miles through the untracked wilderness. To my surprise, after leaving the Makalaka village, the old chief, Khaetsa, overtook us when a few hours ahead.

* A primitive and still common substitute in the south of Europe, as well as in Eastern lands, for casks to hold either water or wine.—ED.

He had run after me to apologise for his non-appearance upon my invitation, informing me that some Bamanwato sent by Sechelli, his paramount lord, were stationed in his village for the express purpose of preventing any intercourse between us, and of course to deprive us of the services of a guide. He said that they nearly beat him to death for having accompanied us the previous year to the elephants' field; finally, by way of atonement for his seeming neglect, presenting me with two tusks of ivory, for which I, of course, made an equivalent return. In the evening we outspanned at a place called Thamkerrie, where one of my men shot a large ketloa (*Rhinoceros Africanus*) by the water at night.

Next day we left Thamkerrie. After travelling for six or seven hours in a north-westerly direction, we came to a small pool of muddy rain-water, a few miles westward of a grove of palmyras. The sight of this little pool of rain-water within the first fifteen miles of the desert encouraged my companions, who now felt certain that the whole desert would be full of water, as Bechuana, who had visited us before leaving Zoutharra, had falsely told us; but, doubting this, I ordered about eight gallons to be stored in the wagon for greater safety.

We now continued travelling all day, as fast as our oxen could draw the wagons, cutting our way through dense bush. For a short distance we followed a spoor of the preceding year of some wagons, which must have been those of Messrs. Oswell and Livingstone, in their journey to Sebetoane's country; but as we proceeded, finding these tracks obliterated by the rains of summer, we were obliged, having no guide, to cut a road for ourselves, and slept at night in a dense field of low flat-crowned mimosa trees, without any signs of water. Next day we started early, and travelled with only one hour's intermission till dark night, when I was obliged to put the men, as well as ourselves, on short allowances of water, a pint per day. The oxen began to show great signs of distress, owing to the intense heat, their excessive thirst, and the heavy sand of the desert, in which more than once they came to a stand. We now began to feel how grievously the Bamañwato had deceived us, and the more we reflected the more we became convinced that this was a stratagem emanating from Sekomi, or his representatives here, with the view of thwarting our designs and effecting our injury, owing to his jealousy of our opening up communications with tribes beyond.

During the night, some of our oxen went off in search of water; but they were found and brought back safe in the afternoon, having strayed some 18 miles from the outspan. At the same time we overtook Mr. Campbell, from whom we had procured the loan of two oxen in the absence of our own. Hourly the sand became heavier, the bush more dense, and the heat of the sun more intense; and at last the oxen became so fagged that, several of them refusing to draw, and lying down in their yoke, we were compelled to resort to all manner of cruelties to get them on.

Travelling again till night, this being the third day, our oxen felt their work severely—the rate at which we had travelled having brought us over 75 miles of desert country. Their distress became pitiable: they crawled round the wagons, inserting their noses into every opening, and trying to butt us away from our only cup of coffee, or to upset the boiling kettle from the fire. Next day, the poor animals having been somewhat refreshed by a grateful dew, we started early; but by 9 o'clock it became intensely hot. Falling in with a rhinoceros' spoor, it gave us some hope it might lead to water, and a trustworthy servant was sent on the spoor while we still pursued our way, not wishing to lose time, in case of disappointment. The man returned with the intelligence that the spoor only led to a lick, or *brak* place, to which, as appeared from the spoors, the animal was in the habit of resorting every other day—a thing I was not acquainted with before, and still doubt. This intelligence was very disappointing to our people, limited to a pint of water per day under such toil as driving a jaded team of oxen with a heavy wagon through an African desert, and they showed strong symptoms of exhaustion in their husky voices and parched lips. All we could do was to allow a small increase of the allowance of water; but this seemed only to have the effect of creating a craving for more.

In this way we struggled till the fourth day, our pace gradually becoming slower, when our wagon fell with the hind wheels into a large wolf-hole, out of which the oxen could not pull it, and obstinately refused to make further effort. However, having unloaded the wagon, we put our shoulders to the wheels and, with the aid of a lever, lifted it out; and having with much difficulty got the oxen on their legs again, we moved, or rather staggered, forward for another mile, and found it impossible to proceed any farther. On this, the driver Abraham and the rest of my people volunteered to go forward with the cattle and horses to some spot where water might be found. Thompson and I resolved on remaining by the wagon, in order to protect it from being plundered by the Bamañwato, as they doubtless reckoned on doing. This being settled, we shared our small stock of water with the servants, and killed and divided our pet milk-goat, which had followed us like a dog through all our troubles, and had become as attached to us as a friend.

Upon the state of our feelings at this time my pen is too feeble to dilate; the reader can better imagine what they must have been. All left us, and we knew neither what would be their fate nor our own. Our conjectures were sad and gloomy for the future. We had one resource, however, which is always a poor mortal's prop—Providence; and we sincerely trusted ourselves to its mercy and care. A circumstance that tended much to increase our anxiety was, that we had with us two or three guns, which had been the property of persons whose fate had proved worse than ours.*

In the morning, wishing to spare our water, we adopted a plan of cooking maize in boiling fat, which caused three-fourths of the seeds to burst open and become four times the usual size, soft, rich, and beautifully white, and, having mixed a little coarse meal with fat, we made a few cakes for our supper; but this rich food only increased our thirst. We lay down at night with our rifles in our hands, and, owing to a variety of conflicting feelings, anxiety, hopes, fears, doubts, and an occasional touch of despair, had no sleep till near morning. The following day our sole employment was to measure and re-measure our water; the evening came without any signs of our men. We heard many birds, whose chirping at any other time would have been musical. These birds evidently lived without water, or drank the morning dew.

We parted with our men on the understanding that, as soon as they found water, they should send us a supply and set our minds at ease as to themselves; but the whole of the next day passed without any tidings of them. Our little stock of water was now nearly exhausted, and we discussed the propriety of abandoning the wagon to its fate, and steering in such a direction as would most likely lead us to some spring or vley; and the fourth day, the 13th of the month, was decided upon for commencing our march, as by that time, our stock of water would be reduced to three half pints each. Many were our doubts whether we could exist on half a pint a day whilst travelling; but, hard as it

* Mr. Dolman and his servant, who died of thirst in the Bakwain country. These articles, continually staring us in the face, looked ominous —seeming as though mocking us and threatening us with the same fate.

was, this was the only conclusion we could come to; and calculating that the Tamalukan river could not be distant more than 100 or 120 miles in a north-westerly direction, we determined upon taking that course—not indeed with much prospect of our accomplishing, weak and feeble as we were, what it might be doubtful the strongest of men can do, but in reliance upon Divine aid, and the hope of finding water midway in the desert.

We remained thus on our small allowance, in a state of extreme anxiety, when on Tuesday night, just as we had made every preparation for starting early on the morrow, and had laid ourselves down to take what we thought our last slumber at the wagon, we were roused by the barking of a dog, and what I considered the smack of a whip applied to a jaded steed. But my friend Thompson, misinterpreting the sound, and grasping his rifle, thinking that the natives had come to plunder the wagon, very fiercely called out, "Who comes there?" In another minute my boy Adonis came up with two horses, one carrying two kegs of water. Here was joy! I mentally returned thanks to Providence, our Preserver; we flew to our old water-cask, and drained it of its contents at one draught.

It appeared from the boy's account, that they had been indiscreet enough to entrust the carrying of the water to one of their party, a most consummate scoundrel and thief, who robbed them of the whole of it the first night. They rode the horses and drove the oxen as fast as they could during the day, but at night the men, overcome by fatigue, fell asleep, and the cattle decamped. It was not till the morning of the third day that they found a vley of dirty rain-water at the distance of 45 miles from our position. They lost, however, only one ox, which died of thirst before reaching it. Campbell had been more fortunate, reaching it the night before, having been six days without water for cattle, and making at least 25 miles a day. Owing to his kindness, we were in receipt of some ready-made bread; and if before we could not sleep for anxiety, we now could not sleep for joy.

On the 13th, Thompson and myself taking the horses, and leaving the lad Adonis alone in charge of the wagon, we started for the vley ahead, which, owing to the state of the horses, we did not reach till night. On our way we fell in with a party of Bushmen, who sustained existence by sucking water (a painful process before described). We should have liked to have quenched our thirst, but the water was dyed red with the blood of their gums; so we delayed till our arrival at the vley, when we had a joyful meeting with Campbell; and my man Abraham having slain a few elands, we regaled on fine steaks for supper. We found a number of Bushmen had collected at Campbell's wagon, begging of us to shoot some elephants, which they informed us were plentiful ahead a short distance; but we had not yet recovered from the shock of our last predicament, and felt that our cattle and horses required rest. My oxen had still to return to the spot at which we left the wagon, a distance of 45 miles, and draw it to the vley all the way without water. This fatiguing journey took four whole days, with some failures, during which other oxen had to be sent to their relief.

Here I first experienced a new kind of nuisance, in the shape of a kind of grass, which, coming in contact with the body, is irritating in the extreme; creeping through your clothes, it twists and screws itself into your body, and generally forms a small pustule. A kind of burr, like the beard of barley, is likewise troublesome, and often prevented me from sleeping; and another tall weed was covered with innumerable downy-looking thorns, almost imperceptible to the eye, but when it is disturbed by the passing traveller the thorns are scattered on the wind, and then, settling on your clothes and body, become so irritable and itching to the skin as to put one into a regular fever; and often the skin of my body has been in a complete state of excoriation.

The Bushmen here, finding that we were in no great hurry to commence operations with the elephants, were obliged to gather roots and berries for their food. I noticed a pretty bean growing on a large shady tree, drooping something like a willow, with a beautiful bright green foliage, like the apricot tree, with a small tapering double leaf. The pod hangs in thick clusters of cabbage-green tinted with vermilion, which when ripe turns black, then bursts, and exposes to view a very pretty bright scarlet bean. These beans are eaten either raw or boiled.

As the vley was fast drying up, we left it on the 18th of July, and the same day reached a river-bed, in rather a narrow valley, which the Bushmen called Dum, probably signifying river, or equivalent to the Bechuana Molapo, a nullah or ravine. It was impossible to discover its direction; but it probably joins the Tamalukan river westward, and seems to have been the channel of a running stream, to judge from the appearance of its banks lower down. A great deal of ironstone abounds in this neighbourhood, and limestone in other parts.

The next morning we arrived at two conical hills of ironstone, called Chenamba, when, finding several large vleys of water, we proposed recruiting the strength of our oxen by a few days' rest. At sunset the Bushmen, whose ears are as acute as their eyes are keen, were seen in a listening attitude, and the next moment came running up in great excitement, informing us that some elephants were at the water. In an instant our horses were saddled, and we soon fell in with three of the herd, making off as fast as they could.

When on the point of firing at them, I was compelled to remount and make way for a troop that emerged from the bush behind, and came thundering and crashing down upon me. I fired at one of the largest bulls, and, having spent three shots on the beast, and brought it to a stand, it was claimed by one of my companions, who finished it with another in the head. The rest had now escaped, except one, attacked by Mr. Campbell, whose gun I heard at intervals of a few minutes. Riding in his direction, I found he had expended all his ammunition, eighteen bullets, and wanted my assistance; so, having headed the animal after a smart gallop, one bullet from my double-barreled Blissett brought him to a stand, and three more to the ground, just as the moon's beams began to penetrate through the foliage of the trees, which all around for several yards were besprinkled with blood dashed from his trunk. After receiving the last bullet, the poor beast twisted his trunk around a very frail tree, by which he tried to support his ponderous frame; but the tree as well as his legs soon gave way, and he sank expiring to the ground.

During the night three bull-elephants drank at our encampment, and next morning we followed on their spoors for a long way. At noon one of these elephants was pointed out by our Bushman guide, whose keen eyes soon detected him at a distance of eighty yards. My companions seeming to prefer the other two that were not yet in sight, I spurred forward, and, at a distance of forty yards, dismounting, I put a hasty shot behind his shoulder, rather too high, and another in the flank, as he wheeled off at a rapid pace, with trunk and tail erect. My friends Campbell and Thompson, failing in their own object, rode up and offered me their assistance, which was not refused; but before they could render me any service, a large white dog of Thompson's rushed in and set up a terrible barking under the elephant's trunk, mistaking it for his tail. The enraged beast turned upon him like a whirlwind, and I had just got time to come up and give him another shot, when, with extended ears and outstretched trunk, he turned upon the dog, and trumpeted with a vehemence that made the forests ring and seem to tremble around us. The dog, seized with a panic, fled towards his master, who, not knowing my position, and finding no time to pull a trigger, made straight towards me. The elephant, seeing us, made us his mark, and on looking round for Thompson, who but a second before was close behind my heels, I missed him from his horse, which was still following mine, with the elephant's trunk over his back; but the sagacious beast, missing the rider, allowed the horse to move out of his way, and when within five paces of the tail of my horse, who was galloping and leaping at his best speed, the forest became so dense that I was despairing. Making one last desperate effort, just as the elephant's trunk breathed against the back of my neck, I drove the spurs into my favourite and willing Dreadnaught's side, who flew over the hook-thorn bushes and fallen logs with the agility and grace of a bird; but it was of no avail. The trumpeting in my ear was deafening. Dreadnaught flew, and broke through a cluster of mogonono spars, when I was hooked off and rolled under the elephant's feet, fortunately escaping his heavy tread only by a few inches. Here I lay as still as death, until he had passed over me, when, missing me from the horse, he stopped and became furious; raising his trunk aloft, he worked it about like a monster pump-handle, snuffing the air to find out my whereabouts. Taking advantage of his turning his head in the opposite direction for a moment, I rolled under a small green bush which favoured me, and was effectually screened from his view. I now raised my gun at his head, but being nervous, and still breathless with excitement, and knowing that unless I killed him on the spot I or my friend who was crouching near me would rue it, I refrained from pulling the trigger, and was glad to see our friend march away up wind, carrying his trunk aloft.

As soon as our enemy was out of sight, Thompson and myself emerged from our hiding-places, and searched for Campbell and my servant Abraham. The former had thrown away his gun to lighten his weight, and so escaped unscathed. Abraham, seeing the danger Thompson and myself were in, owing to the denseness of the bush, abandoned his horse and wisely took to his heels, concealing himself under the wind.*

The first thing now to be done was to search for our lost horses, which in the course of the first hour we found; and while my servant still assisted Campbell in his search for his gun, Thompson and I made a fruitless effort to pursue our retreating foe. Some time after our return, Campbell's gun was found buried under the sand and bushes by the elephant's foot. I had lost part of my bridle and both my stirrup-irons and leathers: these articles I recovered, but returned from the fray minus the tail of my coat, which was left, fluttering in the breeze, on one of the hook-thorns with which the locality abounded.

Another adventure, which bid fair to be equally stirring, soon followed. On the 21st Thompson and myself took up the spoor of a solitary bull-elephant, which in the course of an hour we fell in with on beautiful ground, standing fast asleep. I gave him, as I thought, a good shot, but it only had the effect of bringing him round very slowly, eyeing us with his head on one side, evidently lost in amazement, or else contemplating mischief. As we generally do for safety's sake after discharging our guns, I remounted, and the fact of the elephant turning towards us so alarmed my little mare (who was well up to elephants) that she expected a charge and bolted; the

* It is, or ought to be, always the hunter's first thought to get under the wind of dangerous animals.

curb-chain broke, and for two miles I could not hold her, and was compelled to throw myself from the saddle in order to stop her. Having adjusted the curb, I returned, and found Thompson still peppering at the elephant, who was yet in the same attitude, motionless, at a distance of about forty yards. On my arrival he seemed to have made up his mind for a start: never have I seen an elephant run with speed at all approximating to this, and our endeavours for a long time to catch him were unavailing. Thompson's horse giving up, and he having broken his gun over the horse's head, I alone followed, and succeeded in putting nine unavailing shots into his body, until at length I found myself in an extensive mogonono field, swarming with elephants in every direction, crushing down the bushes with a sound resembling the roaring of waves. Not knowing in which direction to turn, I climbed a mokala (camel-thorn) tree,* of which there were a few about, and which seemed to be the standing-place† of numerous elephants every day. On reaching the top of this tree I found myself surrounded by elephants, mostly cows, in every direction, in groups closer than I ever dreamt of. My position not being very enviable, I sat in breathless suspense observing the movements of the numbers of gigantic animals round me on every side. At the distance of about 120 yards to the eastward of my position, I soon discovered the old bull which I had followed, surrounded by a group of about a dozen cows, caressing and fondling him, some of them dashing him with water from their trunks, others with sand. Those elephants below his wind, probably scenting his blood, lifted their trunks, and after smelling a moment gradually moved off. The elephants below me were working their trunks about, but made no effort to move out of the bush, in which they pro-

---

* *Acacia giraffe.*

† In localities where elephants abound, they generally frequent the same forest at the same time every day, to stand (sleep) in the shade at noon.

bably considered themselves concealed, while groups in other directions, that seemed to have got over the alarm caused by the reports of our guns, were moving slowly, followed by their calves, breaking down the branches, and pulling up young shoots and grubbing roots, which they strewed before their young with an air of the most maternal solicitude. A shot fired on the outskirts by my servant Abraham drove a troop towards my position, and another and another shot seemed to be bringing masses of cow-elephants from every direction round me; but, to my great relief, as if actuated by one impulse, the foremost began to move into two parallel files, one on each side of me. I fired a snap-shot at the file above wind when the other file had passed, and then, rounding the latter, rode parallel for some distance, thinking to get the bull, but, failing, fired at a cow, which ran off in the opposite direction, trumpeting pitifully with pain, and setting the whole herd soon in commotion.

Having hunted near the wagons for some time, on the 25th of July we collected a number of Bushmen, and saddled our horses; and taking sixty men with us, while their wives remained feasting on the flesh of the elephants we had killed, we proceeded on our excursion eastward, leaving the wagon in charge of our people. Our party formed rather an imposing procession, a single file of about 300 or 400 yards long. Two Bushmen took the lead as guides, bearing our guns; then came ourselves on horseback, our servants behind the horse carrying pots, kettles, provisions, guns, &c., and a blanket or two. The rest carried skins and calabashes of water on their spears, with a small skin blanket slung over it behind their shoulders. They had rubbed themselves well with the fat of the elephant, an operation in which they particularly delight, as it softens and cleanses the skin, cures all scrofulous diseases, and it is, moreover, a non-conductor of the heat of the sun. They wear few ornaments or trinkets, save a round plume of black ostrich feathers, closely clipped, and tied with a string on their foreheads.

At night we bivouacked by some vleys of water, and slept under a dense kind of fibrous bushy tree, with small roundish leaves, a white flower smelling like jessamine, and the same in appearance, only much larger. This tree (*Gardenia Thunbergia*) bears a poisonous fruit, about the size and shape of the prickly pear, having a most luscious smell, tempting one to eat of it. It is called by the natives *Moralla*, and affords good shelter from the attacks of elephants, being so densely matted with its fibrous wide-spread branches as to resist the entrance of an elephant. During the night we heard elephants drinking at the distance of two or three miles, and the next day followed in their spoor, and, falling in with a young one, I succeeded, with the assistance of Thompson and my servant, in slaying it after a very long chase, and I spoilt a very good horse by frequent discharges from a very short-barreled gun, close to his face. We also fell in with numerous cow spoors, but declined hunting them.

We now divided our party, and spread over a large tract of country, Campbell steering eastward, and Thompson and myself to the south. In about an hour we fell in with a small troop of cows, and, contrary to the rule on which we had hitherto acted, determined to attack them. We commenced driving them about, shouting and yelling at their heels, until one of the larger cows, on our following them into a bush, made a succession of charges upon us, which caused us to lose sight of the whole. While renewing our search for them we were nearly overrun by a small troop of seven or eight, that came thundering down from behind us; these we also commenced driving about, but, Thompson's horse knocking up, I was left to myself to continue the chase. Giving the best cow a shot, which soon brought her to a stand on the brink of a forest, where I gave her a second, which broke both her shoulders, I placed myself between the forest and the troop, with the view of driving them back, in which I succeeded, though frequently repulsed by a vicious cow. Another cow I brought down with a single shot; but my companions, now firing from the opposite side, drove the elephants into the bush, followed by the one I had last wounded.

The first charge of the cow-elephant this day drove me incautiously in the direction of a "borele,"* and, escaping from his paroxysm of rage, I started a panther from his lair in a field of very long grass. On our return home we fell in with another young bull-elephant; I sent two shots into him, and, making straight for a neighbouring bush, he was soon out of our sight. We endeavoured to follow, but, owing to the dense underwood and the briars, &c., we found it impossible to make an entrance. In consequence of this remarkable feature in the forests of this country, which are all overgrown, and completely matted with thorns and briars, we were often obliged to abandon a fine elephant, which in the fairer fields of the Limpopo and Bamañwato we might to a certainty have reckoned our own. One satisfaction, however, which I had, was to learn the subsequent year that five of the elephants which I had here wounded had died, and that their tusks were disposed of by the Bushmen to a trader who was travelling with Dr. Livingstone.

On the 1st of August Thompson, myself, and servant rode out early, and soon fell in with a large bull, who seemed inclined to charge us before we had troubled him; so, dismounting, I took a deliberate aim as he approached on the open plain, and fired a bullet between his neck and shoulder, and, as he wheeled round, another behind the shoulder. The blood streamed copiously from the wounds, dyeing the trees

* The smaller species of black rhinoceros.

as he passed, and, telling my servant to follow and keep him in sight, I felt for my powder horn and pouch to reload, but, to my great disappointment, found that I left them at our bivouac. I sent a man for them, and when he was some time gone, becoming impatient at the delay, I followed him, but he had already left with the things in search of me. I raced backwards and forwards for two hours, and then by mere chance fell in with him, and followed the large bloody trail of the elephant; so freely had he bled, that I could gallop on the spoor and keep it. After a pursuit of about six miles, I found him standing, and, leaving my horse, crept in and gave him the contents of both barrels behind the left shoulder, when he ran off at a smart pace. Delaying to reload, I despatched Abraham to keep him in sight, threatening to shoot him if he lost him, as he had often done before through cowardice. On coming up with him, to my question of "Where is he?" I got the usual reply, "He was here just now, sir." I tore along through a terrible jungle, and reined in suddenly within ten paces of an elephant, which I made sure was the same. The elephant turning towards me, I hastily retreated about twenty paces, considering myself too close, and with my first shot brought the animal to the ground. It rose, however, immediately, and a furious charge followed from a short distance beyond, and I soon observed the heads and trunks of ten or twelve elephants in the air, and that neither the one I had fired at nor any of the rest were bulls, but, as I had lost sight of my victim, thought I might as well continue shooting at these, and consequently fired two more shots at the largest cow, which walked away a few paces into the forest, when I heard her fall.

The jungle into which they had retreated was one of those in which these animals might well consider themselves perfectly secure, being matted with the sickle-thorn briars—a most virulent thorn, through which I tried to make my way, but was forced to give it up. The following day, returning to the spot to get the cow and follow the trail of the bull, I encountered three koodoo bulls, the best of which I shot, and returned with the flesh. On Wednesday I sent a Bushman in quest of the dead elephant, who soon returned with the tail and point of the trunk,* and the tusks soon followed.

On the 4th of August Campbell and I returned for the hunting fields. I wounded en route a fine cock ostrich, pursuing which Abraham and I disturbed a family of four ketloas,† which dashed up wind and dispersed us, causing us to lose our feathers. Continuing by myself the search for elephants still farther eastward, I fell in with some Bushmen belonging to a hitherto unknown tribe, called Mashubea, under a chief called Chapatani. From these I learnt that an elephant had died six months previously about 60 miles eastward, and as the tusks were reported good, I despatched two servants on horseback in quest of them, and was rewarded with two tusks of the value of about £25 for my trouble, and then returned towards my wagon, shooting a quagga and a brace of elands.

On our return I discovered that one of our Bushmen guides was either a maniac, or had some disease partaking very much of the nature of madness, was troubled with convulsions, &c., and was held down while the fat of the elephant was anointed on his side, from which they believed hair would make its escape. The manner in which they explained to us that he was mad was that "the devil (parra) was in him," who sometimes led him about into the veld, with great danger to his

---

* They never missed bringing me an *eye* and point of the trunk, with the request that I would burn the former to a cinder, and broil the latter, which would give me full power over the life of any elephant I encountered, and prevent them from charging me.

† The larger kind of black rhinoceros (*Rhinoceros Africanus*).

life from the wild beasts. Having anointed him with lard,* and held some smoking roots under his nose, while the natives meanwhile anointed him, moreover, in sundry places with the spittle of the father, who was also the doctor, and used many unmeaning ceremonies, we left him in their charge, and reached the wagon on the 9th of August.

* Elephants' lard is the most cooling and beneficial ointment for all inflammatory wounds that can be obtained.

CHAPTER VIII.

Travel North for the Chobé—The Commanè River—The Grass on Fire—Moreymi's Town—Sebetoane, Chief of the Makololo: his History and Character—Dr. Livingstone: his Intercourse with the Natives—The Makololos and their Country: Trade with—The Tsetse Fly—The Chief Sekelètu—The Author hears from Dr. Livingstone and endeavours to meet him.

WE had made the camp at the Chenamba Hills our headquarters for upwards of three weeks, while we hunted the country round for elephants with fair success. But a very visible change had now taken place in the season. The vley waters were fast drying up, and the north-east winds, which prevail at this time of the year, had set in, stripping the seringa and mopani trees of their golden foliage to give way to a young growth of fresher hue. The elephants, also, had become so alarmed by the havoc we had made, that they were fast trekking eastward to some large river understood to lie in that direction. Our own course was planned for Sebetoane's country on the Chobé, and we resolved on pursuing it without delay. Parting on the best terms with the Bushmen, to whose services we were much indebted for assistance in our hunting, we distributed among them some beads and trinkets, and left them in the full enjoyment of the flesh of six elephants, besides others that I afterwards learnt had died of their wounds, an abundance they had not enjoyed for many a day or year. I also distributed among them some tobacco, with which they were smoking themselves into fits, dancing, feasting, revelling, and making the wildest demonstrations of joy.

On the 11th of August, starting from Chenamba, we travelled for five hours on the western edge of a patch of tsetse, the poison-fly which Livingstone was the first to make familiar to English readers, and employed the rest of the day in cutting a road through the forest ahead. In two days more, travelling north-west, we reached a river called Commanè. Passing through more open forests, we regretted that, as the waters were so fast drying up, we dared not delay to hunt, although elephant spoors were numerous, and we heard them trumpeting and breaking the boughs in every direction. But as we intended resting the Sabbath at this river,* I despatched Abraham in search of an eland, and he returned in an hour, having killed one that weighed 1000 lbs. or more.

On the 15th we travelled all day west by south, following the course of the Commanè river, and slept by a Bushman village. The next morning continued for three hours farther down the river, and then crossing it, and steering north, came to a village of Bushmen under a chief called Dorila, whence we observed a range of hills, not very distant, and well known as the boundary of the fly country, bordering the plains south of the Chobé river. Having travelled all night, to escape the attacks of the fly, next morning early we were passing through a forest of tall straight trees, something resembling the poplar; when, as the sun rose, we were met by a Bushman, who reported that he had been sent by Moreymi, a petty chief of the Makololo nation, living on the Chobé, about 30 miles distant, to warn us of the fly and to direct us safely through it. A delay of three minutes on his part would have brought us within the range of this pest, which was just

* This was the first water that might be considered permanent after leaving Thamkerrie; and had it not been for the providential supply of rain-water which we got, I fear we must have perished before we could have reached it.

ahead of us; and we were obliged to remain all day, though our oxen had not slaked their thirst for two days. They became very restless, probably smelling the water of the river some 20 miles ahead.*

Soon after we were startled at the sound of a rumbling noise like distant thunder. On looking back, we observed a dense cloud of smoke between the trunks of the distant forest trees, and we soon discovered that the forest was on fire, and was fast spreading; and the dry, rank grass, feeding the crimson flames, sent innumerable forked tongues into the dry foliage of the tall trees, and escaping above their tops some eighty feet from the ground. The wind, springing up from the north, increased its virulence and speed, as well as the ominous sound attending it. Without thinking of any other resource, we unlashed our spades and reaping-hooks, and cleared away the grass to windward as fast as we could; we only succeeded in clearing away a few paces around the wagon, and, having secured the oxen to it, covered it as well as we could, and then laid ourselves under it, burying our faces as close to the ground as possible, for fear of being smothered by the heat of the flames and smoke of the passing fire; and, after a few moments of intense anxiety and partial suffocation, we crept unscathed from our concealment, and set to work extinguishing the flames around us. Some of the oxen were slightly singed, and some of the harness was also damaged, but nothing serious. Until night the burning logs continued falling around us in every direction, resounding loudly as they fell.

At night, as the moon rose, we started to get through the fly,† but, on entering it, our wagon struck against a large tree, and we were compelled to put the oxen to the back of

* Cattle and game smell water and young grass a very long distance.

† A narrow strip just here, not more than a mile broad, was infested by this terrible insect.

it. We now travelled only by night, when the cold prevents the fly from biting the cattle. On the second morning we struck a spruit flowing into the Chobé, which latter river flows into the Shesheke or Zambesi, a couple of days to the eastward of this. Next day we travelled about 18 miles farther eastward, and reached at last the banks of the Chobé, opposite Moreymi's Town. We crossed four other streams, all, like those of the previous day, flowing into the Chobé, and generally four to five feet deep at the drifts, and 100 to 200 and 300 yards broad, the surface being generally covered with reeds, bulrushes, or different kinds of lotuses.

For about fifteen or more miles on both banks of the Chobé the plains are dotted over with little mounds, all bearing palmyras, date-palms, wild cotton,* indigo, and other indigenous tropical productions. The plains are periodically flooded, and become inundated for several months in the year, so that instead of the river, which is scarcely more than 200 yards broad, the natives at those times navigate sheets of water many miles in extent, in search of the animals which resort to these mounds. The soil in the vicinity and for several miles seems most fertile, as appears from the great abundance of grain of different kinds in the possession of the natives. They cultivate also several kinds of beans and vegetables, which, with the pistachio nut, form their principal diet.†

* An indigenous short-stapled cotton, converted by the natives into rugs, which they sometimes sell to the neighbouring tribes.

† I have since planted some of these nuts and beans in Cape Town, and found them to thrive well, as also the lebèlèbèlè, a small kind of grain or millet, something resembling rye, but scarcely larger than canary seed, for which the prize was obtained at the Cape show. This is a very prolific grain, yielding a return of some thousand for one; it came originally from the *Batawka*. Another plant which I introduced at the Cape is a large flat-grained and very white species of Kaffir corn, resembling lentils in shape.

On our arrival on the banks of the Chobé we encamped on a large mound, prettily overgrown with evergreens, among which clustered a variety of creepers in full blossom, and of a delightful fragrance. Another kind of plant twined round the trunks of the larger trees, climbing from branch to branch, and forming a complete network at the top, with clusters of sweet-scented flowers drooping towards the ground. Under the shade of these trees we had a spot cleared for our wagons; a carpet of mats was spread on the ground, and, with an awning over our heads, we squatted like Turks, with our pipes and coffee, and received the natives in state. The river, which washed the base of the mound, seemed to have nowhere less than ten or twelve feet of water, and generally much more, even at the banks, undermined as they were by the continuous flow of the stream, running at the rate of three knots per hour. It swarmed with waterfowl, consisting chiefly of varieties of ducks and geese, with herons and cormorants of every colour, pelicans, and other aquatic birds too many to enumerate. Crocodiles were said to be very numerous, and hippopotami could be seen or heard from the wagons. We were in full expectation of receiving a visit from some of these animals in their nocturnal rambles, as was the case with the first party of travellers who reached this spot last year, when the visitor was slain over the pole of the wagon. Besides these, and all the well-known animals of these parts, I noticed here, for the first time since leaving the Sovereignty, the Black Gnu (*Connochetes gnu*), lashing their white tails across the plains, and capering among vast herds of quaggas and luchees; we heard here also of the Nakoñ or Setertunka, the only antelope that seems worthy of the name of a waterbuck, to which its aquatic habits fully entitle it. After examining the hides and horns of these animals, I have come to the conclusion that Water-koodoo is the most appropriate name for them, as they resemble the koodoo in colour, mark, shape of horns, action, and everything except their aquatic habits, and a very long coat of liver-coloured hair, nearly three inches in length. This animal is so shy that no white man has as yet got a glimpse of it, except Mr. Oswell, who, I was told, shot a very young one, in fact a kid. They never venture out of the reeds or rushes, lying all day on islands that are densely overgrown, with their bodies generally under water. They are seen singly or in pairs, and are captured by the natives at seasons when the reeds are dry, by setting fire to them on the islands, which they surround with boats, and spearing the animals when they take to the water. Another unknown buck which I heard of here is the Poku, an animal as large as the luchee, of a pale or bright red colour, and said to be gregarious.

Moreymi, the chief of this district, declined, as usual, being seen, and sent word by his son that he was "not at home." This is what the traveller must expect when coming for the first time among a strange tribe in these parts; and Moreymi, taking advantage of the absence of the great chief Sekelòtu (who shortly after accompanied Dr. Livingstone to the Barotse country, the land of cattle, a long way north) from his town of Linyanti, situated some three days to the westward, took the opportunity of secreting himself, the object being to obtain a preliminary present; but, failing in this, he made his appearance in the evening, after our arrival.

I informed Moreymi that we had come a long way to see Sekelètu, his paramount chief, and obtain his permission to shoot elephants in his country, in return for which we would be glad to supply him with any articles we had to spare in exchange for ivory. I then gave him an account of our perils and adventures, with a present of beads, brass wire, knives, handkerchiefs, &c., with which the chief professed his heart was so touched, that he felt for us, and assured us we should never be hungry while in his territory.

I had brought up some boxes and parcels for Dr. Livingstone, and I now handed them over to Moreymi, to be sent to the doctor's servant, a coloured man named George Fleming, who officiated as the doctor's cook, besides being a trader on his own account, and who, by the way, went by the name of the doctor's wife; for, according to their rules, this branch of service devolves upon a wife alone. These goods were forwarded at once to Sekelètu's Town, Linyanti, where, in his absence, Poonoani and Mahali, two rival under-chiefs, had the management of affairs. From our camp on the Chobé we amused ourselves by navigating the river in canoes, shooting wildfowl, hunting the hippopotami, or stalking or coursing the luchee.

As the people amongst whom we were now sojourning fill an important place in the native records of the South African interior, I believe that a few particulars respecting them and their famous ruler will not be unacceptable to the reader.

Sebetoane, a Basuto, one of the greatest chiefs that South Africa has known, the conqueror of many nations,' came originally from a hill in the Orange River Sovereignty, situated between the Sand and Vet rivers, near Winburg. It appears that his people had been living in the present Trans-Vaal country, becoming increasingly powerful, and subduing many of the small tribes in their neighbourhood, until they at length encountered Moselikatze, "the great lion," who drove them back. They had afterwards, in 1824, an engagement with some Griquas from the Kuruman,* who, possessing guns and horses, of which Sebetoane's followers had hitherto no experience, succeeded in driving them northward to the Molopo river. Malitsani, Sebetoane's brother, remained in that district, but Sebetoane himself, sweeping all before him, found his way to the Botletlie or Lake river,

* For an account of this engagement, see Moffat's or Thompson's Travels.

after suffering great loss from thirst while crossing the desert, as well as subsequently from fever. He and his followers subdued the Bakwains and other tribes dwelling eastward of Sekomi's present site, and, finding that the Makobas, or Bayèyè, on the lake and its rivers, were living in small clans, without any paramount chief, it was an easy matter to subjugate them. With his forces thus augmented, and the aid of their boats, Sebetoane reached the country on the Chobé, in which he settled. Exaggerated reports of his successes, his numbers, and prowess, had preceded him, and so terrified the greatest of the Barotse chiefs, Santhuro, that he fled into Central Africa, without waiting to strike a single blow, or even to look upon his enemy.

Sebetoane died during Dr. Livingstone's first visit to Linyanti. The chief had treated the doctor and his companion, Mr. Oswell, with great kindness and hospitality, inviting them to return next year, and to encourage traders to bring merchandise for ivory, of which there was abundance.*

Sebetoane was succeeded in the chieftainship by his daughter Ma-matsasani, but it shortly afterwards devolved on Sekelètu, a son, about whose legitimacy there was some question. Ma-matsasani retired to the great town of Naliéri, situated ten days up the banks of the Shesheke. Meanwhile Sekelètu put to death, one by one, the councillors and friends of Sebetoane, and nearly daily murders or massacres, disguised as executions, took place by his command.

Sebetoane was not only one of the greatest warriors of his

* Dr. Livingstone gives a very interesting account of his reception at Linyanti, with a valuable notice of the history and character of Sebetoane. He says that this chief was decidedly the best specimen of a native chief that he had ever met. ("Missionary Travels," pp. 90, 178, *et seq.*) The year following Dr. Livingstone revisited Linyanti, then under the dominion of Sekelètu.—ED.

nation, but his name is held in great respect for his liberal government, and generosity to his enemies. He had subjected a great many tribes in these parts, fifteen of which I have heard enumerated. His policy was generally to spare life as much as possible ; but the conquered chief he would either kill, or, separating him from the rest, would place him in a tract of country where he would be always in his power. He would return them their cattle to live on, give them a daughter or relative to wife, and administer his own laws. This liberal plan, unlike that adopted by other tribes, combined with a judicious and uniform treatment, inspired the conquered people with such confidence in and devotion and reverence for their new chief, that they generally soon preferred his government to their former. In this manner amalgamation took place, and the original tribe of Basutos are now, perhaps, the least of the whole population ; and the climate not being congenial to their former habits, they have become the most effeminate of the races under Sekelètu's sway.

The Barotse, who seem to comprise the majority of the Makololo natives, are an active, tall, and athletic people, broad-chested, and, owing to their usual occupations, boating, &c., are very muscular and powerful in the arms. They are very expert in killing the sea-cow and elephant, which they delight in attacking when in the water, and generally take the opportunity when they are fording a river or come to drink. Thus every day elephants are killed by these people, and ivory and flesh conveyed to town in their boats. The Basutos are generally decorated with a white ostrich feather on their head, their bodies greased with butter, with scarcely any covering, except when they happen to be the possessors of European clothing, of which they are very fond. Their shields are generally borne by a slave behind them, and they are seldom seen carrying more than a single battle-axe. They have rather an imposing carriage, and walk with a free and easy, confident air, which yet cannot be called swaggering. They have the reputation of being very proud, and of scorning a dirty action or mean advantages, but we were led to form a different opinion in this regard before we parted company with them.

From the banks of the Chobé the Makololo country extends probably northward for 300 miles or more up the Shesheke river, which runs all the way down nearly from the centre of Africa. On the west it is bounded by the Bavicko nation, under Lebébé; north-east by the Bashukulumpe (a people wearing their hair plaited like a sugar-loaf on their heads, and no clothing whatever on their bodies); east by the Batawka, and south by Sekomi's Bamañwato. The Sheshcke, which is joined by the Chobé, about 50 or 60 miles east of our position, is said to be about a mile in breadth, and sometimes assumes more the appearance of a lake than a river.

The different conquered tribes under Makololo dominion are generally esteemed according to their power, or the difficulty they gave their capturers. Amongst these the Zulus rank high; few of these, indeed, were actually taken in war, for the Makololo were never a match for the Zulus; but a number of them were once taken by stratagem, and are now settled amongst the Makololo, who esteem them highly.

Of the religion of the Makololo it would not, perhaps, become me to judge on so short an acquaintance; but during my stay I gathered that they implicitly believe that when they die their souls enter other bodies and live again. Their belief in the immortality of the soul was further corroborated by a circumstance which then occurred. A young man being taken ill, the doctors held a council, at which it was affirmed that the young man's father, who is long since dead, was haunting him, being hungry. Thereon a fat ox being doomed to slaughter, a portion of it was thrown over the hedge of the kraal in which the father was buried, to appease his ghost, the doctors sitting down to feast on the remainder of the sacrifice thus offered to the dead.

Of Dr. Livingstone's labours we hear but poor success. Previous to his last arrival amongst them, when told that he was coming, the first question they asked was, "What is he coming to do?—to bring guns?"—"No; the Book." "Well, then, he had better stay away; his God has killed us." Sebetoane's doctors attribute the chief's death to the white men coming amongst them, and whenever Dr. Livingstone preaches in the presence of or visits a chief, the doctors burn something as a charm to protect them from his witchcraft. Being, as they find, a doctor, he has also the reputation of being a wizard. This makes him either feared or admired, and gives him a certain influence. They give him credit for being a good doctor, and say he has cured many, but killed some natives. They do not believe in natural deaths; when a man dies he has been killed. By all accounts the doctor's preaching is barely tolerated by the chief, who is at heart highly displeased at his doctrines concerning rain and polygamy.* The people say that Dr. Livingstone has promised them all the good things of the earth, rain, corn, cattle, &c., if they would believe in God and refrain from polygamy, slavery, and other malpractices; that they have waited a long time for these good things; and that they would wait another year to see if the Good Man he talked about helped them nicely (tusa sintle). While they were relating these things, and conversation grew slack, the councillor Poonoani was observed

* Dr. Livingstone's servant, Suyman, a man of colour, told us that when his master preached an eloquent sermon against polygamy, ending with a hymn, Sekelètu would collect all the young girls in the town, and make them wind up with a dance.

sitting with a piece of a newspaper upside down, mimicking the doctor singing a hymn,* and, observing that he had attracted our attention, he rolled over on his back, threw his feet into the air, and exclaimed, bursting out into a loud laugh of ridicule, "Minari" (a corruption of the Dutch mynheer, generally applied to missionaries). Such is the sort of impression as yet made on these barbarians. It is to be hoped that in time better success will attend missionary efforts.

The slaves of the Makololo seem to be as industrious as their masters are idle. They work in iron and copper; of the former they smelt abundance, of a very superior quality, and manufacture it into implements, which, considering their rude materials, would astonish any European. Hoes, spears, lances, battle-axes, bows and arrows, are abundant and cheap, of excellent workmanship, and are even chased and ornamented. They work in wood also, making stools, vases, &c., out of solid pieces of wood, on which they carve and decorate very tastefully. Their earthen beer-pots are also superior to those of all the other tribes; and they make nice baskets and mat-works of every description, in which they show a degree of skill, taste, and ingenuity far above any of the other tribes that I know, which skill extends even to the culinary art. Their canoes are large, well-shaped, and often ornamented, made out of a solid log with small adzes only an inch broad. Some of these are rowed by sixteen men, paddling at the rate of ten knots an hour. Their fishing-nets are similarly constructed to our own, saving the floats, which are made of rushes instead of cork, and the twine of a kind of flax (*Sansevicra angolensis*), growing in great abundance on all the mounds. They are expert in rope-making of every description.

The Makololo proper are never active but in war or the

* The natives call psalm-singing *Bokolèlla* (to bellow like a bull).

chase, though too idle for the latter. The slaves in general hunt, as well as plough (hoe) and reap for their masters; and even all the domestic departments seem to be conducted by male and female slaves, under the control of the wife or wives; and to the number of slaves, and the abundance of food and rain, I attribute the very limited extent of polygamy in this part of the country, where even the greatest man has very few wives, or only one. The slaves hunt and fish, and do everything, even nursing the children. They are very expert in killing the hippopotami, which are numerous, in the same manner as we kill whales, and are quite as or more dangerous, especially when they have young. Fishing is made easy from the variety of modes they have of capturing them with nets, baskets, fences, &c., as well as a small kind of turtle that abounds. Otters are also killed in great numbers, and their beautiful black furs sold to traders for a trifle. Jackals, not being hunted so much here as in other parts, are exceedingly numerous, as well as the panther, the skin of which is so much valued all over the world; and the elephants are so numerous and troublesome, that in the grain season the natives have to battle with them to drive them from their gardens.

The population of the Makololo country is very large, owing to the fruitfulness of the soil, which is capable of supporting any amount of population. The country is full of people everywhere, and where you would least expect it. I cannot form any correct idea of the numbers. I do not think 300,000 souls would be over the mark, perhaps much under it. The town of Linyanti must contain, from all accounts, more than 15,000 souls alone, and that is not the largest town. In the northern part the population is denser, and fever, which is fatal to them, sweeps off numbers every year.

The climate of the Makololo country is said to be unhealthy, and the malaria arising from the decayed vegetation, after the receding of the yearly floods, is very productive of fever. By a letter received from Dr. Livingstone after his return from the Barotse, I learnt that he ·had suffered from several attacks while travelling. We kept ourselves very quiet in order to preserve our health, and during a stay of six weeks cannot complain of anything, saving a few rather severe headaches, and slight accompanying fevers, though this, I must admit, was considered the healthiest season.

Besides various kinds of grain, beans, earth-nuts, water-melons, pumpkins, calabashes, and sweet-canes, the natives described the sugar-cane as abounding in the north, together with a vegetable resembling yams, and other well-known kinds of tropical vegetables. They possess abundance of cattle, goats, and sheep. Their cattle are of two peculiar breeds. The Barotse, very diminutive cows, standing very little higher than three feet, or as large as a year-old European calf, but exceedingly good milkers. The others are the Bakalihari cattle, immensely tall and lanky cows, rather higher than an Africander ox, with immense horns.* Their sheep are the Egyptian breed, with large tails, but their goats are exceedingly small, and generally only marked black and white. Their cattle are carefully kept in the plains, or in parts known to be perfectly free of fly. The only remedy they use, they say, is to administer the fly in milk to a fly-bitten calf, but they do not seem to be very sanguine about the cure. Besides cattle, goats, and sheep, the only domestic animals they have are dogs, and an inferior breed of fowls, of which we bought several for trifles of beads.

Now, a word with regard to that insignificant-looking insect, the Tsetse, or Poison-fly. This great barrier to African

* Compare with Livingstone's account, "Missionary Travels," p. 192.—ED.

travelling was first met by the Boers and other travellers on the Limpopo; and though most people on their first encounters felt doubts regarding its repute of the sting being fatal to horses and cattle, too painful experience of its ravages has left no doubt on the subject. We find again the insect rising here, after we have completed more than a thousand miles of our explorations towards the unknown interior of Central Africa, crossing our path and stopping our progress in every direction. The tsetse is, in extreme length, half an inch, or very little more, and has very much the appearance of a young bee just escaped from its cell, or a bee half-drowned in honey, the wings being always closed when stationary. The body is not quite so long as that of a bee, and much more slender. It is marked with alternate stripes of yellow and dark chestnut in transverse bands on the back of the abdomen, which, fading towards the centre of the back, gives it the appearance of having a longitudinal stroke of yellow immediately down the vertebra, the belly livid-white, glossy dusky-brown wings, folding over each other, eyes purple, six long legs, and its proboscis from one-sixth to one-eighth of an inch long; it has tufts of hair over the body, which are longest at the mouth, on the back of the thorax, which is brown with black spots, and at the extremity of the abdomen. It is extremely quick of sight and keen of scent; its flight is rapid and straight.

The bite of the tsetse is something like that of the mosquito, but the pain not so lasting. It assails different animals in their most defenceless parts: a man behind the back between the shoulders, and an ox on the back or under the belly; a horse in the same places, and inside the nostrils; and a dog on the forehead, &c. With the proboscis they penetrate a pilot cloth coat and whole suit of under-clothes. The bite of this insect is fatal to cattle, horses, sheep, and dogs; but there is a peculiar breed of the latter known as *Makoba* dogs, which are exempt from the effects of its poison, the breed having from time immemorial been reared in the "fly" country, and escaped a *cow milk diet*, as the natives say. It has no ill-effects whatever on game or upon men, except that the being bitten by numbers is likely to induce headache, as with the irritation of mosquitoes. The symptoms, as I have observed them, are, first, in the ox, a swelling under the throat, which if lanced emits a yellowish fluid. The hair stands on end, or is reversed. The animals become debilitated; and, though the herbage be ever so luxuriant, refuse to eat their fill, and become thin. The eyes water, and at length, when the end is approaching, a continual rattling in the throat may be heard at a few paces' distance. It sometimes happens that a fly-bitten ox will live, but very rarely, and only when it has no work to perform. Work and rain are great precipitators of their end. In horses the symptoms are swelling about the eyes, nostrils, testes, the hair is reversed, and, though they have the best of food, they become thin, sleepy, and, pining gradually, at length die.

Both cattle and horses live from fourteen days to six months after having been bitten by tsetse, but they generally die after the first rain has fallen. A dog dies in ten or twelve days, or two or three weeks at latest. It is perceptible in the eyes, which are swollen and protruding. After death the heart of an ox is generally incased in a yellowish glutinous substance, which might be mistaken for fat. The flesh is full of little bladders of fluid, and the blood also is half fluid, which becomes congealed on cooling. The vitals are of a livid colour.

The tsetse fly is generally found within a few miles of water, in rich sandy ridges near marshy spots, and generally in mopani or mimosa forests. I have known them to shift their positions, or encroach on new ground, or leave parts where fire-arms have driven the game out of a district. They are mostly only found within a certain range from water. To the buffalo in particular the insect is more attached, and often moves about with them in the rainy season.

The thorax of the tsetse is brown, covered with hairs, and is divided into *three* sections, by two transverse indentations across the middle, the posterior or triangular one having a white margin like the tail of a soldier's coat, being edged with dull white, and a stripe down the middle. The middle section has four oblique oblong dark spots, and the front section one or two irregular spots. These markings are, however, very capricious, perhaps differing in male and female.

We had been a fortnight or three weeks on the Chobé when Poonoani and Mahali arrived, bringing with them for each of us a present of six baskets of native corn, a large pot of honey, and a valuable tusk of ivory, and placing them at our feet said, "Sebetoane's children greet you with a tusk. Our king is absent with Dr. Livingstone, and we have taken these things at his house, knowing it is his custom." I thanked Mahali, and gave him in return a present of 15 lbs. weight of regal beads, an equal quantity of brass wire, 5 lbs. of powder, and 10 lbs. of lead, with some knives and forks, handkerchiefs, &c. &c. He was mightily pleased, but, placing greater value upon a common musket, I readily supplied some in exchange. They said it was useless for us to think of hunting elephants, and that we had better barter our goods with them; they had abundance of ivory, and we might easily fill our wagons and return home.

Since the discovery of Sebetoane's country by Dr. Livingstone and Mr. Oswell, some Bechuanas had come here, and so duped the natives, that they, not knowing the value of ivory, readily gave them 80lbs. of ivory for about twenty leaden bullets, or something equivalent. Their visitors had thus driven an exceedingly profitable trade. But the natives, having since heard that they had been overreached, had become exceedingly cautious, and now, knowing the real value of ivory, adopted the plan of asking most exorbitant prices, such as ten valuable guns for a tusk, gradually decreasing their demands according to the effect produced on the trader's countenance, the only indication from which they could at all guess when they were near the mark; and in this they were generally very acute. But they carried this too far with us, endeavouring to obtain everything we had at almost nominal prices; and when we informed them that they could not procure the articles in England for the price they offered, they left us for a few days, and then, returning again, renewed the attempt to traffic with us on their own terms, spending a whole day in haggling for them, and only towards evening, perhaps, offering a slight advance. Finding that they were not disposed to give a fair equivalent for our goods, and that they seemed more bent upon having our horses than anything else, we could not succeed in our traffic. Perhaps if the chief himself had been present business could have been done to mutual satisfaction.

A further stay of a week which I made amongst the Makololo was employed chiefly in boating and shooting—several exciting attempts which I made to add the hippopotamus to my list of trophies being rendered fruitless by the excessive care of my life taken by the boatmen, who refused to go within a nearer distance than 150 yards of that animal, at which distance I found it impossible to strike their snouts, when they were allowed, for a moment only, to protrude above the water for the purpose of breathing. I vainly entreated the boatmen to bring me at least within ten or fifteen yards, but they positively refused, saying that if any accident should happen to me their own lives would be the forfeit.

At length, having been promised boats for a visit to the falls, described by the natives as being only three days to the eastward, we prepared for the excursion, hoping to enjoy grand sport with the numerous sea-cows inhabiting the river, and the elephants that come to drink at night. But when it came to the point, we found that the promised boats were not to be had. I subsequently learnt from Dr. Livingstone the probable reason of this, viz. that Moselikatze, having lately cut off a tribe of Banabea, under one Juankie, a petty chief to whom he was paramount, for having had intercourse with the Makololo people, had caused Juankie to be slain. Afterwards, however, when in Moselikatze's country, I learnt that Juankie and his tribe were massacred for no other reason than that he was a brave and formidable warrior, and withal so crafty that Moselikatze had never been able to get him into his power. Although acknowledging Moselikatze's authority, he had frequently defied his power with impunity. The Matabele, however, believing that the body of a brave warrior, when eaten, acts as a powerful medicine, imparting courage and skill proportionate to that which he possessed when living, long sought an opportunity for entrapping him; and at length succeeding in their object, they conveyed him to Moselikatze's Town, where he was killed and skinned, and his head, heart, and sundry other parts cooked, and distributed amongst the renowned warriors of the Matabele to be eaten.

On the 1st of October, receiving a letter from Dr. Livingstone, from which it appeared that Poonoani had informed Sekelètu that we had already left, I determined on paying the former a hasty visit, and therefore engaged a boat, started, and travelled about 25 miles, passing many troops of sea-cows, at which I occasionally fired a snap-shot, as they put up their noses to breathe. In the evening I slept on a mat stretched on the damp ground, and suffered acutely from a severe headache, accompanied by fever, which was aggravated by the torments inflicted by swarms of mosquitoes. A sea-cow grazed near us all night, but it was too dark for us to find it. Next morning, starting early, we travelled another 20 miles, when I fell in with Poonoani in a canoe paddled by sixteen men, and shooting down the stream at a prodigious rate.* On a nearer view I found that it contained also one of my own men, who had journeyed up from the lake in quest of me—a report that we had all perished in the desert, or been murdered by the people about, having reached the lake and created great alarm amongst my followers. I now learnt that Mr. Andersson had arrived at the lake, on pack-oxen, from Walvisch Bay, and that he had gone a short distance up the Teoghe, or Teougé, river; and, further, that our oxen and horses were fly-bitten, two or three of the former having already died. This disastrous intelligence induced me to abandon with regret all idea of seeing either the chief Sekelètu or Dr. Livingstone, and we immediately put about to return towards the lake.

Our cattle and horses having been bitten by the fly, their speedy death was inevitable; but my object now being to get to the lake, I thought they might hold out for a two days' journey, in which time the wagons might reach the Tamalukan river, where, embarking my ivory and other property in boats, we might pull down its stream till it joins the Botletlie, as the latter river does the lake.

On leaving Linyanti, Dr. Livingstone, who was now accompanied by Sekelètu, visiting the northern part of the chief's territories, kindly sent a messenger to me with a small stock

---

* These men, standing upright in the boat, eight paddle dexterously on either side, lifting the boat at every stroke. After six strokes, they change simultaneously to the other side of the boat, and so cleverly, that it can hardly be observed. This change is in order to avoid injury to the chest.

of sago, rice, and other things which might be useful in case of sickness. He had already informed me of his intention of proceeding through an unknown country to the Portuguese settlements on the west coast, and I was sorry that circumstances did not permit my having the pleasure of at least offering to bear him company in that perilous enterprise.

CHAPTER IX.

Travel South—Mahabe Flats—Interior Watercourses—The Tamalukau: embark on it—Native Roguery—Crocodiles in the Tamalukan—The Botletlie River—Makobas—The Chief Lechulatèbe: Trade with him—Visit to Lake Ngami—Lechulatèbe's Town—Journey to Chapo's: Disasters on the Way—Journey to Sechelli's Town—Shock of an Earthquake—Reach Kuruman.

WE parted company with our fellow-traveller, Mr. Campbell, and leaving Linyanti on the 10th of October, 1853, after crossing two considerable spruits of the Chobé, we halted for a day near Moreymi's village, which, with all his people's stock of grain, and the ivory they would not sell me, we found in flames as we approached it. On the 13th, having disposed of one of my fly-bitten horses to Dr. Livingstone's man, who wished to have it, thinking it could be cured, we started again, struggling along as best we could with our sick oxen, and, having been refused guides through the country infested by the fly, we pressed into the service some Bushmen who passed us. At night they bade us stop, and next morning before daylight I sent the oxen back to the plain for fear of fly, although we were told there were none here. Thompson and myself walking out after sunrise, we fell in with a flight of these poisonous insects, which alighted on our backs and bit us severely.

Having thus far, contrary to our expectations, been more fortunate than Mr. Campbell in not losing any more of our oxen after leaving the Chobé, we thought it advisable to treat the oxen kindly, and rest them occasionally, as we did this day. Thus we struggled on for several days, Thompson and I riding forwards all night to save our horses from the attacks of the fly.

On the 19th of October, a journey of about 10 miles south-west through a very fine country, and its continuance on the morrow 14 miles farther, brought us to the edge of the Mahabe flats, a plain about 12 miles broad, covered with a tall reed-like grass, generally growing to the height of twelve or fifteen feet. Fearing the grass might accidentally take fire, we travelled fast to get over the plain, and by 4 o'clock in the afternoon, having outspanned at the Banabea village of Borogo, we sent the oxen forward a great distance to the water. The people of this tribe, which came originally from the southern bank of the Zambesi, have their two front teeth filed so as to leave a notch between them. They pay tribute to Lechulatèbe, the chief of the lake, as an acknowledgment for living within his territory; but they seem to be on friendly terms with the Makololo also, probably fearing the one as much as the other.

On the 21st, 10 miles more brought us to another such village, under a petty chief named Murimajani, a great rascal. Towards evening three Mashunas, lately escaped from the tyranny of Moselikatze, made a dance for us, and sang a dirge on their conquest by Moselikatze. These people have a greater notion of music and song than any other tribe I know of; they played with their thumbs on a musical instrument fixed inside a calabash, and made of slender bars of flexible iron, strung in a manner to produce the different notes. Unlike all the other tribes, they have two-storied huts, and sleep in the upper story while a fire smoulders all night beneath, and sends up the smoke to protect them from the annoyance of mosquitoes. The Mashunas dress very like the Damaras, and from the great affinity of the languages, and other circumstances, it appears that both people must have originally inhabited one country near the east coast.

I had flattered myself with the hope that I should be able to embark my goods in boats here, and glide merrily down the Tamalukan river; but these hopes were disappointed, for there was no such thing as a boat to be had, and if it had been otherwise, the water itself was too shallow to float a boat. It sometimes happens, however, that when the Chobé and the Tso, which flows into the Tamalukan out of the Teougé, are full, the water runs up the Tamalukan, and the overflowing streams, meeting here, flood the Mahabe flats for many miles eastward, forming a sheet of water nearly 20 miles in breadth, where at other seasons not a drop can be found. At such times the river is navigable from the lake to Sebetoane's, and one might then travel in a canoe from Chapo's, at the terminus of the Botletlie, to the mouth of the Zambesi on the east coast; or for several hundred miles north-west of the lake from a very long way beyond Lebébé's in the same direction. And if, as the natives assert, the Teougé branches off from another river, beyond Lebébé's, flowing to the west coast, then the continent of Africa is probably navigable for boats right across from east to west.

I had sent one of my servants to Linyanti to fetch some ivory I had purchased, and, as he had not yet returned, I was obliged to wait for him. Thompson meanwhile having determined on starting for the lake, to set the people there at ease respecting us, I gave him my best horse; and on the 24th he started, under the guidance of three natives, who were to direct him through the country so as to avoid the fly. Feeling lonely after his departure, and hearing that buffaloes were to be found close by, I walked out, accompanied by a number of the Banabea. Soon falling in with a troop in marshy ground, I found them exceedingly tame, having probably never been fired at before, so I easily succeeded in shooting two, which, although aged, I was surprised to find differed materially in the size of both body and horns from those I had seen in other parts. Their bodies are short, their horns very much so, and twisting very abruptly. Gnus, of which I shot a few, were plentiful. Indeed, during the rainy season, the Mahabe flats are said to be swarming with game of every description, in great multitudes, including elands, buffaloes, rhinoceroses, tsèsèbies, &c. Their resort hither I conclude must be to escape the bite of the various species of fly which infest the forests during the summer.

Some of the natives arriving from Linyanti on the 27th, I was enabled to start after sun-down, and, journeying on for several days, exposed to all sorts of depredations from the petty chiefs, on the 1st of November found myself travelling along the bank of the Tamalukan, making inquiries everywhere for boats, as the river was gradually getting deeper and broader—but without success. The oxen were by this time in such a wretched state of debility from the venom of the fly, that I feared every moment they would drop down dead. The poor creatures still struggled on, falling at every quarter of a mile, till by dint of torture they got up again, and then, after dragging on a mile or two farther, the whole team tumbling over each other, we were obliged to let them out of the yoke till they recovered themselves a little, and then goad them on again. I could endure the sight of this no longer, and every day when these cruelties commenced I left Abraham to come on the best way he could, and started ahead, shooting a wild duck, or muscovy, for my dinner, hardly caring whether I got a bed in the wagon at night or under the shelter of a tree. However, guided by a Makoba chief whom I fell in with on the road, I reached, about 10 o'clock at night, his village, which stood on the left bank of the river, here 200 yards broad; but though my Bushman guide shouted for several hours for a boat to be brought for the white man, it was near morning before they came across, pretending to be very much frightened. Next day I bargained with the chief for the conveyance of my ivory down the river in boats, and immediately unloading the wagon, sent it on empty, instructing Abraham to drive hard for the lake before the last ox dropped, as there were now only eight left, and even these were failing fast. Having allowed Abraham a sufficient start, I left the village, where I had again been robbed of everything the natives could lay hands on, even to all my clothes, with the exception of those on my back, and was paddled down the river for 12 miles, sleeping at a place well known as the Sea-cow Holes, a series of small lakes parallel with the river, and full of sea-cows. Just as we had got out of the boat, and were looking at a suitable spot to make a fire, and spread my bedding, we heard a loud snort a few feet behind us, followed by a violent splashing of the water, and, looking round, perceived that our boat was in great jeopardy from the attacks of a sea-cow, in whose path we had unfortunately moored the boat. Observing us, the brute instantly plunged into the water and disappeared. I had the annoyance of finding the boat half filled with water, and my blankets and gun completely saturated. A fire was soon made, and, having scratched a hole in the ground alongside of it, I lay down for the night, while the two boatmen made their fire a dozen yards off, and roasted and feasted on the spoils of the day, consisting of some large muscovy ducks (*Anser melangoster*), the largest and heaviest waterfowl we have here. Waking in the night, I was astonished to find that the two boatmen had disappeared, and, shouldering my gun, strolled down to the river-bank to see if the boat was also gone. Finding the boat and ivory as I left them, I lay down again, hoping that the men's absence was but temporary, but, daylight coming without them, my perplexity increased.

If ever I felt utterly at a loss during my travels it was at this moment, when I was left in the wilds by these two savages. Here I was, with a boat containing 1200 lbs. weight of ivory, worth about £300—a very serious object in my circumstances, especially after such a disastrous trip, during which I had lost all my cattle and horses—and I would now have to abandon a wagon. What my feelings were may be easily imagined; but in this worst of plights I did not abandon myself to despair. I felt that no time was to be lost in the endeavour to extricate myself from my forlorn position. Having happily found a long pole and a paddle in the boat, I resolved on trying to get to the lake, or overtake my wagon by managing the boat myself, until I should fall in with or surprise some native, who by compulsion, if other means should fail, would paddle me down the river.

With this resolution I started on my voyage; but having had no experience in the use of a paddle or pole, my first efforts might stand comparison with the flounderings of a youth first learning to swim, and instead of making a forward movement, the boat continued for a long time describing a circle on the surface of the water. At length, however, I discovered the secret, and pushed along at the rate of about two and a half miles an hour, and succeeded in making about three miles, when I came to a stand, the channel becoming suddenly closed by a dense growth of reeds and bulrushes. With great toil I at last picked my way through; but, finding a bend to the north, I concluded that this must be the junction with the Tso, which branches from the Teougé north of the lake, and, not wishing to steer in that direction, I determined upon hiding the boat in the rushes and running forward to overtake the wagon. Fortunately, at this moment, on looking back over the plain, I observed a number of natives at the spot from where I started, but on the opposite side of the river. Quickly, therefore, crossing the stream, I ran towards them, with the view of getting their assistance and obtaining information about the course of the river. But when I got within hailing distance, the cowardly wretches, observing a white man with a gun, fled precipitately. In vain I called out to them, laying down my gun, or, to intimidate them, levelled it at them. They fled all the faster, and after a chase of two miles farther I gave up the fruitless pursuit, sick at heart, and returned to the boat: I refilled my pouch and powder horn, and then, concealing the boat, started forward again with greater resignation. Having travelled five miles along the river, and without falling in with anything but the spoor of Campbell's wagon, I struck out to the desert, in order to meet the spoor in case the wagons should have made a short cut. Unsuccessful in this, and baffled again, I came to the melancholy conclusion that the wagon must have broken down and was still behind. So away I ran back as fast as I could, half mad with suspense, and just as the sun was setting I fell in with my wagon, only a few miles from the village of the rascally old chief last referred to, who, having received his payment in advance, and also robbed me, crowned his villany by sending a messenger to Abraham, in my name, to stop him because I was still behind!

Starting at once with the empty wagon, I reached my depôt of the ivory the same evening, and, having transferred it to the wagon, felt inclined to sink the boat. But, on second thoughts, I left it uninjured, and proceeding down the banks of the river, fired all night at the troops of buffaloes that were feeding in my path. In the afternoon of the next day I reached the village of Lingalo, one of the principal men of Lechulatèbe's tribe, in charge of all the Makobas in his country.

Lingalo, perceiving on our arrival that five of the eight oxen that were drawing fell down before we outspanned, raised many obstacles to my being supplied with boats, informing me also that the junction with the Botletlie river was still ten days distant. This I knew to be incorrect, and gave him the proper distance from a rough map, which seemed to justify their belief that "the Book" tells us everything, and that any attempt to cheat us was useless. On my part, knowing that Lechulatèbe was very anxious to purchase from me everything I was disposed to sell, I resisted his most exorbitant demands for conveying my ivory to the Botletlie, informing him that if he refused to carry it for me, though I offered double the payment he had demanded from travellers the year before, his chief would get nothing from me. This settled the dispute, and having, according to the custom of the country, paid the freight in advance, I counted the ivory given into Lingalo's charge. Abraham I left with only four oxen to come on the best way he could without guides, which were refused us, while I got into another boat and started ahead to avoid sight of the sufferings of the poor oxen, and to obtain assistance from the wagon and people whom I had long since sent up the Botletlie to the lake, to trade there and wait for us.

During the night we saw many troops of buffaloes, upon one of which I expended nearly all my ammunition, and wounding three or four. Next morning we observed occasionally the backs of huge crocodiles floating on the surface, or were aroused at intervals by a splash from those that plunged from the banks where they had been basking. While these were in sight, we perceived an old woman standing up to her neck in the river, now and then diving for the root of a lotus, which, with fish, constitute their daily food. Having plucked a root, she held it aloft, and, divesting it of its jacket, threw it into a dish which floated by her side. On expressing my astonishment at her rashness, and asking my boatmen whether crocodiles never devoured human beings, they said, Yes, often; but gave me to understand that they left it entirely to chance: sometimes God allowed them to be eaten, and sometimes He did not. I kept my eyes upon the woman till the boat glided out of sight, expecting every minute to see her disappear. I spent the next night at the village of one of the petty chiefs, and on the morrow, leaving the boat moored to the bank of the river, and marching across the country, under the guidance of some natives, thereby cutting off a considerable angle, in 12 miles reached Makato's Town, on the Botletlie river.

On applying to Makato for a boat to convey me to the lake, and making inquiries about Thompson, I was stunned at hearing that he had already left for Chapo's, a distance of more than 300 miles, and also that my man January, who was stationed at the lake to wait for me, had become alarmed at not hearing about us, and, giving us up as murdered, or having died of thirst, had left the country immediately after —despatching a messenger in search of us—without even waiting to hear tidings of our fate. There seemed now to be no end to my misfortunes. I could only hope that Thompson would overtake my wagon before it left Chapo's, and send me some assistance. For the present, instead of going on to the lake, I returned to my wagon, to render what assistance I could to Abraham, but I was so famished, that I was compelled by hunger to eat a mash of white ants (*Termites*). On my return I found Lingalo still in charge of my ivory, but he refused to send it on unless I added to the payment he had already received a gun, 50 lbs. of powder, and twenty bars of lead. Considering, however, that sufficient advantage had already been taken of my misfortunes, I determined there should be an end of this system of extortion, and made up my mind to stay where I was, if I had remained alone for six weeks, while Abraham went to Chapo's for the oxen I had left in his care; and I gave Lingalo to understand that his chief would get nothing from me. This threat had the desired effect, Lingalo promising my ivory should be sent forward immediately. But although in this matter he kept his word, the pilfering commenced again, and I found that I had not a single plate, knife, fork, spoon, or basin left. I was obliged to cut my meat with an assagai, and drink my water by lying on my stomach, like a beast, at the stream.

Having now come to an understanding with Lingalo, I shot several buffaloes and water-ducks for the natives, and at last they started with my ivory in boats. My wagon, which started at the same time, reached the lake on the 16th, and, having been taken to pieces, was conveyed across the river in two large canoes. One horse that same night fell into a game-pit; another was taken by the crocodiles; two oxen also fell into pitfalls: so that I had but two left, and my favourite pony Dreadnaught died the previous day. After following the wagon like a lamb, the good beast staggered for a moment and fell down dead. This faithful animal had been my companion for several years, and had borne me safely through the most exciting chases, and carried me out of many difficult positions; and now that he was so suddenly snatched from me, I am not ashamed to own that I was almost weak enough to shed tears over his lifeless corpse.

At the time of their greatness, and before any dissensions occurred among the Bamañwato to create civil wars and divisions of the tribe, the great chief Matebe ruled over them, and the nation, then in their full power, feared none of the surrounding tribes. They then possessed the same country they now inhabit, with considerably more territory to the north-east, their chief town being situated somewhere north of the Shua river. This Matebe had two sons, who divided between them their father's territories, the half tribe, called Batuana, under Toncana, the eldest, removing to and settling at this lake. He also had two sons. Moreme, the eldest, was conquered and slain by Sebetoane, and his son Lechulatèbe, the present "chief of the lake," was carried into captivity. After a time—to cut short a somewhat romantic story—his uncle Magalakoè, undertaking a long and perilous journey in disguise, managed to redeem his young nephew from some of Sebetoane's boatmen, to whose lot he had fallen, and, having reared the lad to mature age, handed over to him his people and government, and is now acting as his chief councillor, Lechulatèbe acknowledging him as his preserver and father. Yet, notwithstanding this noble conduct of Magalakoè, of the like of which not many instances are to be found in the traditions of barbarous tribes, there are still times when Lechulatèbe regards him with the suspicion natural to savage minds, and desires to be rid of him.

The Bayèyè, or Makoba, comprising by far the greater portion of the people under Lechulatèbe's rule, are scattered along the banks of the Botletlie, the Tamalukan, the Teougé, and the many minor streams which form a complete network east and north-east of the lake, nearly as far as the Chobé. They number, probably, 200,000 souls; and over these, and several thousand Bushmen, scattered throughout the surrounding deserts, Lechulatèbe holds his arbitrary sway.

The Makobas always build near the rivers, or on islands in the streams. The marshy ground on which they build and sleep seems not to affect their constitutions, though their habitations are generally all but in the water. Owing to their aquatic habits, the soles of their feet are very white and tender, and they find it a most painful toil to walk any distance from the river. Take them away from it, they become perfectly useless. They are, however, very expert fishermen, and skilful in making twine, cord, or cables of a kind of flax (*Sansevicra angolensis*), growing so densely on the banks of the rivers as to be an obstacle and barrier to one's progress.

They also make fine nets and baskets, and construct weirs for fishing, and frequently capture the hippopotamus by spearing it from their canoes, as we do the whale. The modes in which they take the numerous kinds of fish, of which I have seen twenty-one different species, are various. They set a net across the stream by night when travelling; and when hunting during the day two boats go together, and each taking the end of a seine, many fathoms in length, they separate, and, forming a semicircle, drag the net for some distance; then closing, they draw in both ends in the water until the fish flounder about between the boats, and they kill them with their paddles. Also a kind of hedge, or weir, is set constantly across the shallow part of a river, with many intricate chambers and trap-doors, admitting the fish, but allowing no exit. This gives them nightly a plentiful supply of fish, their principal food. But, besides the river, they have other sources from which to obtain it. Along the banks, for some 20 miles either way from their habitations, they have pitfalls, which they visit every morning to look after the game that may have fallen into them during the night in coming to the river to drink, and on these excursions they prefer the toil of rowing themselves in a boat to that of walking. There are also many kinds of roots in the water, on which they subsist. The lotus they eat, root, stem, leaf, and flower, raw or boiled: the flower and root of the bulrush, as well as the tsetla root, or palmiet (*Juncas serratus*), forms the main article of the diet of the Makobas, as well as of the poorer Bechuanas, or even the better sort, when their corn fails. This tsetla, when dry, is pounded between stones, and the pulp separated from the fibre, if well cleaned, makes a capital pottage.

The Makobas are the most superstitious of all the tribes; at least they have a great many ceremonies and observances, which some people are disposed to construe as religious acts.

Probably they may at one time have had that meaning, but they are now so corrupted, that one cannot trace many of them to their origin, though with regard to some we can entertain no doubt. such as circumcision, which they practise like the other natives. They scarcely do anything without accompanying it with some extravagant or unmeaning ceremonies. In each village there is a chief, to whom the inhabitants seem to pay a sort of worship. This high priest, as he may be termed, is generally found sitting under the shade of a tree, which travellers have learnt to designate as the altar, from the circumstance of the practising something at it which might be construed into the offering of sacrifices to their god to propitiate his favour. On this tree they hang the horns, heads, or tails, or some more insignificant part, of the different animals they kill; otherwise they believe they would have no more luck. Of reptiles that are not eaten, such as lizards, they hang up the whole. All that is killed in the shape of game, as well as fruit and vegetables, are brought here, and, having been subjected to some unmeaning ceremony from the chief, who generally mutters something to the Invisible One, while kissing or spitting upon it, the article is considered blessed and fit for use. The fortunate young man who kills an animal, finds a bee-hive, or anything else, is led to this place by the chief; a portion of the honey, or whatever it may be, is rubbed into his arms and wrists, accompanied by the muttering of the chief, who generally spits all over the poor fellow, and then dismisses him with a blessing, and he departs proud and contented, though he gets none himself, it being *miela* to those who find it. Thus are the people duped by the old men, who persuade them that if they eat of anything of their own finding or killing, instead of bringing it to the chief priest, they will surely die. In telling the tale of their achievements, they adopt the same sort of rhythmical cadence already mentioned as belonging to the Bushmen, and which is very musical. The language of the Makobas is at all times soft and musical, seeming as well adapted to the water as that of the Bushmen is to the forest; they make themselves audible to their own people at a great distance without any effort.

The manner of greeting each other among the Makobas is sometimes highly ridiculous. For instance, Makato, meeting his father-in-law, took a mouthful of water, and, running up to him, spat it all in his face; then, grasping his hands, he kissed them most rapturously. At other times, when friends have long been absent, on meeting they rush upon each other, and wrestle to determine which of the two is the superior, or has become so since last they met; and after the trial, the weaker exhibits signs of the greatest respect towards the other. Some of the Makobas, on finding a boa-constrictor, will make a kraal round it, and they are said to worship the reptile, for what reason they do not know. It seems, however, sufficiently obvious, since, when they find that the boa has swallowed an antelope, they jump upon it and make it disgorge the animal, and then set it free to kill another for them. Some clans, however, eat the boa; and all the Bushmen hunt it for food, and relish it exceedingly, especially when fat.

After the discovery of Lake Ngami by Dr. Livingstone and Mr. Oswell, in 1849, it was frequented by a crowd of traders, who for the first few years carried on a very lucrative traffic with the natives, some of them bringing home, perhaps, ivory to the value of £1000 for goods which had cost only £200. But the reader will not think the trader over-recompensed by this or any profit, however large, when he takes into account the risks, dangers, and anxieties which have to be encountered in reaching this place. At the present time, after five or six years' intercourse with the numerous traders who have been at the lake, Lechulatèbe had gained sufficient experience to be able to make a satisfactory bargain on his own account. On the present occasion he had taken the trouble of going a distance of 60 miles from the lake to meet me, with a number of his followers carrying tusks upon their shoulders. Care was taken, however, to conceal this from me, and the chief, having seated himself, in a long discourse set forth what handsome prices he had been in the habit of getting for his ivory. Trading with natives being a most tedious and vexatious affair, to cut it short I at once gave him to understand that I had but "one mouth," and that it would be useless to try to make me alter the price I offered. Still the whole day was spent in higgling for more, while tusk after tusk was brought from the bush, and carried away to be concealed again. Thus the first day passed off without trading.

I took now the opportunity of complaining to Lechulatèbe of the trouble his Makobas had given me, and the manner in which they had robbed and deceived me. His eye sparkled, and he seemed to gloat on the details of my sufferings, and, turning to his people, exclaimed, "*Oa ootla banna?*" ("Do you hear, men?") and then turning to me, said, "Can I help it? what did you want at Sebetoane's? I cannot help it; they are Sebetoane's people, and act under his orders." I signified that when a great man ceased to speak the truth he was no chief. Finding, however, that I had broached a subject which gave them as much delight as it had caused me pain, and as they continued to annoy me with their sneering and heartless remarks, I gave it up. The chief, then turning to his people, and saying, "What think you, my men? will he go there again?" to which they replied, "He lies; he won't."

The next day, finding me still resolute, Lechulatèbe brought over some larger tusks, and, after a deal of chaffering over the bargain, he seemed disposed to accept the terms I offered, asking, however, the opinions of his followers, each of whom made a long dissertation on the respective merits of my wares and the ivory, and, having keenly examined the articles I produced, gave his opinion. This occupied a weary length of time, and when I flattered myself the bargain was concluded, some insignificant-looking fellow would start up and give an opinion adverse to the arrangement. However, it was all to no purpose; I would not yield, impressing upon them that I was no liar, and had but "one mouth;" so at last the chief, finding me resolute, and being anxious to return, traded with me in one day for about 1500 lbs. of ivory, being all he had, and which just cleared my wagon of all the goods intended for barter. A blue and pink bead was very much in demand, and for any nice articles, either of clothing or otherwise, that he had set his heart upon possessing, he would give a good price. I found, however, that the palmy days for trade were gone; and although I received a very remunerating profit for everything I sold, the ivory which I got, together with that obtained by my man January, whom I had sent here expressly to trade, and what was collected at Sebetoane's, and shot by myself, did not cover my expenses and losses in cattle, horses, &c., and a wagon which I was obliged to abandon afterwards.

The loss of so many of my oxen now placed me in a dilemma as to the conveyance of the ivory. I had been trying hard to get boats from the chief to carry it to Chapo's, but no reward would induce him to assist me. He reminded me again that on any future occasion I must not go to Sebetoane's, that his own country abounded in elephants, and that, if I came back, he would give me guides who would show me more than I could kill. Fortunately, however, my messenger had overtaken January at Chapo's, and brought up eleven out of the fourteen oxen I had placed there. This relay, though somewhat short of what I expected, coming just at the time I had failed to procure the assistance of boats, relieved me from much anxiety.

Having now satisfactorily finished my trading, and made a new axle for my wagon, the greatest piece of work I had ever accomplished, and being, moreover, importuned and worried, as usual, by the natives, I at length grew impatient, and was glad to get away and indulge my desire of visiting the lake. This I did on a pack-ox, brought by Mr. Andersson from Damara Land, and sold to one of my men.

Ascending the Botletlie river for a short distance, I was surprised to find, on a tree overhanging the stream, a fruit very much resembling the loquat of the Cape, and, having plucked and tasted it, I was delighted to find it had the same flavour. The seeds are exactly similar, the only difference between the two being that the leaf of the one I found here is much smaller and narrower, and the fruit not quite so sweet and luscious as the Cape loquat, though I have no doubt a little cultivation would bring it to the same perfection. Wild medlars are also found in abundance.

We travelled by the wagon-road along the river, and seldom out of sight of it. For about twenty or thirty yards on either side the banks are open, admitting of easy traffic for a bullock wagon; but the road is often abruptly terminated by the steep limestone cliffs occurring on the very brink of the dense jungle and forest which border its valley on either hand. This wooded region extends a few miles in breadth, till its outskirts become less tangled with creepers and underwood, though seldom free from the briars of sickle-thorn and the flat-crowned mimosa. Many of the trees near the water's edge are evergreens—some of immense height and most majestic appearance. One of them is the "Moku-choñ," male and female, the latter bearing a very delicious fruit, converted by the natives into a pleasant drink. A dense belt of reeds overgrown with convolvulus, yellow, pink, and blue, fringes the water on both banks. In this belt there occurs an occasional gap, made by the troops of animals resorting to the river to drink in the night, the spoors of which we passed all along in great numbers, the chief of them being those of buffaloes.

During the day, having ascertained that Lechulatèbe remained behind to take his dinner, and having nothing provided for mine own, I took possession of a pot of *L intloa* or white ants, which was prepared for the chief; and, having dined upon a portion, I gave the rest to the slaves. Hunger will create an appetite for strange viands; and it was after suffering, on another occasion, three days of starvation, that, driven by necessity, I learnt to eat and to relish the insect, and to be grateful to Providence for having made it a means of human subsistence. Those that have eaten the locust prefer the white ant. The mode of catching these insects is curious. I observed it a short time after this, when travelling in advance of my wagon, and, being overtaken by a storm, had taken refuge in a Bushman's hut. During the day I had noticed the different members of the family tying together bundles of dry bark six inches in diameter, and smaller bundles of a kind of plant, which they had at first split in two, or others made of reeds, but had no idea for what purpose they were intended. The rain ceasing, I again started, travelling till dark, when we made a fire, upon either side of which my Bushmen guides and myself lay down to rest. It was dark and cloudy when, about 9 o'clock, happening to be moving, I saw a great many bright lights dancing about in every direction, and could not conceive what caused it. I thought that, perhaps, some natives had lost a child, and were seeking it with lighted torches. But asking my guide, who spoke afew words of Sechuana, what it meant, he replied, " *Ba chuma L intloa* " ("They are hunting ants"). I could not understand how they could hunt them at night, and, to satisfy my curiosity, followed the light. On arriving at a large ant-hill, fifteen feet in height, I found against the slope of it a hole about two feet deep and one wide. Just below it was planted one of the large bundles of dry bark, in splendid blaze; the Makobas, all expectation, made small holes in different parts of the ant-hill, and, having inserted in each some buffalo-dung, a ceremony which they called "medicine," they slit one of the smaller bundles, and, using it for a torch, proceeded in search of another ant-hill, while I remained to watch progress. For a few minutes the ants, whose wings are twice the length of their bodies, attracted by the fire, came floundering out of the hill, and so continued down the slope, till they lodged in the holes. The stream increased until they literally rolled over each other and filled the hole, many of them losing their wings. One of the Makobas now stirred them up, breaking their wings and disabling them; he then put them into a bag, and waited for the hole to fill again. Two or three large pailsful were generally obtained from each hill. Having then planted a torch beneath the hole by other ant-hills, they left them to be filled as before, and went in search of more. By this simple process they obtained myriads of these insects in a very short time. Having first dried them in a kiln and winnowed them, they are again dried in the sun, after which they are fit for use. The white ants are only taken at one particular season, the beginning of the summer rains: and after the earth has been once wetted, they make their appearance when first attracted by the light. If the natives are aware that the season has arrived, and no rains fall, they may be seen all day carrying buckets of water to wet the ant-hills. These insects eat only grass. I have frequently been an interested observer of their sagacity, and particularly of the manner in which they assist each other at their labours.

Before rains, they are most sedulous in their exertions to lay in a stock of food. In the neighbourhood of dwellings they are considered a scourge, destroying all wood-work, papers, &c.; but the queen, having an abdomen of immense proportions—two inches long and half an inch thick—in comparison to the thorax, which is like that of another ant only, is unable to move herself, and is located in a cell, having many entrances, by which the rest feed her. The queen-ant once killed, together with the king, who is generally found near her, the rest all die.

Out of business hours, I found Lechulatèbe an exceedingly pleasant and jocular fellow. He was at this time about twenty-three or twenty-four years of age, stands about 5 ft. 7½ in., has rather a lighter complexion than is common, very small and neat feet, hands delicate as a lady's, and his figure is very good. With all his pleasantry he can be very cruel, and is said to have cut the throat of one of his own wives, for having given the Makololo the medicine with which to overcome the Batawana; though it is more than likely it was in a fit of the jealousy he is known to indulge in to such an unlimited degree that he even keeps a number of fierce dogs trained to bite any intruder into his harem. One poor fellow, who had in some manner displeased him, was tied to a tree for a target, while the chief and his followers practised upon the unfortunate victim with their guns, at a long distance, uttering cruel taunts, and wild with delight at his sufferings as he writhed under every wound. About a month before my arrival, Lechulatèbe had bought a screw of a horse from one of the old soldiers who travelled with us from Kuruman. This horse he handed over to two Makobas to graze. By some accident the horse, being knee-haltered, fell into a bog and was smothered in the mire. Fearing that they would be put to death, they endeavoured to conceal the manner of the accident; but the chief, having made the discovery, took them over to the carcase, and had them tied by the neck to it, telling them to remain there and take better care of the horse for the future. In a few minutes, of course, they were smothered to death, in spite of the intercessions of all the traders that happened to be there, and who offered compensation for the loss sustained through the poor sufferers' crime. The chief replied that they were his dogs, and he would do as he liked with them.

On approaching the lake, I was astonished to see the number of waterfowl that swarmed over the surface of the river, or moved overhead, in clouds that for some moments literally obscured the sun. As the lake is receding, the fish become plentiful, and waterfowl of every description flock hither from great distances northward, to regale themselves on the fish. Ducks, geese, muscovies, pelicans, flamingoes, herons, cranes, gulls, water-hens, rails, snipe, a species of fish-eagle, and many other kinds, crossed each other on the surface, making a deafening noise; and immense crocodiles lay floating within a few feet of the women, who dip their vessels for water in the river. One of these animals had just killed a large ox, and the chief, desirous of having a piece of crocodile-skin, as an antidote against death, requested me to shoot one. Having also been requested by one of the chief men of the Bakwains to confer a like favour on him, I did my best to fulfill their wishes, but having no elevation to fire from but the boat, I was unsuccessful. The women, at first believing I was shooting birds, were delighted, and wished to have some; but, becoming acquainted with the fact that I was shooting at the crocodiles, I received such a torrent of abuse from both sides of the river as I have never been subjected to before or since — such original, scornful, and malicious curses as the Bechuana women (who are well-known adepts in this line) could alone invent. They swore that if I molested the crocodiles, which they call Morimo (God) or the "Lion of the waters," the river would dry up, and they wished me all manner of evil deaths for attacking the brutes.

In the evening I travelled up the margin of the lake, outside of the dense growth of reeds and rushes which here skirt the bank, and, running out into the water for a mile or two, effectually conceal its broad expanse of water from view. Round the lake is a belt of plain covered with a short prickly grass, which generally grows on soil emitting a saline efflorescence. This plain was formerly covered with water, but during the last twenty years the waters of the lake have receded very much, generally for two or three miles all round; many of the people, or rather most of them, remember the time that they used to navigate their canoes amongst the dark evergreens that skirt its shores.

Turning westward, the opposite shore was not visible, the length of the lake, which ranges from north-east to south-east, being at least 50 miles; the width does not appear to exceed 10 or 12 miles. The depth I have since had an opportunity of testing, and found it not to be more than about twelve feet in the deepest part; but the water, for the most part, is so shallow, that the average depth may not be more than six feet. Even now the lake is fuller some years than others; old Magalakoè remembers the time when the waves were so high as to throw the hippopotami on shore, and describes them as roaring like thunder. A few of those animals are still found in the lake, which abounds with many kinds of edible fish, and is infested with crocodiles. A great many canoes are out on the lake every day employed in fishing, but generally keeping close to the bank, for fear of the wind springing up, which would certainly sink most of these boats, as they rise only a few inches above the surface of the water, and some of them not more than an inch. The reeds and marshes teem with waterfowl that hide therein, but are generally very clamorous.

I remained on this occasion only a single day at Lake Ngami, returning on the 24th of November to my wagon, and reaching it at night, having travelled 50 miles on a pack-ox. The wagon, which stood very close to the river's bank, was visited every night, it appeared, by a crocodile, which stole from underneath it all the pieces of hide and riems it could find, as well as a very fine koodoo's head, with its horns, which I had preserved as a trophy. Our men, fearing a nocturnal visit from these reptiles, always made their beds far above the banks, where they lay closer round the fire.

The crocodile, which infests these rivers in such numbers as to make it extremely unsafe for people to venture in the water at any time, or even near it at night, is, when fullgrown, about eighteen feet long. The bulk of the body is equal to that of a large bullock, and they weigh, probably, about 1500 or 2000 lbs. Round the neck a large crocodile measures 5 ft. 6 in., and the jaws, containing sixty-six fangs (thirty-six in the upper and thirty in the lower), are twenty-four inches in length. During the heat of the day, especially when the water becomes stagnant, they delight to bask in the sun, lying like logs of wood against the sloping banks, and then may be easily shot. But they are seldom killed instantaneously, unless hit in the brain or neck. To hit the brain is a difficult shot, attainable only from a certain elevation, and through the eye, which, instead of being at the side, is on the top of the head in a slanting direction backwards. The fore part of the skull, unless fired upon from an elevation of more than 45°, is always impervious, as well as the hard and scaly substance of the back. Though always prompt to attack, the crocodiles chiefly seek their prey at night, lying in wait for the animals as they come to quench their thirst, when they either draw them into the water by the nose, or kill them with a blow from the tail. Unless very hungry, they do not devour their prey at once, but, like the tiger, conceal and watch it till it is partly decomposed. During the winter months they lie in a state of torpor, which gives rise to a notion entertained by the natives that their teeth are blunt in the winter, at which time, they say, you may risk a plunge into the water with impunity. But that this is an error was shown by the sad fate of a Mr. Robinson, who lost his life at this season, and just at this spot, a year or two ago, while waist-deep in pursuit of a couple of ducks he had shot. He was snapped in the presence of his companion, a Mr. Moyle, who saw him disappear with a fearful shriek, without being able to render any assistance. Crocodiles, like wolves, are said to note the flight of vultures during the day, and at night they crawl out in pursuit of an animal, perhaps killed by a lion, and the king of beasts, when vindicating his right to his lawful prize, generally pays the forfeit of his life in the encounter. I am also told by the natives, that when game is scarce the half-famished reptile will leave the river in quest of food, and, getting into a field of "dubbleje thorns," will so lacerate his tender feet that he lies down in despair. It also sometimes happens in the summer that, straying too far from the river, they are disabled by the hot sand, and the soles of their feet peeling off, they die. For the truth of these stories I do not vouch, though they seem to me quite probable.

Lechulatebe's Town stands in an angle between a bend of the lake and the Botletlie river. It contains about 400 huts, with the usual khotla in the middle, near the chief's residence. Near his seat was placed a staff, with the wings of a bird tied to it, intended as a charm to ward off the witchcraft of the white men. The Makololo also, who live so close on their borders, are a constant source of anxiety to the Batawana. The huts in the town are built on the same fashion and principle as all Bechuana huts, but from the abundance of grass and reeds easily procured, they have a warmer, cleaner, and more cheerful appearance. Instead of a wickerwork of bushes, their kraals are surrounded with a fence of reeds, clipped even at the top. The inside floor is, like the walls of their houses, composed of a hard clay,

LECHULATÈBE'S TOWN.

made of ant-heaps, which is smeared with dung every other day.

In their habits and customs the Batawana resemble the Bamañwato and other Bechuanas. On days when the rite of circumcision is performed, and on occasions of the marriages or deaths of persons of importance, the people refrain from work. The women toil most sedulously in their gardens, singing in chorus to every stroke of the hoe. They grow maize, several kinds of beans and grain, the pistachio nut, and pumpkins, calabashes, and water-melons.

My wagon was now perfectly repaired, after much labour
bestowed upon it, and a thick iron axle that had been bent
was restored to its former position with the aid of a Makoba's
hammer and bellows. This latter article, something resembling a Jew's harp in shape, is excavated at the base out of
the solid wood, and covered over with prepared skin; in the
narrow part there are bored two tubes, from the muzzle
leading to the base. The muzzle of the wooden bellows is
put into an earthen one, which goes into the fire, and then a
native, with one of the little sticks sewn to the skin in each
hand, pulls it up and down to a tune something like a horn-
pipe or a jig. Their hammers resemble our own, but are of
very rude construction.

We started from Lechulatèbe's Town on the 28th of November, and rattled along the banks of the river on the hard
ground, now at a gallop and then at a trot, with difficulty
stopping the oxen occasionally for a rest. But in the afternoon the road turned away from the river through the bush,
climbing the steep sand-bult (rising bank), which sloped towards the water. Here we came to a stand, and were obliged
to outspan the oxen and carry the ivory on our backs to the
top of the hill. This we accomplished by sunset, and then
drew the empty wagon up.

The oxen having failed in this first effort, and there being
many steeper hills to be ascended daily till within a couple
of days of Chapo's, it became evident that they would
never get on with such a load as we had over a piece of road
which all travellers dread to face under the best of circumstances. Consequently, after vain attempts to procure boats,
and having been unable to accomplish more than six miles
in two days, we reached, on the 5th of November, the
Makoba village belonging to Palami, a chieftain of some
importance.

While resting, on the way thither, at a small intervening

village, I had for a meal the choice between egoanas, frogs,* and white ants. My efforts to enlist the sympathy of Palami, or to procure any aid from him, were utterly fruitless. I accordingly sent Abraham forward with half of our load, ordering him to take it on for about 30 miles, and then return for the other half. Meanwhile I had despatched Andries to Thompson, at Chapo's, requesting him to send up the remainder of my oxen as quickly as possible.

On the 12th Abraham returned, and informed me that some of Palami's men overtook him where he had unloaded the ivory according to orders, about 30 miles east, and falsely told him they were sent expressly by me to say he must go on for three days' journey farther! However, he only went half that distance, and returned two days over his time. What could be the object of this abominable attempt at deception it is not easy to make out, but probably the hope of plunder in some shape or other.

We now hastily loaded the wagon with the remainder of the ivory, and were glad to get away from the neighbourhood of these thieves. Passing the one half of my property, stored on the north side of the Ngabisani Flats, on the 15th we arrived on the south side, and found the names of Bushe, Shelly, Green, and others, carved on the trees.

Left alone on the 16th of December at Ngabisani, in

---

* These are an exceedingly large kind of bull-frog (*Pyrecephalus adspersus*), eaten and much prized by all the natives. They are of a greenish-yellow colour, and often attain a length of twelve inches, are very fleshy, and, when boiled, look very like chicken. They feed upon beetles and other insects, and burrow in the ground during the whole of the dry season. As soon as the ground is sufficiently moistened by the rains to allow of their escape, they make their exit, and find their way to the pools, where they deposit their spawn, and where they may be heard at a great distance by their peculiar noise, resembling the distant roar of a lion. After every shower the Bushmen make regular excursions in pursuit of these animals, and procure great numbers of them. Their size and weight in general equal those of a full-grown chicken.

charge of a portion of my property, while Abraham and the other men were engaged in bringing up the remainder, I packed up the tusks, one above the other, under a tree, to the height of about three feet, and made my bed on the top of the heap, and, having kindled a fire, went with a tin pail to the river, a quarter of a mile off, for water. While strolling about on the banks of the river, I gathered a quantity of a kind of wild plum, called *morutoloogoë*, growing on a large bush. The colour was red, the taste a combination of sweet, sour, and bitter, with a slight astringency. The bush had small, dull green, oval leaves, guarded by thorns above and against it. I gathered the fruit, to make a sour drink, by steeping and bruising them in the water, as the natives do with the wild loquat. But on my way to the fire-place I was seized with a strange sensation of giddiness and stupor, and great debility, and had a feeling of nausea growing upon me, accompanied by headache and pain in the neck and back. I felt at once that I was attacked by the fever prevalent in this country, particularly in the summer, and wondered only that I had so long escaped it. While still able, I looked out the few small papers of medicines I possessed, and took from them ten grains of jalap and ten of rhubarb, with three of quinine, and longed for four grains of calomel to put with it—the usual prescription, I believe, of Dr. Livingstone for this fever. Having taken this mixture, I laid a large log of wood on the fire, which would burn for several days, and, having placed the bucket of water by my side, lay down on the heap of ivory, panting with fever, and soon fell into a state of semi-oblivion, haunted by the most frightful dreams. The next morning found me in a state of burning fever. I took large draughts of the cold water that stood beside me, for which I felt an incessant craving, returning to it at intervals of every few minutes, and refrain from it I could not. In the evening the expected wagon had not arrived, and I fell into the same state of dreamy obliviousness, losing sight of everything around me, and fancying myself transported into indescribable regions, fighting all sorts of visionary battles, and seriously imagining myself, at times, to be a many-headed monster, a hideous serpent, or other reptile.

The third morning after Abraham's absence I was able to take another dose of medicine, and fell again into my former state. But the fourth day, having a considerably greater degree of consciousness, I began to reflect upon my situation, and felt no regret, pain, or anxiety at the prospect of death. For this I was probably indebted to the reduced state of the body and enfeebled state of the senses. Death, indeed, seemed quite welcome; the only thing that I remember having given a thought to was, whether anyone would find and bury me, or whether, as soon as my spirit fled, my body would be devoured by the troops of vultures that sat on the trees all around. They had already robbed me of all the meat, which I was unable to protect, and now seemed only awaiting my decease. Whenever one of these ill-omened birds flapped his wings, or hobbled from the trees, I longed to shoot it; this feeling, however, did not last long without reproducing the fever which had already consumed me. I again had recourse to my can of water, which I drank to the last drop, and then, craving for more, crawled out of bed, and, with a small kettle, staggered towards the river. I soon fell to the ground, but, the fever and thirst increasing, I renewed my efforts from time to time. Crawling on my hands and knees, it took me two or three hours (though the distance was only 400 yards) to reach the water; I quenched my maddening thirst, and, having rested and revived, I managed to get back to my fire-place with my little kettle of water. There I lay till about 5 o'clock in the afternoon, when Providence sent Thompson's man to my relief. The messenger I had sent from Palami's had at length arrived at Chapo's, and given Thompson my letter, who despatched January with the oxen, as I requested. He was as much astonished to find me in this plight as I, to all appearance, was indifferent about his arrival. I had, however, enough of remaining consciousness to tell him that Abraham had been away four days instead of two, and that I imagined the party must also be sick, and I advised him to go in search of them. This he did, after handing me over to the care of some Bushmen, who came with him. He returned two days after, bringing up the wagon, with Abraham and the rest huddled together in it. They were scarcely able to recognise me, being in a sort of delirium from the effects of the fever, which attacked them the same day that it did me, and but for the timely arrival of January from Chapo's, who brought us again together, we all, probably, and I at least, must have perished within a few days.

The ivory was now all put into the wagon, and the oxen, now fourteen, had the additional weight of five human beings. On the 23rd of December I was put into the wagon amongst the rest of the sick, and tossed about by the jolting vehicle upon a hard bed of ivory. I scarcely remembered anything till the evening of the 25th (Christmas Day), when the hind part of the wagon came down with a crash, pitching out some of the sick, and laying me in a very uncomfortable position, with my heels elevated to a height of about forty or fifty degrees. Being now on the plains near Chapo's, January went forward to fetch the other wagon, and next day Mr. Thompson arrived with it and took me on to Chapo's Town, leaving January to bring up the broken wagon in the best way he could.

I now determined to abandon my wagon and put all my effects into Mr. Thompson's, feeling satisfied that with the number of oxen now at command we could get on safely, though slowly. We placed the weaker two of the other sick men in the wagon, on the top of upwards of 3000 lbs. of ivory, and were ready to start, having purchased thirty-seven goats for slaughter.

During our short stay at Chapo's a shocking occurrence took place—the putting to death of a child born with some deformity. I was too sick to pay any attention to what was going on outside the wagon; but our people saw them smother the child under skin blankets, singing, groaning, and wailing over it, and heard the cries of the dying infant. Among all the native tribes deformed children are not suffered to live; that is probably the reason why we saw no deformities amongst them. Besides, the way in which they bring up their children, without too much pampering, lacing, and bandaging, has the effect of giving them their full and natural development. From the earliest age they are exposed to all inclemencies of weather, their little shaved heads laid bare to the heat of the sun, and they grow up hardy and able to endure anything.

On the 1st of January, 1854, departing from Chapo's Town in the middle of the reedy swamp, which was now dry, we reached Shogotsa (incorrectly called Inchokotsa, as Chapo is by some called Tsapo). On the night of the 2nd, travelling over the salt-pans, two feet deep in the water, we next day reached Kokonyani, where, for the first time, I managed to get out of the wagon and sit in its shade. Travelling by the usual stages, it raining almost every day, and the country being full of water, we reached Sechelli's Town on the 24th of January. The chief had returned from the Cape, whither he had gone to seek protection for the natives from the oppression of the Boers, and with the intention, in the last resort, of proceeding to England, and imploring redress from the Queen. He returned disappointed, having proceeded no farther than the Cape. He received us kindly. He had built a small chapel, in which a native teacher, paid by him, officiated and taught his children.

In the afternoon of the 25th, it being a very fine day, we were startled by a most unusual sound, and a loud rumbling, like subterranean thunder; it seemed approaching nearer and nearer, and we saw everything visibly shake before our eyes: we felt it was the shock of an earthquake, unusually severe. In an instant the whole town was in alarm, the women rushing out of their huts, with clubs and hoes in their hands, holding them up at the sky, and cursing God with most awful imprecations. After an instant's cessation there was another shock, accompanied by a crash which was even louder than the first. Some of the women lost their footing as they stood rending the air with their demoniacal yells. The two successive shocks must have lasted more than a minute, and Sechelli declared there was a great chief dying in another country, and we were to let him know when we got home who it was.

On the 26th, in the evening, we left the town, and on the 27th I fell in with Mr. Samuel Edwards, who will be no stranger to the reader of the succeeding chapters. He, also, had felt the shock of the earthquake, though at the Bawankitze Town of Santhoohie's at the time. I remained in his company three days to talk over our adventures, having both travelled over most part of the same country. After our parting, I pushed on at an unusually fast rate, and reached Kuruman on Saturday night, the 11th of February, where Mr. and Mrs. Moffat, commiserating my reduced state of health, turned me daily into the garden to feed on the grapes and abundant other fruits, the best medicines I have ever taken. I soon gained strength to walk about again, though still exceedingly emaciated, and frequently troubled with attacks of ague. Amongst the variety of fruits we enjoyed were pears, apples, figs, grapes, almonds, and walnuts.

Many itinerant traders, called Smouses, visit this country yearly to barter for slaughtered oxen and sheep for the Cape Town market; but latterly, since the Sand River Treaty, the disappointment of Sechelli, and the rumoured intention of the British government to give up the Sovereignty, traders are very cautious about entering the country. The natives seem all to apprehend attacks from the Boers, and hold themselves in a constant state of preparation for war.

Mr. Edwards having now returned from Sechelli's, we were again thrown into one another's company, and had abundant opportunity of comparing our adventures in the far interior. This intercourse led to my entertaining a growing desire to undertake a renewed expedition, in spite of the determination, so often mentally resolved on, of avoiding a recurrence of the perils and hardships to which I had been exposed. But the charm which attaches to a life of adventure is irresistible. My friend and I had not often met, before we came to an agreement to make joint-stock of our means, and, with new wagon, cattle, and horses, start for another hunting expedition in Moselikatze's country.

Having settled all accounts with Mr. Thompson, that gentleman set off to join his family in Natal, and a few days later Mr. Edwards and I proceeded towards Fauresmith, in the Sovereignty.

CHAPTER X.

A new Expedition—Fauresmith—Cession of the Sovereignty—At Phillipolis—Back to Kuruman—Rev. Mr. Moffat—A Stalactite Cavern—Reach Sechelli's—The Question of Polygamy—Arrival at Sekomi's—Plans of Procedure—The Author again sets out for the Lake—Reach Chapo's—Buffalo-hunting—Crocodiles in the Botletlie River—A rare Antelope Shot—Arrival at Lake Ngami.

ON the 21st of February, 1854, we proceeded on our journey to Fauresmith, by way of several well-known farm stations of the Dutch settlers. At one of these, called by the name of Saul's Kuil, on the margin of a basin which seemed to have been once a small lake, I observed a large and deep hole bearing the appearance of a volcanic crater. The surrounding rocks, sloping towards the crater, were composed of a conglomerate, with innumerable agates embedded in it—presenting the appearance of having formerly been in a fluid state. At midnight, on the 24th, we reached Campbell Town, having travelled twelve hours, or about 30 miles, during the day. Campbell Town is a mere farm, with a small missionary chapel, and a couple of dwelling-houses, one of which is occupied by the Griqua chief, Cornelius Kok, a man entitled to a wider command than he holds at present; his influence, however, is great amongst the Griqua people, as well as amongst a large number of the Korannas. He led us next day across the ford of the Vaal river, which was at this time rather deep—so much so, that the water washed through our wagon, and the leader and oxen had to swim. On the southern bank of the stream we found the mission station of Backhouse, where I met my kind friends, the Rev.

Mr. and Mrs. Edwards, again. The next day Mr. Edwards preached in Dutch and Sechuana to a mixed congregation.

On the 1st of March we travelled in a south and sometimes east direction through a fine sheep country, passing some farms and small fountains, and next day continued our journey for eleven hours in a rather easterly direction, through the same sort of country—long level plains here and there, broken by tabular ranges, rising abruptly on the horizon. We passed during the day some farms well-stocked with cattle, horses, and large flocks of sheep, for the rearing of which latter the whole of this district beyond the Vaal river seems particularly well adapted. One farm that we passed belonged to an old emancipated slave, of negro extraction, who, by his industry and well-known integrity, has accumulated large flocks of sheep, to the number of several thousands, as well as a great number of cattle and horses. He is the proprietor of several farms. His daughters are said to be respectably married to Dutch farmers. On the 4th we reached Fauresmith, a rising little village with several fine stores, and about 100 neat flat-roofed dwelling-houses.

At Fauresmith we found the majority of the people, and even some of the Dutch farmers, exceedingly indignant at their loyal allegiance to the British government being discarded, and the country on the point of being transferred to what is called the Orange River Free State, in spite of their remonstrances. A few days later, we learnt that the British troops had already left Bloemfontein, which place was described as being in consequence like a city of the dead, and Sir George Clerk, to whom the settlement of affairs was entrusted, had taken up his quarters at Fauresmith. Hearing that I had recently returned from the lake, Sir George desired an interview with me, with the purpose of making inquiry respecting the condition of the interior. I did not withhold from him my opinion of the ill consequences which would attend the policy pursued by the British government, and subsequent events have fully justified my anticipations. The respect and esteem which the natives once entertained for the British government, as their avowed protector, no longer exists.

On the 15th of March Sir Geo. R. Clerk left Fauresmith with a train of mule wagons, proceeding to join the troops at Phillipolis, where he had still questions to settle with the Griquas. Thither Mr. Dickson, Mr. Edwards, and myself followed next day, and arrived in time to see the entrance of the troops, accompanied by a train of more than a hundred bullock wagons.

On the 17th Sir George made liberal offers to Adam Kok, on his being released from his allegiance to the Queen; but Kok refused to accept of any pecuniary recompense, or to give up his country to the new Free State government on any terms. Sir George admired his motives, which, he said, did him great credit; but, under the circumstances, he would be obliged to withdraw British protection, and cancel all treaties, as there was now no longer a British government on this side the Orange river. Kok, however, still persisted in rejecting all offers of land in the colony, or any other recompense for surrendering his rights.

While staying at Phillipolis, I enjoyed the hospitality of the Rev. E. Solomon and his kind lady; and returning to Fauresmith, Mr. Edwards and I effected in the course of a few days the purchase of two new Paarl wagons, for each of which we paid £112 10s., with a number of fine strong oxen and several horses. We also made purchases of additional guns, ammunition, stores, merchandise, beads, &c., amounting in all to about £900; when, leaving my fellow-traveller to make some further purchases of horses, &c., I set out at midnight, on the 23rd of March, making direct for the Vaal river. I forded the Vaal river, then just passable, on the 4th of April, and proceeded to Campbell-dorp, there to await Edwards's arrival.

I found the chief, Cornelius Kok, that "fine old *black* gentleman," fat and hearty, keeping, as usual, his tribe in good order, and bringing them to punishment regularly for misconduct on the Sabbath. This was formerly a very wretched tribe, but missionary labour has had a great influence on them, and they are now much advanced in civilization, though I believe it was only introduced among them subsequent to 1815.

Old Kok was surrounded by a number of Bushmen, who were making active preparations to gather the locusts that passed over in clouds, partly obscuring the light of the sun, and, settling on the ground, covered every blade of grass. Dogs were running about feasting on them, and the horses also stood quietly feeding on them: as for the Bushmen, it was quite a happy time; they were watching the flight to see where they would settle for the night, and then repairing thither to gather them in their bags and baskets.

On the 10th of April Edwards arrived from the south, and news of a rather unpleasant nature came at the same time from the north: that the natives had taken it in their heads to plunder all white men, believing that the British government had, through fear of the Boers, given up the Sovereignty to them; and our countrymen being now completely vanquished by the Boers, our friendship was now not worth counting. It was also reported that Waterboer had unpacked the wagon of some Cape traders to search for gunpowder, and they said that we should be stopped and plundered at Daniel's Kuil. To satisfy ourselves of the truth of these statements, we rode over to Griqua Town, where we were kindly entertained by the Rev. Mr. Hughes, and learnt that, to a certain extent, the information we had received was correct. Our wagons were to be examined, and we should be compelled to sell anything they required upon their own terms. This we determined should not be the case; and though the Griquas everywhere were assuming a hostile bearing towards Englishmen, we determined to go forward on our course, and trusted to our address and experiences to make our way through the country.

On the 14th we left Campbell-dorp, and at midnight on the 15th reached Daniel's Kuil, where we remained over the Sabbath, and were visited by a field-cornet of Klaas Waterboer's, who had been sent to unload our wagons; but his good sense prevailed, and he did not molest us, Matsabon, the field-cornet of Kuruman, informing him that he had had sufficient experience of me to know that it could not be done without a scuffle. We left at sunrise on Monday morning, and reached Kuruman on the 18th.

At Kuruman we found Mr. Moffat in rather indifferent health, suffering from a peculiar affection of the head, brought on by close application to his laborious studies in translating the Bible. On learning that Moselikatze's was our destination, Mr. Moffat determined to accompany us. He had previously decided on accepting an invitation which that chieftain had given him, but until our arrival had given up the idea of going this year. During the brief delay which Mr. Moffat's preparations involved, we visited, in the company of the Rev. Mr. Ashton, a large cave in the mountains at the back of Kuruman. Its roof was encrusted with stalactites, and the crevices of the rock swarmed with bees. On the way back we met with numbers of guinea-fowl, as well as a few pheasants and partridges.

We left Kuruman, with Mr. Moffat in company, on the 22nd of May, and, travelling the old road by easy stages, getting some exercise for our guns by the way, reached Sechelli's on the 10th of June. One night during the journey our rest was disturbed by the roars and growls of a troop of lions, who kept prowling about our wagon until daybreak. We found Sechelli in high good humour. He supplied us with abundance of beef, corn, milk, and tamanies—the last a kind of large bean, growing on a creeping plant, with an edible root resembling a yam, and called morama, but more juicy. The bean, when roasted, is a very good substitute for cocoa. We remained here over Sunday, when Sechelli, who had been reading the first portion of the translation of the Old Testament, gave Mr. Moffat a text, and asked him to preach upon it, as his people, to whom he had been reading, were puzzled to know why the missionaries had made him discard all but one of his wives, while Solomon and David had so many wives and concubines, and were still " men after God's own heart."

Under the influence of Dr. Livingstone's teaching, Sechelli had turned off all his wives excepting the one he married first; and was then baptized, and admitted into the Christian communion.* It must, I believe, be admitted that, however contrary the institutions of polygamy may be to the established principles and usages of the Christian churches, its abolition is the great stumbling-block to the conversion of the heathen. The chief, Sekomi, who is still a heathen, once said to me, " I should like to have a missionary, and to become a Christian, if I could be allowed to keep my wives. *I don't want any more.* I have transgressed, and

* It is but fair to quote Dr. Livingstone's own account of this matter. " Perceiving at last," he says, "some of the difficulties of the case, and also feeling compassion for the poor women, who were by far the best of our scholars, I had no desire that he (Sechelli) should be in a hurry to make a full profession by baptism and putting away all his wives but one. His principal wife, too, was about the most unlikely subject in the tribe ever to become anything else than an out-and-out prosy disciple of the old school. She has since become greatly altered, I hear, for the better. . . . .
" When the chief at last applied for baptism, I simply asked him how he, having the Bible in his hand, and able to read it, thought he ought to act. He went home, gave each of his superfluous wives new clothing and all his goods in their huts, and sent them to their parents, with an intimation that he had no fault to find with them, but that in parting with them he wished to follow the will of God."—*Missionary Travels,* p. 18. [ED.]

nothing can ever undo that which has once been done; but I cannot turn my wives and children out. All men's hearts will be against me; I shall be alone on the earth. To have my wives disgraced, and my legitimate children branded with a false and ignominious name, would bring overwhelming ruin and trouble without end upon me." If it were possible to admit such men, with their wives and families, into the church, without degrading them in their own or their people's estimation, the labours of the missionary in South Africa would lead, not to those meagre results with which we are but too well acquainted, but to the conversion and civilization of whole tribes, as the people would all follow the example of their chiefs. In the case of Sechelli, the great sacrifices he has had to make have, unfortunately, terrified all the other tribes, who now hold that these sacrifices and his other troubles are the fruits of his newly-adopted faith. It was very natural that Sechelli, perplexed by the questions which some of his people who had been reading, or had heard read, Mr. Moffat's translation of the Old Testament, now put into their hands for the first time, should ask the missionary for explanations on so important a subject. I regret to say, that, notwithstanding Mr. Moffat's able discourse, the native mind was far from satisfied.

It is all very well for those whose consciences are enlightened, but it seems a hard case to persuade a man at a stroke to put away his wives against their will, when he and they have contracted together in good faith and according to old-established custom. They have known no other law; and we know that under the old dispensation polygamy was permitted and practised by pious men, without the Almighty's intimating any displeasure or reproof. Hence Mr. Moffat failed to make any impression; for, notwithstanding that he adduced David and Abraham's after-troubles as the consequences of their polygamy, and said we had every reason to believe that David and Abraham might be now undergoing the awful penalties of this their sin, still the natives could not get over the idea that those men were men after God's own heart, though they had many wives, besides concubines. Although we knew very well that polygamy is not in keeping with the spirit of Christianity, we could give them no better elucidation than Mr. Moffat had done on this and many other points; and it certainly seems to savour of cruelty and injustice to require the natives to break off these marriages, and put away all but the first wife and her children against their wills. It may not be so hard in some respects, in the case of a wealthy chief who can still support them, but the evil and the responsibility incurred in other cases appear to be very great. In their own estimation they are not adulterers. The people were married according to the rules of the land, and to brand them as adulterous will turn all heathens against the faith which we are so anxious they should embrace. It is a pity that some middle course could not be found to remedy the evil.

While at Sechelli's we learnt that he had been trying to unite the tribes against their common enemy, the Boers. But he has not altogether succeeded in this effort, since an impression appears to prevail that his former career has been in too many instances marked with treachery to allow of his being trusted.

We left Sechelli's on the 12th of June, and—getting between us abundance of sport on the way, bringing down some fine giraffes, and also a gnu, and some smaller game— reached Sekomi's Town, a distance of 145 miles, on Monday, the 19th. At Sekomi's Edwards and I held a consultation, at which we decided on a change in our plans. Our object was to secure as much of the trade of the country at first as would insure us against loss, or at least leave us the means for a future undertaking; and, as a wagon of mine was left on the road to the lake, and Moselikatze's country was quite an uncertain field of operations, we resolved on separating at this point—one of us proceeding towards the lake to trade there, while the other would visit Moselikatze, form an alliance with that chief, and trade for as much ivory as he could furnish. Upon the successful execution of this project, which would, we hoped, secure us against any pecuniary loss, we looked forward with sanguine expectation to the selection of some field for the enjoyment of the exciting pursuits of the chase alone, undisturbed by any other considerations. Having determined upon this plan, we drew lots for the choice, when it fell to Edwards to proceed to Moselikatze's, my lot being for the lake. Making an equal division of our property, we equipped ourselves for a long journey in each direction, Edwards a distance of 270 miles north-east, and myself 450 miles north-west of Sekomi's.

On the 21st of June Edwards, accompanied by Mr. Moffat, started for Moselikatze's country, while I steered in the opposite direction for the lake. The finding a number of scorpions, which were driven out of the burning logs one night, was among the least pleasant incidents of my progress. One of these reptiles—an exceedingly large black one, nearly six inches long—found its way into my bed, which the coldness of the air had led me to have located near the fire, and bit my leg. Another was found in the bed next morning, and, while sitting reading after breakfast, a third leaped on to my breast; of course I soon dislodged it, but it fell in the wagon between my boxes, and lay concealed there for several weeks. On the 30th of June I fell in, at a place called Kokonyani, with Messrs. Dyme and Moyle, travellers who had left the Cape two years previously, and had encountered great disasters in attempting to reach the lake by a road parallel to that followed by M'Cabe through the Bakalihari desert, losing all their cattle and horses from thirst. When reduced to almost the last extremity, they had providentially received assistance from Mr. Andersson, who had fallen in with them at Riet Fontein. By means of oxen, with which he supplied them, they struggled on to the lake, and afterwards with great difficulty had got as far on their homeward course as the place where I met them. But it would have been impossible for them to have got through the desert without additional help. I accordingly let them have eight of my oxen, and willingly gave them a small portion of my coffee, sugar, tea, and flour, luxuries which they had not tasted for more than a year, having subsisted during that period almost entirely upon flesh, with sometimes wild roots.

On the evening of the 5th of July we arrived at the edge of the reeds in Chapo's marsh, which, being on fire, our progress was stopped, but next day we reached his town. At the time of my abandoning my wagon here I had instructed him to keep it safe, thinking I might fall in with some friend or traveller to whom it might prove of service, little imagining at the time that it would be of any use to me again, as now turned out to be the case. A nineteen-gallon cask left in it was now full of honeycomb, a swarm of bees having taken up their quarters in it during my absence. Finding that, contrary to expectation, the wagon had been well cared for, and not robbed of any portion of the iron-work, I rewarded Chapo with a cow, which was of more value to him than half a dozen wagons. The chief kissed my hands rapturously on becoming the owner of his first cow, bidding me "*Gorogo Ka pula Ra!*" ("Arrive with rain, father!")—the most hearty welcome a native can give you; for with them no blessing is so great as rain, which makes the grass spring up and kine fatten, the grain to grow and man to thrive, with many other good things of which it is productive, making it, in their estimation, the source of all earthly happiness.

On the 10th, having completed the repairs of my wagon,

we were in trekking order, and left Chapo's for the lake, with our two wagons. We saw large herds of gnus, springboks, and luchees, two of which latter were caught by our dogs and speared by the Makalaka followers whom we had engaged at Chapo's. The beautiful guinea-fowls made a harsh and grating noise, and thousands of pheasants ran all day long fluttering in our track, scared by the unwonted appearance of the wagons. The plain along the river-side, which is covered with a short and prickly grass, seen only where a nitrous substance forms a portion of the soil, abounds with hares and korhaans, of which we saw a great many.

On the 13th I rode out hunting parallel with the wagons, accompanied by my achter-rider, Molihie. After passing among numerous herds of quaggas and springboks, we fell in with a troop of about seventy buffaloes. Molihie wounded a calf and the mother. A smart and well-conditioned young cow remained to protect her offspring, and, having made several active but ineffectual charges, she fell, pierced with three of my large bullets, dying the victim of maternal affection. Falling in with buffaloes again, I bagged another very heavy bull, which I rolled over at full gallop at the distance of 200 yards; rising again, he charged furiously, and received a broken leg, but still continued to charge on three legs until we despatched him. An exceedingly large and fat jackal, anxiously awaiting my departure, at the distance of about 300 yards I knocked over, smashing the vertebræ; and, reaching the wagons, I played havoc amongst the pheasants and guinea-fowl. After waiting in vain for the flesh of the fat cow-buffalo, I started next morning at about 9 o'clock. Being in advance with my dogs, they scented some buffaloes that had taken up a position in a jungle, into which I found it impossible to penetrate, so dense was the underwood and briars, and was obliged at last to call off my dogs, and give them up. I next fell in with three fine koodoo-bulls, and while giving chase fell in with Chapo's son, at the head of a party of hunters. Marching at the distance of 300 yards apart, their line extended over the mañana plain for two miles. The hunters, each accompanied by a few small curs, and armed with a club and spear, marched westward, making a semicircle to enclose the game. I came up in time to be in at the death of one of the kosi (*Megalotis capensis*) kind, the fur of which they value more, as being warmer, though not so handsome. Of these skins they generally take great care; the moment the animal is killed they rub the skin dexterously with sand to remove the saliva left by the mouths of their dogs, and at their first resting-place they skin it, drying it carefully, and when they have collected a number, convert them into karosses or blankets for the use of the wealthier amongst them, or dispose of them as tribute to Sekomi.

In the afternoon we travelled on, and slept in the middle of the bend of the Botletlie river, called Thamesan, our cattle suffering dreadfully from the sickle-thorns, which drew blood from their noses and continually brought them to a stand. Horses and cattle having once suffered from this virulent thorn are afraid to face it, and show symptoms of fear at every tree or bush bearing leaves resembling the sickle-thorn.

We started again early on the 17th. Learning from the natives that crocodiles were abundant, in the afternoon I strolled ahead with Molihie, who soon discovered and pointed out a large cow, basking in the sun about thirty feet below me. I carefully levelled my gun and fired a bullet through the vertebræ from above, near the region of the head: she tumbled hastily into the stream, and, having gained her element, struggled for half an hour in the agony of death, and, dyeing the water with her blood, sank beyond my reach. Observing another equally large cow swimming to our side of the river at some distance ahead, I approached within twenty yards unperceived, put a bullet into the centre above the two shoulders, and, like the last, she struggled to the stream, when, after writhing and lashing the water with her ponderous tail for a long time, now shooting into the air and diving again to the bottom, and performing all kinds of tortuous and painful evolutions, sickening to behold, she also sank in the blood-stained element.

A little farther on I killed another, rather larger than the two former, which only measured, I should say, about thirteen feet. But just before leaving the river-side I fell in with a troop of about twenty, amongst which I observed a wary old patriarch, probably eighteen feet long, and bulky as a buffalo-bull. This old one, possessing a pair of intensely acute nostrils, became aware of my approach while I was still very distant, and, swinging his head suddenly from side to side, turned it towards the stream and glided into it, reluctantly followed by the younger members of his family, who seemed not half so wary or sagacious. Coming immediately opposite to where he floated, I fired at the distance of 200 yards: the bullet took effect in his back, and, without making any such lively demonstrations as the preceding ones I had shot, he sank with one splash of his enormous tail, followed by the rest, leaving the broad basin of the river in a state of commotion for several minutes after.

The next morning I again fell in with crocodiles, and, shooting another in a shallow stream, had an opportunity of judging of their tenacity of life, while the Bushmen that accompanied me, and who have a great antipathy for these reptiles, were in roars of laughter at its extravagant evolutions.

Tired of the tameness of this kind of sport, I rode out again with my achter-rider, bagging a buffalo and wounding some pallahs, and on my return shot a few brace of ducks and guinea-fowls. These latter birds are so numerous that we sometimes fell in with flocks extending over a quarter of a mile, often making a dismal and grating noise. They are heavier than a large fowl, and are very good eating, especially when plain boiled. At one season of the year their eggs are found in great numbers by the natives, who subsist for days and weeks on them alone. Each bird lays between sixty and eighty eggs, speckled like a turkey's, the size of a fowl's, and exceedingly delicate. They are often hunted by the natives, with the aid of their dogs, and, taking refuge in trees, are killed with short clubs thrown at them. Birds of every description are plentiful on the river-banks, but were only slain by us when in want of other flesh. We saw every day numerous monkeys, also paroquets, and a species of lory, as well as many kinds of waterfowl.

Continuing, on the 19th, my course up the river, and making havoc amongst the crocodiles on the way, at night I mortally wounded one of three great Kring-gat waterbucks (*Aigocerus ellipsiprymnus*), as they pranced like untamed horses before me; taking to the river, it expired on the opposite side of the stream, where, as I learnt a few days afterwards, the Bushmen had a feast upon it.

On the 20th, still advancing along the Botletlie, we observed a number of bull-elephant spoors: I followed these for 20 miles without success, and, returning, shot a fine bustard, measuring nine and a half feet from tip to tip, also three or four brace of pheasants for our larder.

On reaching Makato's village, I heard from him the story of his recent misfortunes, and the whole particulars of Sekeletu's attack—his loss of all his wives, and the sweeping off of all the Makobas, excepting the few that fled to Chapo's, and are gradually returning, or who sheltered themselves in the desert till the storm was over. I passed on for the junction of the Tamalukan with the Botletlie river, and slept there that night.

Being feverish and restless, and strolling out early, ahead of the wagon, to enjoy the balmy freshness of the morning in the groves of the golden-blossomed mimosa, I had not proceeded far when I fell in with the loveliest animal that, as it seemed to me, was ever created. It seemed to partake of the appearance of the bushbuck, the koodoo, and the domestic goat. On first sighting it, I observed it with its front feet climbing the branches of a thorny bush, on the young leaves of which it was browsing. Its colour is a bright red, with transverse stripes of round snow-white dots, which become larger and bolder towards the tail. A streak of longer and darker hair runs down the whole length of the back; the front part of the body is rather of a neutral tint, the face marked like a koodoo, and the horns resembling those of a bushbuck. It stood about as high as a goat, which it resembled also in its shape and its actions.

Hearing my approach, it started from its occupation, and, looking back over its shoulder, eyed me with the same degree of astonishment that I must have done, for I entirely forgot my vocation, and the pretty animal bounded off towards the desert, but, having gained the height of the sand-bult, it halted for a moment to get a better look at me. In an instant my gun was at my shoulder, and, with a quick but certain aim, I fired, and the animal fell to the ground, pierced with a large bullet through the upper parts of the two shoulder blades. Before, however, I could reach the spot it was up and away again, much to my mortification. I hastily loaded my gun, and, having found that the creature bled profusely, I determined to have it, if it cost me a week's delay and trouble; and I never had two hours' more exciting chase nor was ever rewarded with a more magnificent trophy in the many years I have shouldered a gun. Taking my guide with me, I followed the trail back to the dense bush on the river-bank, and, having crept through the underwood on the bloody trail for some distance, found that the beautiful creature had taken up its position in the dark shade of a moralla tree, which was so well fenced in on all sides with briars that I could not penetrate them. My guide daring to raise his arm to throw his spear at the animal, was nearly frightened out of his wits at my rage, for I would not have had the skin spoilt for any consideration; but I soon repented, for the next instant away went the buck, and I feared greatly that I should not see it again; however, I had better luck, putting it up repeatedly, but unwilling to fire for fear of spoiling the skin, until at last it escaped me for a long distance, seeming to show signs of recovery. The next glimpse I got of it, it bounded off in so lively a manner that I was only too eager for a shot, and the animal halting at the distance of 100 yards to look back, I sent a hasty shot through its heart, and in the agony of death it ran directly towards me, and, while I seized hold of one of its legs, it literally expired in my arms. In a few minutes my servants and about fifty natives arrived, greatly admiring the beauty of the animal, and, congratulating me upon my success, rent the air with deafening yells and plaudits. I must say that I never before or since felt so pleased and proud of a trophy won on the field. The animal seemed quite a stranger here, and I had some difficulty in finding out its name. By the Batawana it is called thamma, and by the Bayèyò or Makoba, tugwumgo. I presented the skin of this animal to Mr. Moffat, who intended sending it to the British Museum.

In the afternoon I left the wagon, and, accompanied by my servant, rode on horseback to the lake, which I reached the same day.

On the arrival of my other wagon the next day, I observed a great commotion among the chief and his followers, who all took their departure in a high state of alarm. To my astonishment, I found that the cause of this commotion was the circumstance of my people having brought the crocodiles' feet into the town, and coolly skinning them by my orders. The natives here have a great aversion to the sight of a crocodile, and cannot even be persuaded to look upon a dead one. They worship it, as do the Bakwains, who are called after it—that is to say, they biena * it—fear it, and believe that if they were to kill a crocodile, or look upon one that is dead, woe would betide them, or they would die. I always observed that they made a great circuit to avoid the sight of a dead crocodile, which they also say is Morimo (God). Nearly all the tribes, excepting a large portion of the Bushmen and the Makobas, have this same superstitious fear in a more or less exaggerated form. However, the doctors or wizards, who believe that this reptile furnishes powerful *molemo* (charms or medicines), are often anxious to obtain a portion of its carcase. But so far do the people carry their belief in the evil of coming in contact with or even seeing one, that had any misfortune happened to the tribe while I was in the town, it would have been ascribed to me. A special embassy was sent me from the town to beg that I would remove the obnoxious article, but I did not think it right to lend any sanction to their superstitious fears, and I sent a message to the chief, telling him that "God's medicine was stronger than the crocodiles," and he need fear nothing. In reply, being very anxious to see me, he begged that I would sprinkle the white man's medicine about to keep the evil spirit from his person when he came. I did nothing of the kind; but I observed that he took the precaution to bring his doctors with him, bearing medicines as antidotes against the devil.

* Biena, to dance before, reverence, worship, &c.

CHAPTER XI.

Return to Chapo's—Start for the North-eastern Interior—Alarms by Natives—The Ntwetwe Salt-pan—Encounter with Lions—Abundance of Game—Reach the Shua River—An Elephant-hunt—Makarikari Salt Lake—Confluence of Rivers—Borders of Moselikatze's Country—Dealings with some petty Chiefs—A Night Encounter with Makalakas—Bushmen massacred—The Simonani River.

My hopes of falling in with elephants having hitherto failed of realisation, I now meditated a long excursion in an eastwardly direction towards Moselikatze's country, and through a region as yet quite unexplored. After consulting one of my Makalaka servants, whom I had engaged at Chapo's, as to the practicability of such an undertaking, I determined (although with small encouragement on his part) to make the attempt, and accordingly hastened back to Chapo's, which was to be my starting-point, for the purpose. I already knew the country for nine days in a north-eastwardly direction, and had no doubt of my ability to get through it; but I did not feel equally confident as to the treatment I should meet with at the hands of the native tribes on the way, who had never seen a white man, a horse, or a wagon, and whose minds had been poisoned against the whites, from interested motives, by Sekomi's tax-gatherers.

Having fully determined, however, on the venture, I arranged the wagons in such a manner as would enable me, on arriving at Chapo's, to despatch one of them homewards, while the other was filled with the luggage and supplies necessary for my long journey. I left the lake on the 3rd of August, 1854, and in doing so gratified the secret wishes of the chief, who, fearing another attack from Sekelètu, felt concerned lest he should incur blame from the head of the white nation in the event of any accident happening to us. Some of his sage councillors, however, had already concluded that we carried a "medicine," and that our lives were charmed, as otherwise we should not dare to force our way into the presence of great chiefs, through their multitudes of people, without betraying any signs of fear.

On the way to Chapo's Town, which we reached on the evening of the 24th, I had a good deal of sport, with the details of which I will not fatigue the reader. The crocodiles, in particular, I had a feeling of satisfaction, amounting almost to a sense of duty, in destroying, and the numbers of those reptiles with which this river Botletlie abounded gave plenty of employment to the rifles of my servants as well as myself. On one occasion as many as fifty-seven crocodiles were seen huddled together on the margin of a pool to which the women resorted for water. One of those that I killed in the Botletlie river was of enormous size, its extreme length being upwards of sixteen feet, the circumference of the neck more than five, and the jaws rather more than two feet long. The body of this brute was as bulky as that of a buffalo, and the united efforts of eight men failed to drag it out of the water. Its legs, as thick as those of a rhinoceros, were however secured and skinned, and now, I believe, adorn the shelves of a "Children's Mission Museum" in London.

A few days sufficed to complete my arrangements, and guides were promised me for a start to the Shua river, where, judging by the favourable reports of the natives, I hoped to shoot at least a few elephants, and to secure some fine long-tailed rhinoceroses. For these latter the Shua, hitherto unknown to white men, is famous. When on the point of setting out, however, news came which rather damped the spirits of my people, and nearly prevented the prosecution of my plans.

Some Bamañwato arrived with the tidings that two regiments (*mepato*) of Sekomi's men had departed for the Shua, to attack there the outposts of Moselikatze and carry off his cattle. This news, coming as it did at a time when my followers had hardly recovered from the perils and fatigues of a long spell of work in the desert, made them hesitate to accompany me on an expedition beset, as it seemed, with dangers and difficulties still more serious. I thought it highly probable that the story was an invention of Sekomi's to deter me from exploring those parts of the country which contributed most to his wealth as a preserve of elephants, and were hence best suited to my purposes. I endeavoured to instil this persuasion into the minds of my men, but to little purpose, until I assured them of my full resolve to undertake the journey, in spite of the report in question, when they at length reluctantly consented to bear me company. On the 28th of August I took my leave of Chapo, or *Masella*, as the natives also entitle him, after the large kind of otter which is the worshipful patron of his tribe, as the crocodile is of the Bakwains and the young lion of the Batawanas.

I had already sent on Meroepie with one of the wagons to wait at Lutlochoé either for my own or Edwards's arrival at Sekomi's. I now struck away to the northward, having with me the other wagon, which was nearly empty, in order that I might travel with rapidity, in case I should presently find myself in a wider tract of desert than I reckoned upon. I knew that the oxen would hold out for five days without water, and calculated that within that time, with a light wagon, I should be able to get over 180 miles of ground. Moreover, I greatly doubted the probability of finding so extensive a tract of country altogether destitute of water, although the natives, both at Chapo's and along the road, swore by their chief that we should die of thirst, and declared that we should not reach the Shua until we were grey-headed—or for six months at the very least. They resorted also to their usual tactics, endeavouring to mislead and intimidate my servants, and attempting to frighten me by affirming that the Makalakas dwelling on and beyond the Shua, as well as the Matabele and Mashona, were cannibals. This last piece of most appalling intelligence I met, much to the confusion of my bewildered informants, by professing great delight at the prospect of having at last an opportunity of falling in with that singular race, the Man-eaters! These and other such marvellous legends, however, could not fail of having some effect on the minds of my credulous followers, and I found no small difficulty in allaying their fears and inducing them to continue the journey.

Our course lay for a time along the same track that I had followed on a previous journey. We passed Koobye a few miles north of the Botletlie river, and afterwards crossed the great Ntwetwe salt-pan, there about 18 or 20 miles broad, reaching its opposite shore at midnight on the 2nd of September. In the absence of water at Gootsa, we were obliged to push on at once, passing various stations with no more than the ordinary incidents of travel in the African wilderness.

One morning, while inspanning, I shot a beautiful bird, called the Kaffir crane (*Balearica regulorum*). It is of a bluish slate colour, pale about the head, with a long black tail, and six black feathers in each wing. These are much prized by the native warriors, the plume being stuck on their heads, in the manner practised by the North American Indians.

Throughout this portion of our journey I had been under the impression that we should yet feel the effects of stratagems practised to obstruct our march, and had prepared my people that they must expect to hear some startling news concocted by the natives to frighten us back, as all their other machinations had failed. My anticipations were shortly realised. On the 6th of September we had outspanned at a spring-pond called Tsagoobyabsa, when, at the dead hour of midnight, a file of meagre Bushmen joined us from the east, and, squatting round the fires, were soon engaged in earnest and mysterious conversation with our pretended guide, Goroge, who seemed to have expected their arrival. The next moment he woke Molihie and begged him to rouse me and bid me fly. There was no time to be lost, as the Matabele warriors, having overrun the country eastward, would be upon us in the morning, being in search of a tribe of Makalakas who had fled past us. I now found that all my previous admonitions and warnings were lost on my people, who joined their entreaties to those of the Bamañwato that I would return at once. But all this I was prepared to hear, without the slightest intention of being influenced by it, though there seemed till daylight a great likelihood of my being deserted by all my people at this junction. My remonstrations on their undutiful and cowardly conduct at last produced a good effect, and with daylight their courage returned.

Goroge and his gang, who had accompanied us for some days, feigning great alarm, now abruptly left us, taking every member of their party with them, as well as every particle of flesh they could lay their hands on, and prophesying that we should all die of thirst, if we were not killed by the Matabele. I had been so accustomed to these manœuvres during my travels, that I had become callous to all such reports, and, bidding Goroge and his followers "God-speed," felt almost thankful at being rid of their company, though I must acknowledge that I was rather at a loss for want of a guide. I had formerly reached, a few days farther to the north-east, a chain of springs called Motlamaganyani (a succession), but my present destination, the Shua river, lay about south-east, or east by south, and I determined that if I could not find

water ahead during the next few hours I would make a push
and ride night and day until I should find some. Consequently, on the 7th, accompanied by one of my men, I started
on foot eastward, over the same kind of country we had
traversed for the last few days, consisting alternately of long
grassy lawns, and groves of the golden-leaved mopani trees,
growing generally in a sandy soil, and abounding with beds
of calcareous and siliceous pebbles; and after a walk of nine
miles east I captured a few Bushmen, grubbing for the kind
of bulbs well known at the Cape as lunches. These miserable
wretches, never having seen white men, were terribly frightened; but, with some difficulty, tobacco and beads eventually
reconciled them to the strange apparition. We found, however, that they had been informed of our being in the neighbourhood, and warned by Goroge to avoid us, as, if successful
in our object, we should make a road which would lay open
Sekomi's country, his cattle, and his Bushmen, to the ravages
of Moselikatze's people. This injunction had the desired
effect on these Bushmen, who swore with trembling that there
was no water in the land, though I soon found a beautiful
spring within 100 yards from the spot where we stood.

We now travelled eastward, and soon emerged on an open
plain, the margin of the Ntwetwe salt-pan, which we had
crossed before considerably to the westward, and which runs
parallel with our present road, as it also does with the Botletlie.
This river expands gradually, a few miles east of Koobye, into
a large salt lake, which increases in breadth till it is met by
the Shua river flowing from the north-east. In four hours
from our last bivouac we reached a spring called Mananyna,
or by the Bushmen *Chua Katsa*, on the opposite or southern
bank of the Ntwetwe, which stretches still farther north-east,
until near the springs called Motlamaganyani.

The gnus, quaggas, and springboks were so tame here,
owing to their ignorance of the gun, and, besides, so abun-

dant, that we did not care to shoot them; but during our journey in the afternoon in the south-east direction, across what is called the Bontveld, I was diverted from stalking a drove of the finest buck pallahs I had ever seen, by the sight of a troop of about thirty elands, fast asleep, a short distance before me. These I stalked in great haste, as the wagon was coming up, and shot the largest cow; but she proving poor, we abandoned the greater part of the flesh, to the no small astonishment of the poor Bushmen, who nearly wept for grief at such waste.

The track we pursued now gradually inclined more to the south. I was not quite satisfied with this change, but was unwilling to interfere with our guide, lest he should desert us altogether. I felt sure, moreover, that when we had once struck the course of the Shua, which could not be far off, I should be able to advance without any apprehension of want of water. One spring to the north-east, and several south and south-west, were pointed out to us, but only one, called Mamtsoe, lay in our course, about south-east. This we reached the following day, after five or six hours of hard travelling over a wide plain.

We passed the spring, trekking three or four miles farther to be near the wood, and encamped under one of the giant moana trees, of which there are many about. Here I found the bush folk more numerous than I had yet seen them, owing probably to the abundance of game. Their astonishment at the sight of white men was prodigious; they did not forget, however, to warn us of the presence of their *boleo* traps, which were hung on the neighbouring trees, and would certainly have proved fatal to some of our oxen, if not to ourselves.

A number of lions being in the habit of resorting nightly to the spring, either to drink or to waylay the game, I went out early in the evening to kill some animal for a bait to attract them to some spot convenient to my skaarm. I knocked over two quaggas at my first shot, one of which I had dragged up to within five or six yards of my shooting-box, when my man Molihie declared his intention of occupying another position, where there seemed little likelihood of his being visited by lions; but he completely outwitted himself, for, soon after dark, and before the moon rose above the horizon, I observed some object crouching towards his post, and my cry of warning was lost in the report made by his large rifle. The animal he fired at did not move; and five other lions, which were lying within a few yards of his position, now got on their legs, while two others were observed moving round the pond above the wind. My faculties being now also quickened, I observed two lionesses facing me within twenty feet from another direction, at which I instantaneously fired, and probably prevented their springing into my skaarm. Abraham, who was with me, now discharged both barrels of his gun, and while I held my rifle pointed at them he reloaded his, when I again fired with the same bad success. The lions not seeming inclined to beat a retreat, Molihie's courage was well tested, and he cried lustily for his master to come to his aid. Of course I could not comply with this demand, being myself placed in a similar position. Having continued firing till nearly all my ammunition was expended, I became more cautious, and, taking a more deliberate aim, was delighted to find the bullet tell, and to see the lioness bound off with a growl. In this movement she was instantly followed by the rest; the troop, numbering ten in all, having besieged us for a short but anxious period, during which it seemed as if not even thunder nor lightning could terrify them.

After their retreat, we heard them for a long time tearing with tooth and nail at the second quagga, which lay about 200 yards off, but, having only three bullets left, we did not dare to meddle with them, although for the rest of the night we had the advantage of a bright moonlight. Having feasted themselves on the flesh of the quagga, and killed and devoured another, they came again towards the water to quench their thirst, but the recollection of our engagement with them some hours before probably checked their advance, as they halted midway and set up a most terrific roaring. Two others also approached in a different direction with fearful roars, but Molihie having fired a shot which mortally wounded a white rhinoceros, the lions feared to venture near the water. Game of all sorts came and went in vast multitudes all night, many passing within a few feet of us; and I feel no scruple in affirming that, since the preceding evening before sunset, till the next morning after sunrise, except during the time of our being besieged by the lions, no less, at a very moderate computation, than a hundred head of game drank at the spring every five minutes. This in ten hours would make the number 12,000, which, however enormous it may appear, is, I feel confident, far within the mark. The pool, about 400 yards in circumference, was all night kept in commotion, the splashing of water, the din of clattering hoofs, and the lowing and moaning of gnus and their calves, being mingled in discordant notes. The braying of quaggas was terrible, and the pond, excepting at one or two short periods, while we fired, was never clear.

Next morning the wounded rhinoceros was overtaken by the Bushmen and killed, and some of the flesh brought to camp. A few ducks and geese were shot for our own larder, and a vast concourse of cranes strutted about the plain, making a croaking noise all day long. At night multitudes of Namaqua partridges and turtle-doves visited the water; some of these were also shot, which, together with a dozen springboks, two gnus, four quaggas, and the rhinoceros, gave ample employment to the Bushmen in devouring or jerking the flesh. Some of these Bushmen are fine strapping fellows; one old man, who had been crippled by an elephant, wore badges round his arm for eighteen elephants which he had himself speared.

On the evening of the 10th, still hoping to shoot a lion, I lay again by the water, but none came. A multitude of springboks drank breast to breast within two yards of the muzzle of my gun, which, being loaded with a conical bullet weighing five ounces, on a strong charge of powder, passed through the vitals of five and wounded a sixth. This incident considerably enlarged my belief in the powers of a gun. Till this time I had resolutely maintained three animals to be the maximum that could be killed with one bullet. A borele with a very long horn received another bullet from my large rifle through his heart, and ran for more than a thousand yards before he fell.*

Early on the 11th we started east for the Shua river, passing first through a mopani forest, in the midst of which I met with a small spring, and near it a family of Bushmen, shamefully naked, the women as well as men. They were in a complete state of panic at every movement of our own, or of the wagons, having never seen white men, wagons, or horses before. In the afternoon we crossed a level country intersected with shallow watercourses, and plains covered with a short prickly grass, always indicating the approach to a river or watered locality, and at night reached the banks of the Shua, having travelled nearly twelve hours, and killed a borele on the plain, after a sharp contest.

I was much disappointed in the Shua river, of which I had

* I may advert here to the fallacy of the common belief (encouraged, I believe, by anatomists) that death ensues instantly to an animal wounded in the heart. Such a wound is no doubt mortal, but that it does not always terminate life *instantaneously* I can positively affirm, in the case both of the elephant and rhinoceros, which beasts I have known to live, and even move with rapidity, for a quarter of an hour after a bullet had pierced the heart.

heard so much, and had sometimes been almost persuaded to believe, from the accounts of natives, that it must be a navigable river running eastward. The Botletlie too, they had told us for years, became in its lower course a broad and deep river, abounding in hippopotami and crocodiles. The fallacy of these exaggerated reports at once became evident on my visit to the spot, and I regretted that, instead of finding an immense body of water, the Botletlie gradually widened into an extensive salt basin, probably twice the circumference of Lake Ngami, and was met here by the Shua from the north, and four other streams from the eastward, forming by their confluence an immense lake or pan, which is generally dry for half the year or more. From the appearance of the banks, which are abrupt and steep, one would suppose that this immense pan must at one time have been a permanent fresh-water lake, for the older Bushmen assert that about thirty or forty years ago it never dried up, and abounded with hippopotami, crocodiles, and fish. But suddenly, they say, the waters from Lake Ngami ceased to flow; the lake dried up, and the dead fishes and animals were devoured by the vultures. At the present time, this vast expanse, stretching away in every direction farther than the eye can trace, is nothing but a barren plain, level as a plank floor, and covered with a white incrustation or saline excrescence. This level becomes inundated during some months in the year, assuming then a very grand appearance, though only twelve or eighteen inches deep.

The Shua, which here appears to come from the north, is fed by several lesser streams flowing from the north-east, which I shall hereafter have occasion to mention. Though in general about 200 yards broad, and at this time 60 or 70 feet deep, the Shua contained only here and there pools of stagnant and salt water, which when dried up left a crust of good salt on the surface of the white sand. In the summer season, however, this river pours down an immense flood, of which there are strong indications everywhere in the large bogs high up on the banks inclining southwards. The singular fact which surprises me is that, although the soil forming the immediate bed of the river is so strongly impregnated with salt, fresh water may be obtained anywhere by scratching to the depth of nine or ten inches below the surface, as is daily practised by the game, even though within three feet of a pool of brine; so that the water seems only to become salt when acted upon by the rays of the sun, or the influence of the atmosphere generally. The banks of the Shua are studded with the mopani tree for a mile on either side, affording cover to many rhinoceroses, gnus, bucks, bastard-bucks, sassaybies, lions, and other animals. Another strange mimosa attracted my attention here, having never met with it before. It is called *Habbi Khosi* (prick the king) —a very beautiful straight branched mimosa, spreading its branches from the ground.

On moving the wagon next day a few miles south, to the junction of the Shua with the Salt Lake, I observed several of the mañana traps, from which a number of guinea-fowls were, like criminals, suspended by the neck, and also a fine fat pauw. These we left for the owners, and passed on, and, having sought out an eligible position for the wagon, I strolled out with some Bushmen in search of a new buck, called by them *Boo-a-dow*. But after trudging about for several hours in quest of this novelty, the hope of obtaining which I had nourished for more than a month, I found, to my great disappointment, that the boo-a-dow was no other than the common reitbuck. Another animal of which I had heard strange tales still eluded my search; and from the description given of it I felt certain it was not yet known to Europeans. This animal, called by the natives *quelleñ-quellè*, is a beautiful white and spotted antelope. It is said to be some marvellous freak of nature, possessing qualities unknown to the lower animals in general. This strange creature had been in my thoughts ever since I left Lake Ngami, and I determined to satisfy my curiosity respecting its seemingly fabulous character, searching for it whenever opportunity offered. Its peculiarity, as it is represented, renders it unfit to be eaten by females, whose sex it is said in some respects to resemble in its physical formation.

Next day I strolled about again in search of the quelleñ-quellè in a burning sun, beating about through fields of tall waving grass on the margin of the Salt Lake or Makarikari (mirage), but started only some jackals and hares, which generally abound most in this kind of veld. The spoors of rhinoceroses were numerous, but we saw none of the animals themselves. On limping back towards the wagon, with a chafed and blistered foot, I observed my followers suddenly absconding on both sides of the vehicle; and, looking round for some explanation, I received the warning word *Taaw* (lions). This was immediately followed by the angry growl of two large lionesses in the long grass before me, and the yelping and purring of three cubs. Levelling my gun at the nearest, my dog dashed forward, and at once brought them round to face him, with a terrific growl. This drove the Bushman carrying my spare gun 200 yards in a less number of seconds, nor would he again come near; while I, fearful of the odds against me if I commenced an attack, neither dared to turn my back to the enemy, nor to fire, though an old Bushman, who stuck bravely to me, and seemed to have a wonderful craving for lions' flesh, was urging me to do so, thinking, no doubt, that the white man's gun, though a single-barreled one, was always invincible. But, differing with him in this opinion, I was content to secure my dog, preventing him from annoying the lionesses, and alternately levelling my rifle upon the more vicious or bolder of the two, as they lay lashing their tails within thirty yards of me. After about five minutes of rather painful suspense, the cubs moved off, and as the mothers turned to look after them, I squatted down, and, aided by the grass, concealed myself from their view for a few moments. On raising myself again to watch their movements, I saw that they had overtaken their cubs; but observing me, they returned growling for a few paces, and lay down as before. I also resorted to my former manœuvre for concealment till I saw the lions move farther off; and then, jumping down into the bed of a nullah, I ran along it to head them, where a tree on the bank afforded me a safe position for waylaying them; but they had entered a large field of tall tambookie-grass, where they were quite out of sight.

Learning from the Bushmen here, who seemed acquainted with everything that passed at Moselikatze's Town, that the white men, meaning Mr. Moffat and my friend Edwards, had been to that place and were on their return, I determined on a change of plan, feeling that it would be a useless toil to take the wagon farther eastward; but as elephants abounded in that direction, I determined to proceed on horseback a few days farther, to ascertain the character and resources of the country as regarded the objects of the hunter and the trader, and examine some of the outposts at a high range of mountains of which I had heard, called Madumumbela, and, if possible, reach Moselikatze's Town, and pay him a short visit. Consequently, saddling two horses on the 13th, I took with me Molihie, as achter-rider, and three Bushmen, carrying a little ammunition and two spare guns, as well as a tempting collection of beads, &c., intended for an offering to propitiate his highness the great Matabele chief, the terror of the South African tribes, as far as, and even beyond, the Zambesi river.

Shortly after crossing the Shua, I fell in with six strapping Makalakas, all six feet and more high, who, crouching and trembling before me, were in a complete panic at their first sight of a white man. A small troop of elands coming in sight, and approaching within 800 yards, I was tempted to try a very long shot, which, fortunately, passed through the neck of a cow, and killed her on the spot, to the great amazement of our swarthy travellers, who, taking up their shields, ran to the spot, and soon commenced operations for roasting part of the prize. Their admiration of the gun, the horse, and the white man, knew no bounds, and everything belonging to us was " God ! "

Having taken a much smaller portion of eland flesh than my Bushmen were willing to carry, we parted company with the Makalakas, who remained to demolish the rest, and begged me to shoot an ostrich on my way, as they were engaged in an expedition to procure black feathers to adorn the heads of the Zulu warriors. Shortly after leaving them I had the good fortune to fall in with three of these birds, and, still better, to kill one of them, the male, though running, at the distance of full 800 yards. This extraordinary piece of good luck made a deep impression upon our guides, who failed not at every stage to communicate to those we met the wonders they had seen performed by my gun. Having plucked the white feathers from the two wings, forty-six in number (or half a pound in this instance, worth about £6, though generally the feathers only weigh about a quarter of a pound), I left the remainder for the Makalakas, who would be sure to follow the trail of our horses and find it, and moved onwards, while the Bushmen turned back to shower maledictions on the vultures for lighting on the flesh, which I had positively refused them permission to burthen themselves with, as they were already bending under the weight of my packages, besides large junks of eland's flesh.

It is beyond their comprehension how we can be so prodigal as not to remain until we have devoured the flesh we kill.

None of the Bushmen or other natives whom we had met during the last eight or ten days, or might expect to meet farther eastward, had ever seen the face of a white man or anything belonging to him. It was therefore necessary that, whenever we approached a village, one or two of my followers should be sent forward to prepare them for the extraordinary event. Consequently, arriving at a small well, before reaching one of the Bushmen's kraals, some of our party ran ahead to enlighten the simple inhabitants, who would have taken us for *Torrah*, or *Porrah* (God, or the Evil Spirit), while I remained to wash myself, and dress a wound on my heel, which had now perfectly disabled me; I little thought at the time that a party of Moselikatze's warriors were lying in ambush close by, watching my every movement, which, as I afterwards learnt, they faithfully reported, in the presence of my friend Edwards, at the chief's town. Fortunately for me, this party had no hostile feelings towards Englishmen, nor dared to molest them without permission being first obtained from their chief, or I might have been cut off. As I heard afterwards, these Matabele were on the look-out for some Bamanwato who were making a raid on their herds, and, although they thought that I must be acting in concert with Sekomi's people, who had just been making observations on their cattle posts, they were afraid to do me any harm without further watching my movements, for fear of making a mistake, as they knew that their chief was entertaining Messrs. Moffat and Edwards at his town in the most cordial manner.

Next day we were up early, and continued our march, still in a north-east direction, but often inclining eastward, through beautiful groves of the often-recurring mopani; and in the afternoon bivouacked under the delightful shade of a cluster of these trees, enjoying the pleasant interchange of light and shade afforded by their spreading branches.

Falling in with a party of Makalakas belonging to a tribe inhabiting the Madumumbela mountains, about long. 28°, lat. 29°, under a chief called Mènoè, I obtained from them all the information I could; and learning that Messrs. Edwards and Moffat had actually left Moselikatze's Town on their way home, and that the journey to Shunkaan was still a long day's work on foot, I determined to remain here, and invite the chiefs of the Makalaka, who, though now subject to the paramount rule of Moselikatze, had themselves once been chiefs of a powerful independent nation to meet me here. A present of beads, wire, handkerchiefs, and knives was sent with the invitation, and I fully expected the chiefs to make their appearance in a few days, which I was glad to spend in recruiting my horses and resting my leg, which was so disabled that I found it impossible to stand, the tendon above the heel being severely chafed and much inflamed. This may have been owing to the peculiar state of the system when exposed for any length of time to the effects of an exclusively flesh diet.

But soon after my messengers sped on their embassy to the chiefs at Madumumbela, the repose I promised myself was disturbed by the glorious cry of "Elephants near," which has always the wonderful power of dispelling every other care with its all-absorbing excitement. I forgot my pain, and, having rested the horses for the day, was the next morning assisted into my saddle. Followed by a numerous gang of some seventy or eighty armed Bushmen and Makalakas, we started northward, and soon crossed a periodical stream, called Tunkaan, which has its source near the Shashè river, an affluent of the Limpopo, situated about south-east of our course. Another stream which we crossed, called Mietengoè, has its sources in the Madumumbela range, and runs past Shunkaan's

Town, and we bivouacked at a third, which is considered the main stream of the Shua river. This stream, rising beyond the northern extremity of the Madumumbela mountains, which it encircles, waters the heart of Moselikatze's country, passing the town belonging to Impukani, one of Moselikatze's greatest generals, and his uncle.

The sun had just set when we dismounted after our long ride, and, disabled as I was, I lay down by the fire, while Molihie strolled into the bush for a few hundred yards. Soon I heard the report of a gun, and six other shots following in rather quick succession. I felt convinced he had met with elephants. I was however mistaken, for Molihie soon returned, bringing the tails of three rhinoceroses which he had shot, and assuming an air of indifference, as if he was in the habit of slaying a whole herd of rhinoceroses every day. The admiration and amazement of the natives at this feat knew no bounds, and, as we had already made several extraordinary though perhaps merely lucky shots in the earlier part of the day, I advised my man to persevere in taking great pains while shooting in future, and to endeavour to avoid missing a shot if possible, in order to inspire the natives with a proper respect for our guns and our countrymen, and to convince them of the danger of venturing on hostilities with the white men.

A buffalo, a gnu, and a brace of quaggas had been also shot in the course of our ride; and as the following day was Sunday, we informed the natives amongst whom we now found ourselves of our custom not to hunt on that day, and that they would have time to cut up the flesh for themselves. Great was the rejoicing at first, but many the frays that soon ensued between the leading members of the Bushmen families, who each claimed a preference in appropriating the fat. The Makalakas however interposed, and, flourishing their battle-axes, soon settled the Bushmen's quarrels, by forcing them to abandon their fancied rights, taking themselves not only the fat of all the rest of the game, but the fattest rhinoceros into the bargain; and by Sunday evening the surrounding trees were quite concealed by the festoons of flesh on which they were hung.

Early in the morning Molihie was roused from his slumbers by the Bushman who had carried his gun, to partake of a dish of water which he had brought with him. This was such an unusual thing that Molihie wondered what it meant, and, being suspicious enough to imagine that the draught might perhaps contain a dose of poison, he, for a moment, hesitated to drink it; but another Bushman, who, like most of the elder Bushmen hereabouts, spoke Sechuana well,* explained to him the custom that a man who kills an animal is always served early in the morning, by the rest of the party, with a drink of water. I have myself been often since treated with the same ceremony.

Early on the following morning we started in search of the Sibaninne river, on whose banks it was said elephants were to be found in great numbers. Accordingly, at one of the large pools of water in the bed of the river, which abounded with fish, we found the fresh spoor of three bull-elephants. This we followed; but we had not gone far when we discovered by their trail that they had taken to flight, having probably been startled by our wind. Molihie and myself galloped on the spoor, followed by thirty of our armed Bushmen, and in about ten minutes encountered them in a forest, eyeing us as we drew near to them. When within fifty yards I dismounted, and giving the largest tusker a shot, the effect of which was announced by a groan, directed Molihie to follow the second best, and we separated, each pursuing his allotted victim. Closing with mine for a second

* In former times the Bamañwato inhabited these parts of the country. Their chief town was in this neighbourhood, and many of their cattle-wells are still to be seen.

shot, I was nearly unhorsed by a borele, but followed my elephant again, with a number of Bushmen at my heels. I fired several shots, without any other effect than decreasing his speed. At length, having started three buffaloes, they made straight for the elephant, who turned upon them, and, charging them furiously, drove them back towards me. I turned my horse's head towards the armed Bushmen, who, being at a respectable distance behind me, ought at least to have covered my retreat; while their fellows, who had been safe spectators of the fray from the highest branches of the trees, now dropped to the ground like over-ripe fruit; so that the general rout of the white men, the Bushmen, and the buffaloes, all flying pell-mell before the charge of their common foe, the elephant, formed rather a ludicrous spectacle. For the time he certainly had the best of it, but, either his rage or his breath being soon spent, he gave up the chase, and stood sullenly contemplating the retreat of his assailants behind the screen of the forest trees. As for myself, according to the proverb, I recoiled, the better to renew the attack; but while I was endeavouring to drive the elephant towards a more open ground, I heard a loud trumpeting a short distance westward, and, looking round, beheld Molihie in a somewhat similar plight to what had just been my own, being chased by the elephant he had singled out. But the most extraordinary part of the affair was this: the elephant not being able to overtake his enemy, I saw him pull up successively two trees by the roots and cast them after Molihie, nearly striking his horse with one of them. This singular act of sagacity surprised me not a little, being under the impression at the time, as I am to this present day, that the act of thus hurling the trees was not accidental, but intentional. Each of these trees was nearly twenty inches in diameter, and they were thrown twelve or fifteen yards from the spot where they grew, so that, leaving the intent out of the question, it was in itself a prodigious feat of animal strength.

On arriving at the well, where the Makalakas were waiting for us, I was much annoyed at finding, instead of the chiefs, some of their principal men, bringing me indeed very flattering messages from their masters, with a present of some ivory; but regretting they could not conduct me to Moselikatze's Town, as it would endanger their lives. I disbelieved this; but as they begged me to return homewards now, and to revisit them the next year, I came to understand that these petty chiefs had concealed my near approach from head quarters, and would like to carry on a clandestine intercourse with me for their own purposes. They promised to show me, next year, thousands of elephants and other large game; further, that as I had first shot their country, in fact discovered it, they would preserve it for me. Not wishing, however, to confine myself to the acquaintance of these fellows, I threatened to find my way without their assistance. After hearing this the Makalakas disappeared one by one, and at night, the ivory of the elephants we had killed not arriving, I made up my mind to go at once in search of it.

At this moment a Bushman arrived breathless from the east, in search of the Makalakas, with a message from Shunkaan entreating me to fly without delay, as three of Moselikatze's towns were up in arms, and in pursuit of the Bamañwato who had been in my track reconnoitring Moselikatze's cattle, and he feared his troops would fall on us by mistake if I did not get out of the way before they arrived. In the meantime I had discovered from one of my guides that the Bamañwato and Makalakas were leagued together. The latter, wishing to escape from Moselikatze's rule, and place themselves under the protection of Sekomi, and join the tribes he had gathered, had invited the Bamañwato to come and inspect the country, and study the practicability of making off during the rainy season into the Bamañwato country with the whole of the cattle entrusted to their charge by Moselikatze. Becoming aware of this, I refused to fly, being fearful of confirming the suspicion of the Matabele that I, too, was in league with the Bamañwato, and resolved that if they came on I would meet them amicably, hoping, as I still retained a smattering of the Zulu language from my residence at Natal, that I should be able to explain to them how matters really stood.

Having saddled my horses, I left my guides of the Shua in charge of my traps, and, under the guidance of some other natives, sought out the hiding-place of the Makalakas, which I reached about 11 o'clock at night. Contrary to my expectations, I found them quite prepared for my arrival, and drawn up in a formidable array, standing under cover of their long shields, evidently with hostile intentions. Surprised at this, I felt for an instant at a loss how to act; but a moment's reflection satisfied me that it would not do to exhibit any symptoms of alarm, and that my safety depended on the composure and firmness with which I acted in this emergency. Having, therefore, ordered Molihie to stand with his gun ready at full cock, I rode up undauntedly to within a few paces of the dusky crew, whose appearance was doubly horrid, as the bright moonlight did not penetrate the deep shade of the mopani trees under which they stood. At this moment the leader of the gang sprang forward in advance of his comrades in a sort of war-dance, as is their wont when threatening a charge, and rattling the shaft of his spear against his shield, yelling and whistling shrilly the while, he demanded what we wanted there at this hour of the night. Springing from my horse, as did also Molihie, I was perfectly astonished at the effect produced by our assumed composure, when this brave of the braves hastily retreated into the ranks, evidently disappointed at the want of pluck among his followers, who slunk one by one behind the bushes. Following up my advantage, I rated them soundly in Sechuana, in as loud and bold a voice as I could assume, threatening to destroy the whole of them in one instant if they did not instantly lay down their spears. Strange to say, with that superstitious dread which these unsophisticated savages entertain of the white man, but which I had never fully believed in till this day, the whole of the Makalakas laid their spears on the ground, trembling with fear, while nearly all the Bushmen disappeared in a twinkling.

Having brought matters to this satisfactory conclusion, I demanded my ivory, which in another instant was forthcoming. After this I required eight Bushmen to carry the tusks : at first these could not be found, but, stamping impatiently on the ground, I insisted on having them, when the Makalakas set up a shouting, and, explaining to the Bushmen that the danger of hostilities was passed, they one by one emerged from their concealment. I then demanded a bundle of dried flesh, by way of compensation for the trouble they had given me, with another Bushman to carry it; and at length, having cautioned the gang to take care that I did not set the ground on fire under their feet, we took our departure, driving the bearers of the tusks before us to our bivouac.

I had been so often deceived by false reports, that if real danger was threatened it had become highly probable that I should disbelieve in it; but disgusted now with the conduct of these savages, and having nothing further to wait for here, I returned to my camp next morning, attended by eight Bushmen, each bearing a large tusk of ivory, and preceded by a few others carrying our spare guns, blankets, &c.

On the 25th we were up early and marching towards the wagon, which we neared in the afternoon. I shot a giraffe, a koodoo, and a brace of springboks. I found all right at the wagon, and my driver, Abraham, entertaining a jolly circle of Bushmen and their families, who, much to his annoyance, sought, from the colour of his skin, to claim relationship with him; they were astonished that he could not speak their language, and believed he only affected ignorance of it. Amongst the game which he had shot for them were two or three rhinoceroses. These he reported to be very numerous, as well as lions, which regularly killed and devoured every night at least half a dozen quaggas within hearing of the wagon.

The next morning, while working at a *katel* * for my own use, three Bushmen came flying from the east, and reported that a commando of Moselikatze's was within a few miles south-east of us, and would be here the same day. These men were in such a state of alarm that they would not stop even to partake of some of the flesh which our Bushmen were boiling, but hurried on. Shortly afterwards 200 or 300 Bushmen, women, and children, came flying in breathless haste from the same direction, carrying their pots and other utensils on their heads, and occasionally stopping to look round, evidently in great terror. Our followers, too, having been handsomely rewarded with beads and tobacco, made off; one, an old man, who was under an engagement to conduct me southwards, I fortunately caught in the act of making his escape, and insisted on his fulfilling his agreement. Affairs assumed now a more serious aspect than they had ever done before, and as I stood every chance of being left to the mercy of events, and without a guide if I let this man go, while not possessing on my own part the least knowledge of the country towards Sekomi's, I felt that I must start immediately while I had him in my power.

As soon as we had inspanned, the remnant of a Bushman family which had been massacred the same morning arrived, and reported that the Matabele were on the Simoani river, which lay in our route, and were feasting on the oxen and

* Stretcher.

buffaloes they had killed. One of the fugitive Bushmen, who had been out hunting, having discovered the spoor of the Matabele as he returned home, had concealed himself, and learnt that his wife and parents had been killed, and his children made captives. The poor fellow, who seemed wretchedly forlorn, followed the fortunes of the rest of the Bushmen, who, taking all their dried flesh with them, had gone a few miles into the Karikari salt lake, or pan, where they foolishly flattered themselves they would be in safety; once in any place of refuge, however temporary, and having a full belly, they forget all dangers and troubles, regardless of all prudence. But shortly afterwards the sad intelligence reached me that all the eldest of the clan, to the number of twenty-three, had been massacred that same night, and the younger ones carried into captivity to be made slaves or soldiers of Moselikatze.

On the afternoon of the 26th we travelled about 16 miles south, always keeping on the eastern edge of the Salt Lake, and reached the Simoani river, a small periodical stream with very little water, flowing into it from the eastward. Here we slept, and during the night, being obliged to watch our guide, I heard the cattle breathing very hard—a sure sign of their seeing or smelling danger. My thoughts immediately reverted to the all-engrossing subject of the Matabele, but on putting my head out of the wagon two objects close before me loomed as large as elephants. On a second glance, however, I made out that they were rhinoceroses, and the next moment a bullet left my large rifle from the wagon, smashing the shoulder of one of them. The report of the rifle, the puffing and snorting of the rhinoceroses, and the barking of the dogs, woke my people, frightened not a little. They lost all self-possession, and could not find their guns, though they had been sleeping with them close at hand; at length the wounded beast ran off for a few hundred yards, followed by the dogs, and then fell dead.

Next day we started early, continuing our journey down the banks of the lake in a southerly direction for 24 miles, through the same kind of country, with a mopani forest to our left. During the day my old guide made an ineffectual attempt to decamp, and, happening to fall in with other Bushmen, I was able to secure their services and dismiss him. This old fellow saved his life by accompanying me, for all his relatives were killed, as he would have been had he remained with them. We slept at the Chuani river.

On the 28th we still travelled south in sight of the lake on our right, passing on the way a deep well eight miles south of Chuani, and in about 16 miles came to Qualeba, a small, dry stream near a range of hills running from south-west to north-east. We opened two wells in the bed of this river, our horses, cattle, and dogs being miserably thirsty, having been driven fast over a dry country. Next morning we were surprised to find that a borele had robbed us of a portion of our water; but we still found sufficient for our cattle, and, resuming our march, reached Chuatsa, a mineral spring on the banks of the Meea river, resembling that at Shogotsa. We made a distance of at least 24 miles before sunset, having shot a springbok and sighted several rhinoceroses, elands, and ostriches.

On the 30th we travelled again south for about 12 miles, leaving the immense Salt Lake stretching away to the westward, and came upon two beautifully clear fountains of water gushing from the head of a picturesque valley, with a growth of trees, reeds, and an abundance of verdure quite pleasant to behold. At this fountain, which is called by the natives Morimo tsébé (God's ear), we considered ourselves safe from pursuit by the Matabele, and at night we lay by the water to shoot buffaloes and rhinoceroses, and remained over the following day to give the men and cattle rest.

Having had no coffee or tea for some days, I procured two

different kinds of berries from the Bushmen, the *marétloa* and *mogoana*, which forty-eight hours' fermentation converted into a very pleasant drink, something resembling cider. The marétloa berries are distilled by some of the more civilized tribes residing in Mahura's country, and produce a very strong spirit. The process of distillation is very simple. The still consists of a kettle inverted on a boiler, the spout being inserted into an old gun-barrel, which is led through a cask or tub of water.

Absurd and disgusting as some of their practices are, there are traits in the character of the Bushmen in these parts which are much to be admired. Degraded as they are in the scale of humanity, and even in the eyes of their superiors amongst the native races, their morals are in general far in advance of those that obtain among the more civilized Bechuanas. Although they have a plurality of wives, which they also obtain by purchase, there is still love in all their marriages, and courtship among them is a very formal and, in some respects, a rather punctilious affair. When a young Bushman falls in love, he sends his sister to ask permission to pay his addresses; with becoming modesty, the girl holds off in a playful, yet not scornful or repulsive, manner, if she likes him. The young man next sends his sister with a spear, or some other trifling article, which she leaves at the door of the girl's home. If this be not returned within the three or four days allowed for consideration, the Bushman takes it for granted that he is accepted, and gathering a number of his friends, he makes a grand hunt, generally killing an elephant or some other large animal, and bringing the whole of the flesh to his intended father-in-law. The family now riot in the abundant supply, and, having consumed the flesh and enjoyed themselves with dance and song, send an empty but clean bowl to the young man's friends, who each put in their mite, either an axe or spear, some beads, or trinkets. After this the couple are proclaimed husband and wife, and the man goes to live with his father-in-law for a couple of winters, killing game, and always laying the produce of the chase at his feet as a mark of respect, duty, and gratitude. For the father-in-law a young man always entertains a high regard, but after marriage he shuns his mother-in-law, never perhaps speaking to her again for the whole of his life; and there seems to be a mutual inclination between them to avoid each other. The same feeling exists on the part of the bride towards her father-in-law.

CHAPTER XII.

The Great Salt Lake—Sable Antelopes—Ostriches—Vast Herds of Bucks and Buffaloes—Effects of a Snow-storm—The Bakalihari—The Lotlotlani River—Reach Sekomi's Town again—Charms and Wizards—A Court of Justice—Bamañwato Customs—Native Music—Leave Sekomi's—Native Love of Battle—Start North for the old Hunting-fields.

THE margin of the great Salt Lake,* or Makarikari (mirage), from which we had now deviated, being strewn with large agate pebbles, white and coloured quartz, broken pieces of basalt and felspar, and at intervals with calcareous and siliceous rock, as well as the *lechuko* (iron ore) with pieces of a bluish sandy slate, richly spangled with mica, it was now quite a relief to our oxen to get once more into a sandy road. But, as is generally the case wherever the soil is loose, the woods and forests are so dense, and vegetation is so luxuriant, as to cause as great, if not a much greater, inconvenience than the stony road. It often entailed on us hours of most laborious toil for men and beast to drag the wagons through a forest, or cut a path for their passage.

On the 2nd of October we struggled through eight miles of ground, when we came to three beautiful little fountains, situated near each other, and called Potequanyami and Matabele; and having heard that the black buck of Harris, called potequan by the natives (*Aigoceros niger*), abounded here and resorted nightly to the fountain, I determined to remain until I had captured one of these magnificent antelopes.

Next morning I found that Abraham had the good for-

* This lake is from 60 to 64 miles broad, and probably 100 long.

tune, on his way to the fountain, to fall in with a family of
these sable antelopes, and had shot two females. Envying
him his good luck, I started off at once, without breakfast,
determined to overtake the troop and to shoot one
myself if it were possible. Accompanied by a few Bushmen,
I soon fell in with the trail of another family that
had been frightened away—probably by my shot at a borele,
which I had killed early the same morning. These spoors we
followed for about four miles, through a rather dense field of
sandal-wood, when we heard the whistling noise of some
antelopes near us, and soon found by the spoor that they had
been alarmed by our approach. Again we followed for about
two miles upon rugged bush-clad hills, from the top of one
of which I saw, on the opposite slope of a deep valley before
us, a troop of about thirty magnificent antelopes. Some were
reclining, others browsing, and while one fine old buck was
rubbing himself against the trunk of a tree, another magnificent
fellow seemed to be standing sentinel over the rest,
looking intently in our direction, as if he had seen us. We
kept in our position for some time, and taking advantage of a
moment at which the buck moved nearer to his company,
we sank behind the bushes, and, having crawled back some
distance, made our approach in a different direction. Again
we found the old buck on the look-out in the original
direction, while the rest seemed to possess no fear, having
entrusted their safety to the keeping of their watchful sentry.
But now, by ascending the valley in the bed of the nullah, I
got, unperceived, within 200 yards, and, taking a deliberate
aim at the finest buck, I fired, and was delighted to see the
magnificent animal, when in the act of flying, reel and fall to
the earth. In another moment, however, my delight was
changed to chagrin at beholding my prize regain his feet
and follow the retreating herd. Hastily loading my rifle, I
levelled again, and fired at their diminished forms, some 600

yards off, and was delighted to hear the bullet tell amongst them, and, following on their trail, beheld with pleasure a splendid doe prostrate in the grass. Having instructed the Bushmen in the manner of skinning it, I returned, highly satisfied, though perfectly exhausted, in search of the wagon and my breakfast, after which we travelled 10 miles farther south, by a fair road, to a water-pit, Tlalemabèli, in a limestone bed. Here I picked up several pieces of crystallized felspar—a soft, semi-transparent, glassy substance, splitting into uniformly square scales. It is called by the natives "sulu mèle," and worn by them round their necks as a charm. They assert positively that it is brought down in the lightning, and may be found whenever a flash strikes the earth.

By dint of severe labour, we succeeded next morning in watering ten of our oxen, and travelled all day and part of the night, for thirteen hours, without being able to find water for our cattle. The road, however, was fortunately good, the country being open, and affording a magnificent field for sport. We saw numbers of ostriches, and starting one from her nest found sixteen eggs in it.

It has been asserted that the ostrich's note cannot be distinguished, by the most familiar ear, from the roar of a lion. This is not so. Anybody that has spent but a short apprenticeship in the wilds, and has become at all familiar with its inhabitants, can at once distinguish the booming sound of an ostrich's note from the unmistakable roar of the lion. Others, again, have given us to understand that the male ostrich never sits on the eggs. This is also incorrect, for the male bird relieves the hen every twenty-four hours, for a term of twenty-four hours, during the whole period of hatching, which it takes more than forty days to perfect. In general, ostriches are monogamous, though upon occasions three or four females will lay together in one nest; but this generally only happens to a young brood of birds that have not yet parted company, so that I do not think they can be considered polygamous—this being merely a sociable freak, and the exception, not the rule.

On the 5th we travelled all day in the same direction, steering for a range of hills running from north-west to south-east. At night we reached a Bushman village (under Maletsana) at the base of these hills, near a small well called Mununné. Under the guidance of some of these Bushmen we sent our cattle on three miles farther, to a well or spring called Kolokomme, where they obtained a sufficiency of water, and the next morning, having travelled for 10 miles south by west, reached a fine though unfrequented fountain called Imheesa, south of a tabular hill of the same name. Here we fell in with the fresh spoor of three wagons, which we were glad to follow three hours farther that day, as our course led through sandy forests; and the following morning, after five hours' trek, during which we passed to the westward of a small river called Shashani, reached the fountain and village of Mabèli-a-Puli, and five hours farther brought us to a Bakalihari village and stream called Sokos. Here we fell in with some Griqua and Bechuana hunters, who, having taken fright at sight of a party of Moselikatze's warriors, fled the country, and now consorted with the ignoble race of Bakalihari, with whom they were enjoying themselves, shooting elands and disposing of the fat for corn, beer, &c.

Before my visit to this part of the country, I had no conception of the number of Bakalihari, Makalaka, and Bechuana villages spread over this, the most fruitful part of Sekomi's domains. They live on this eastern side in a much better style, and can generally boast of a plentiful harvest. This year, in particular, they have been able to supply many of the famishing tribes in other parts, as well as their masters at the town, who were also in great want, and had suffered great losses in their flocks and herds from the effects of a very severe snow-storm—a most unusual circumstance in this part of the country, and one which is said to have put the chief Sekomi into very bad humour, so much so, that he has sent out commandos (*legagga*) of warriors to spear the rhinoceros, according to a practice of thus exercising the men to try their courage and keep their spears from rusting. But there were, I was told, other and deep-laid schemes in contemplation, and this expedition was merely a preparatory movement for an attack upon Moselikatze, after or before the return of Messrs. Moffat and Edwards.

According to an old custom of the Bechuanas, after the ceremony of circumcision is performed every man is required to have stabbed a rhinoceros, or at least a buffalo, lion, or human being. Consequently all the men of one age, or of one year's circumcision, go out at times in a body and scour the country for this purpose, and it is considered a disgrace to return from such expeditions without having dipped the point of their spears in the blood of a victim of some sort. Failing this, they are held up to public scorn and execration in the songs and dances at the khotla. All the opprobrious epithets that women can muster are unmercifully heaped upon the heads of the unsuccessful candidates for manhood and glory, as well as upon those who begot them.

The Bakalihari seem to live in perpetual dread of the nightly attacks of lions. They have their villages, which generally contain fifty or a hundred huts or more, well enclosed with strong fences of mañana or hook-thorn branches; but notwithstanding this precaution, they have here very recently suffered a loss of some individual for several successive nights. Now, perhaps, the numerous reports of the guns of the Griquas and Bechuanas had frightened the lions away for a time. While here, we succeeded in obtaining, by barter, a sufficiency of Kaffir corn, morama or tamani, and wild coffee, and a small stock of dried fruits called shesha.

On the 9th of October, resting for a time at a fountain called Moku, while the wagon trekked down a picturesque valley leading past the mountain and fountain of Serñe, we rode out on horseback and fell in with klipspringers, which proved too fleet for us. But while in pursuit of a troop of about 400 pallahs (*Antelope melampus*), we found the spoor of an immense herd of buffaloes. Of these we fell in with a troop of only eight, out of which Molihic and I bagged each one, and my dog bringing another hornless cow to bay I settled her with a single shot. While cutting up the buffalo a rumbling noise like that of an earthquake broke upon our ears, and upon looking round we perceived a dense mass of black living game, like a troubled sea, extending over a large tract of plain, approach us. Moving out of the way, no less than from 800 to 1000 buffaloes passed by us at full speed, and, fearful of being overwhelmed and trodden to death in case of a change in their course, we waited till the whole had passed, and, giving chase at their heels, killed four and wounded several others.

I could not help feeling some regret at the haste with which I was obliged to travel through this delightful country. The scenery equals if it does not surpass any that I have ever seen, and the facilities for either riding or stalking game are all that the sportsman's heart can desire—such as we of the north-western districts never had the good luck to meet with before. There were not, of course, the elephants now that were to be found during the time that Gordon Cumming rambled between these hills, but every other sport could still be found to perfection; occasionally, only, a troop of stray elephants visit these parts, and they are either killed or driven out of the country immediately.

Entering, on the 11th of October, the romantic valley of Lotlotlani, or Reed river, we were four hours in travelling through this winding pass, which intersects a range of lofty and rugged mountains, and is overhung in places with immense granite crags, which seem as if a puff of wind could dislodge masses of the rock and topple them over on the caravan passing slowly below. Meanwhile the hoarse chatterings of troops of baboons seemed to respond to the crack of the whip, and to mimic the gruff shouts of the driver, both being echoed and re-echoed in the narrow gorge.

On emerging from this wild mountain pass, we dug for water in the coarse gravelly bed of the river, and having cleared it away to the depth of eighteen inches, the water percolated fast into a pond. Some miles due east we observed a conical hill called Ma-ma-luki, at the base of which another road passes eastward towards the Matabele country. In the evening we travelled two hours farther, during which one of my oxen knocked up, and the next morning four hours brought us to a small water-pit called Gamuroani. From thence I started ahead on horseback in the hopes of joining my people from the lake, to whom I had sent word to meet me in the pass of Sekomi's mountain. After a ride of 16 miles I struck the lake road, and perceiving by the spoor that the wagon I looked for had already passed, I followed and overtook it six miles farther on, having shot a klipspringer (*Oreotragus saltatrix*) and two conies, or rock rabbits, on my road.

On the following day we moved into Sekomi's Town, only a few miles farther on. The chief's behaviour to me on this occasion was marked by great kindness and civility, notwithstanding my sharp remonstrances against the impediments and snares which he had thrown in my path. I made up my mind to stay at his town and await the arrival of Messrs. Moffat and Edwards, sending my cattle on meanwhile to a better grazing district.

Returning next day from the neighbouring mountain, where I had been shooting conies, I called at the house of old Moitoi, an influential kinsman of Sekomi's, and found him doctoring about twenty men before sending them on a begging expedition to the country of the white men. A large tortoise-shell before him contained a mixture of herbs, flowers, seeds, beans, roots, insects, and other matters, taking a portion of which compound between his fingers, he strewed it in a line from the forehead to the back of the head of each individual, muttering some gibberish, while each in his turn knelt before him; and, finally, the whole were despatched with his blessing.

On the 17th, while at Sekomi's, we had some arrivals from the borders of Moselikatze's country, in the persons of a party of Europeans, and also of some Griqua and Bechuana hunters. The latter, having had but poor sport, were envious of the success (small as it was) which had attended the white men, and hastened on to head them at Griqua Town, with the intention, as we understood, of robbing them. Soon after, the town was visited by a very notable personage, being no less than a rain-maker of high renown, who came from the far east, and belongs to a tribe called, I believe, Bariri, from their peculiar fashion of wearing their hair. This man was the youngest rain-doctor I had ever seen, probably not more than twenty-five or twenty-seven years of age, tall and well made, and of extremely black complexion. His hair was dressed in a very singular manner, being parted transversely into equal divisions, and crossed as often, so as to form a number of equal squares, the hair on each of which, being gathered together, was bound up into little horns or rollers resembling the plumes on a hearse. His coal-black skin being quite innocent of any application of fat or grease, this added much to the effect of his striking appearance. The only thing that gave me a prejudice against him was his avocation; for though all the town were rejoicing at his arrival, and the women were plying their hoes joyfully in expectation of the rain which he was making for them, he knew very well that he was practising deceit, and he could not help betraying, by a certain restlessness, the anxiety he felt at sight of myself—a white man—a being he had never seen before, and of whom he probably had a superstitious dread, believing, like most natives, that the white man is omniscient, if not all-powerful.

According to another superstitious notion, the natives never reap their corn until after the first frost sets in; consequently they suffer great loss by the clouds of birds that

NATIVE WOMEN PREPARING WINTER STORES.

invade their fields, and the quantity of grain which is scattered on the ground and wasted from being over ripe. The remainder, being carefully threshed, is stowed away in granaries contiguous to their dwelling-houses, and well looked after. When their toil in the corn-fields is over, the women set to work making beer for their husbands, who imbibe it in large quantities, and enjoy themselves very much during the spring. In October or November the labour of the women is again required in the corn-fields, and having in the daytime sown their corn, they go far at night to fetch spars or thatch for repairing the roofs of their houses, to protect them against the coming rains. At this time these poor creatures may be seen returning every night, carrying immense bundles of spars or thatch, or otherwise fire-wood, on their heads.

Among the many peculiar customs prevalent amongst the South African tribes, it is the practice of the Bechuanas never to eat of the first fruits of the season out of their gardens until the chief has had what they call "the first bite."

Cattle, as well as sheep and goats, thrive well everywhere in the Bamañwato country, and the long-tailed Egyptian sheep is to be found here larger than anywhere else, their enormous tails hanging nearly to the ground. It is quite painful to see these poor animals shuffling along in the rear of the more unencumbered members of the flock. Several kinds of thorns are a great nuisance to them: among these are the seed-pods of several creeping plants, bearing beautiful little purple-yellow flowers, such as the grapple-plant and the dubblejies—the former having a most formidable defence of thorny hooks all round.

My residence at Sekomi's afforded me many opportunities of observing that the Bamañwato held their trials at the khotla in as orderly and well-conducted a manner as one could wish. Till now I had no idea that any judicial inquiry was ever held on offenders, my impression being that it rested entirely on the judgment or caprice of the chief whether an accused person was found guilty and sentenced to punishment, or should be acquitted. Some cases occurred, chiefly of adultery or immorality on the part of the wives, for which they were duly punished, the woman's father (or, in some cases, the co-offender *) having to refund either the whole or

---

* The "co-respondent" of the English judicial system in similar cases. Human nature, and human sense of retributive justice, are, after all, marvellously alike, whether the skin be white or black.—ED.

some proportion of her purchase-money paid by the husband, and receiving back his frail daughter.

The bogoera, or circumcision ceremonies, were going on as usual. There is a difference between the several tribes connected with these ceremonies, and this is said to have divided the Bechuana family, supposed to have been originally one. Attempts have been made to introduce various innovations, which caused from time to time dissensions amongst them.

It is supposed by some of my friends, who are better acquainted with the language and habits of the natives, that they must have some knowledge of mesmerism among them, as they often speak of some of their baloi, or wizards, putting people into a sound sleep, and then torturing them, and inflicting severe wounds without their waking. These wizards are said to go about at night, stark naked, to dig up dead bodies, and convert them into medicine, to torture and destroy those against whom they have a spite. Things like these, although I put no faith in them, may be inquired into more minutely by future travellers; even if these accounts are not corroborated, the inquiry may, perhaps, lead to a cognisance of other practices or forms of belief interesting to the student of human nature.

Several of Sekomi's people being anxious to go to the colony, a distance of several hundred miles, to work a twelvemonth for a cow or heifer, they begged permission to accompany me. When they go on these long journeys they leave their cattle, grain, and other property, in care of their wives, and, should any accident occur to any portion of them during their absence, it is taken as a proof that the women have been faithless to their husbands.

The Bamañwato, as I have already said, are not so much a warlike as a trading people, and that they are desirous of living at peace with all the surrounding tribes is evident from their preferring to travel so great a distance and toil so long for a single cow rather than plunder and molest their neighbours, many of whom are much their inferiors in strength. I have seen Sechelli's cattle graze in Sekomi's territory, and return from it in perfect safety, or live for months on the borders. The love of their cattle is, by all the native tribes, great indeed. A wife is often maltreated by her husband, and they would be greatly astonished at anyone interfering on her behalf, notwithstanding the most piteous cries; but for their cattle they always exhibit the utmost concern and affection, regularly meeting them in the evening as they come to the fold, and examining their condition. They often give vent to wild exclamations, such as the following, in praise of their cattle, or bewailing their loss:

"My *god* with the wet nose, why hast thou deserted me? Thou wast lost; I sought thee, and was benighted," &c.

Again:

"Why am I so poor to-day? Why have I no cattle? I have devoted my whole life to procuring cattle."

"Thou hast slain my father, and his son wilt thou also slay?" &c.

Or:

"My fathers also lost their lives in pursuit of cattle: They have all died fighting to obtain thee, and I suppose It will be my lot also to die while in quest of thee!"

The Bamañwato are a very cheerful people, especially the women. They dance and sing nearly every night, keeping up their revels till next day; but there is not much music in their songs. The only musical instruments the Bechuanas have are reeds, with which they make a very monotonous and discordant noise at their moonlight dances, and a musical bow, with a hollow calabash attached to the back of one end, on which is stretched a twisted string made of sinews, on which the performer strikes with a thin stick, modifying the tones with his fingers by running them along the string. This is a selfish kind of music, intended more for their own gratification than for an audience, who hear scarcely anything of it, while the performer, having the one end of the bow constantly between his teeth, the sounds vibrate powerfully to his own ears, and are lost on the bystanders. This instrument is a favourite one with Bushmen as well as with Bechuanas: they while away many a tedious hour with it. Other tribes, such as the Mashuna and Banabea, have a much more inventive genius for the manufacture of musical instruments, some of which are very ingeniously contrived, and capable of producing excellent music in the hands of a competent performer. Of these instruments, two which I have seen are worth notice. One, which is fixed inside of a large calabash, has sixteen iron keys, eight long and eight short ones, which are attached by one end to a flat piece of wood, and raised half way with a thin cross-wood, in order that the keys may have free play, the performer touching them with his thumbs. The other instrument is a somewhat similar contrivance, with wooden keys of longer and shorter lengths, resting on a string stretched across hollow gourds. On these keys the performer produces his notes by striking them with a stick in the same manner as tones are produced from musical glasses.

After lingering at Sekomi's Town for several weeks, I determined to move to the fountain of Lupèpe, and wait there the arrival of my friends Edwards and Moffat, as our cattle were now breaking their knees over the rough basaltic rocks in their attempts to get at a very scanty supply of water. In preparation for this, I procured a small quantity of the tamanis as a substitute for coffee, as well as another wild bean (our wild coffee) called muchanchua, which grows on a plant two or three feet high, bearing a large and delicate white flower.

We started from the town on the 23rd of October, and, reaching the large well of Kualebel, watered all our cattle by hoisting up pailsful, and pouring it into a small pond close by. The next morning, by another four hours' trek, we reached Mashooi, where we found some wild tea, an insipid parasite like the mistletoe, called by the natives "pallemella." A little barm, or "moer," obtained from the Kuruman people, enabled us to make, in a few hours, a very nice drink of honey and water, tasting like ginger beer, but intoxicating to some people.

We reached Lupèpè on the morning of the 27th, and, finding a sufficiency of grass and water there, employed ourselves in laying in a stock of dried flesh. The following morning we bagged two giraffes at an early hour. Returning from these, we fell in with and followed the fresh spoor of two fine eland bulls. A variety of game, such as hartebeest, gnus, tsèsèbies, pallahs, and springboks, came within the range of my rifle; but, preferring the most glorious sport in the world, we let them pass, and soon espying two elands flying at a rapid rate, we gave chase, overtook, and shot them both. Though they proved fat, these were the fleetest and best-winded elands I had ever met with, and they completely knocked up our horses.

I remained at Lupèpè until the 12th of November, taking very successful lessons from the Bushmen in honey-hunting and stalking game. When nearly giving up all hope of meeting my friends Moffat and Edwards, I received a letter from the latter, informing me of their arrival at Sekomi's Town. Upon this, I immediately saddled horses and rode to meet them. On arriving at Sekomi's, I found Mr. Moffat, with a patriarchal white beard, working hard at a new axle for his wagon.

Having ascertained from Sekomi and some of his principal men that the Griquas had determined upon robbing us when we reached their country, we resolved not to give them the opportunity. They had plundered other travellers on their way, and, having missed us, were determined to satisfy themselves now, particularly as they were returning homewards from a long and unsuccessful journey. Having arranged with Mr. Edwards to get out of the country, if possible, either by Walvisch Bay, or else by steering for the east coast, we communicated our plans to Mr. Moffat, who kindly undertook to take one of our wagons, with about 2000 lbs. of ivory, under his charge, and to pay with the proceeds some little debts we had left in the colony. As we had still an ample supply of ammunition, though not a grain of coffee, tea, flour, sugar, or any other kind of provision whatever, we determined to explore from the Limpopo northwards through Moselikatze's country, and thence make our exit from these central regions by way either of the east or west coast, whichever should appear the more feasible.

The reception of Messrs. Moffat and Edwards by Moselikatze appeared to have been most cordial and friendly, but much more satisfactory to Mr. Moffat than to my friend Edwards. Although the old chief had slain so many thousands of his own countrymen without the least compunction, he shed a flood of tears on seeing Mr. Moffat again. He extended a generous hospitality to both the travellers, but would not allow Edwards to hunt elephants in his country. To Mr. Moffat he held out some distant hopes of eventually planting a mission amongst his tribe.

On the 15th of November we took a rather melancholy leave of Mr. Moffat, who immediately trekked southward for Kuruman, while, on the following day, we steered back into the wilds, not well knowing our own destination. Mr. Moffat had the night before lost all his sheep by the wild dogs, and we felt we had nothing but our rifles on which we could depend for food. On the 17th we met a man named Frans Jood, who resided near Griqua Town, on his return from Moselikatze's country. This man informed me of the jeopardy I was in beyond the Shua, and related all my doings in that country as accurately as if he had been an eye-witness of them. It appeared that I had been continually beset by spies, who regularly reported all my movements to Moselikatze. From Frans Jood and Sekomi's people I also heard of the massacre, by Moselikatze's troops, of the friendly Bushmen left by us revelling in the fat of the land, who were cut off the night after our parting with them on the Shua.

While at Sekomi's, we visited the khotla, and heard a man hold forth in praise of the chief. Edwards, who is a perfect master of the language, told me that his oration was most eloquent and poetical. He spoke rapidly for several hours, and I expected he would drop with exhaustion. His rapturous adulation of the chief was rewarded by the gift of a cow.

Some of the ironwork of our wagons having been injured, Edwards engaged a native smith next morning, with his bellows, to come and blow for him while he welded some pieces of iron, which set us on our legs again. We also procured a lot of native tobacco, which we knew would have a soothing effect on our servants' spirits when depressed.

In consequence of the unsettled state of the country on the Limpopo, the natives there being in expectation of a commando from the Trans-Vaal Boers, we altered our plans, and finally determined on going northwards to my old hunting-grounds beyond Thatharra, thence to hunt and explore the country to the north-eastward, and await the turn of events before deciding upon our ultimate way of exit. We had hopes, if our horses lived, of killing at least fifty bull-elephants each, the ivory from which would remunerate us well for our time, trouble, and hardships.

One of the young men we engaged among the Bamañwato was named Morimo-Pelu ("God is gracious"). His mother, who, after many years of married life, in her old age, and when nearly despairing of offspring, was at last blessed with this son, appropriately gave him this name—a striking testimony to the fact, elsewhere adverted to, of the native belief in a God. Besides, there are many other names which old men had bestowed upon them in their youth intimately associated with this belief, thus proved to have been entertained long before the introduction of missionary labours among them.

CHAPTER XIII.

Journey Northward from Sekomi's—Altered Plans—Reach Kama-kama—Wild Grape-vines—Beginning of a New Year—Camp at the Chenamba Hills—Mr. Edwards departs for Linyanti—Almond Trees—Food in the Desert—Move Eastward in search of Game—Hunting Elephants—Native Fruits—Various Caterpillars—The Epidemic Fever—Curious Cisterns of Rain-water—Increasing Difficulties.

AFTER waiting in vain until the 23rd of November for the recovery of a lost ox, we started on our journey, sleeping that night at Musaga, without water for our cattle. We were now about to plunge again into the wilderness, and the sense of loneliness inspired by the situation was not relieved by the evident inclination to despondency on the part of many of our men; we did our best, however, to cheer them. We found that all the natives whom we met on the road were well informed as to all that had transpired at Moselikatze's. Some also who passed us brought the same news from the direction of the lake, it having reached that place through the medium of gossiping slaves and Bushmen from the Makololo, to whom it had been communicated by the people on the south side of the Zambesi river. It is a matter of astonishment how soon news travels in this country, as will appear from the following circumstance:—When Moffat and Edwards were at Moselikatze's, all that occurred there was reported from village to village by way of the Makololo country and the lake, and actually reached Sekomi's by this circuitous route nearly as speedily as it could have been brought direct.

Just as we had started for our long journey northward, we were visited by a thunder-storm and several fine showers of rain, the exhilarating effect of which, in such countries as this, both upon man and beast, are really wonderful. Our cattle leapt and gamboled as if they had been a month revelling in clover; our dogs, sportive and playful, chased each other; and our men no longer thought of absconding, but were up to all sorts of games, and full of glee and merriment. The women were already tramping to the fields to hoe them and plant their corn.

We had flattered ourselves that we should make a long circuit from hence eastward to the Shua river, and there enjoy some good sport; but news we received here, that the country ahead was all inundated, compelled us to alter our course. We determined, therefore, on making for my old hunting-grounds in the neighbourhood of the Chenamba hills, following the track described in a former chapter, by Shogotsa and the Ntwetwe salt-pan, to the pond at Kama-kama, which we reached on the 23rd of December. We fell in with numerous herds of gnus, springboks, and giraffes, together with some fine buffaloes and a stray rhinoceros, on the way, and our guns did due execution amongst the former. One morning I killed a large puff-adder; and other members of the reptile kind were by no means uncommon. Snakes are generally found numerously after a few days of rain: they come out to bask in the returning sunshine, and the Bushmen, with their naked legs, often get bitten by them. The weather was generally inclement, the nights windy, with abundance of rain. At Kama-kama, the sun appearing for a few hours through the clouds, we followed a little malona (honey-bird), and found a bee-hive. The heat of the sun was so intense that Edwards, who was in his shirt-sleeves, had his arms blistered, causing the natives to remark, "Those white men are good for nothing; they shed their skin like snakes."

On Christmas Day we found a number of chameleons. Wild vines were growing luxuriantly everywhere. The bush-vine resembles in every respect the cultivated grape-vine, and its fruit is perfectly similar, both in appearance and flavour. I have since sown some of the seeds at Cape Town, and found that they produce bulbs at the root, which shoot up every year. The larger game being scarce, we made our meals, for several days, on plovers, kooshans, steinboks, and other small fare; but on the 27th Edwards shot an ostrich and two pallahs, which latter, although lean, were quite a treat.

The year 1855 opened unpropitiously for us. Our horses had begun to show symptoms of sickness, and two of them died. We failed also in our search for elephants, although their spoors were sometimes seen. Game grew every day scarcer, and for supplies of food we had to look chiefly to the flooded ponds, bagging now and then a few brace of ducks, with various geese, and the large wild muscovies. A few kids were occasionally taken. Sometimes, however, we were driven to the ordinary fare of the Bushmen of the desert—bulbs and various other roots. We continued a northwardly course, toiling through the wilderness, and on the 12th of January reached the vley where, the year before, we had been fortunate enough to find water for our cattle, and next day arrived at the Chenamba hills. Here we obtained a change of diet, having bagged a duiker, while our Bechuana servants were delighted to get for their dinner a fine red cat ('ntapi).

As we were now within 60 or 70 miles of Linyanti, it was determined that my friend Edwards should make a hasty expedition thither, to renew my old connection with the chief Sekelètu, and make an effort to barter for ivory; at the same time fulfilling a promise made to Mr. Moffat, that we would endeavour to obtain tidings of Dr. Livingstone, and of the fate of certain packages sent for him by Moselikatze's people. Meanwhile I would remain to hunt elephants, or to discover their haunts—a task of some difficulty, from the roving habits of the animal, especially at this season, when water is so plentiful in certain parts of the desert.

Having made our encampment under the side of a small conical hill, or pile of loose basaltic stones, and erected a booth—a simple construction of sticks, branches, and leaves, fixed pyramidically against each other, and covered with grass—Edwards left for the Chobé, with one wagon, drawn by fourteen oxen, and accompanied by three men.

On my friend's departure I felt, as was natural, a painful depression of spirits on being left alone in the wilderness, and, climbing the pile of rocks, the view presented of as dreary a tract of country as can well be imagined did not tend to dispel my melancholy. The blue sky above was unbroken by a cloud, and the black forest beneath lay stretched around me in every direction, without a single break or undulation—like a vast ocean, neither a ripple or a break to be seen in the distant horizon, while not a sound or sign of life was to be detected anywhere. From my elevated position I felt like one posted on a mast planted in the middle of the ocean, and looking in every direction for land, but in vain. Finding the sense of loneliness too overpowering, I descended to my wagon, guided to it by the cheering beacon which was presented by the smoke of our fire, curling through the downy foliage of a cluster of almond trees. These trees bear a fruit larger, and quite equal in flavour, to those of Europe. The Bushmen obtain a very fine oil from the almond, superior to salad oil. They anoint their bodies with it, and also use it with food, considering it a great delicacy. In obtaining it, they boil the almonds, and skim the oil from the water.

We had now neither flour, rice, coffee, nor tea—nothing, in fact, of the common necessaries of life: our sole subsistence was flesh and water, and for that we had to hunt constantly to provide our daily meal—a difficult thing in the rainy season, when the game, assembled at other times only near the banks of certain streams or fountains, have the run of several thousands of square miles of desert forests and thorn jungles, with water everywhere.

Our friendly Bushmen, more fortunate than ourselves, are never at a loss at this season to provide a meal with astonishing facility, at a few moments' notice. All the different esculent roots known to them have already sprung out of the moist ground, and meet them at every step. It is true that these are sometimes harsh and unpalatable; still, abiding by the principle that "what won't kill will help to fill," they need not starve. The large bull-frogs, already noticed, are also plentiful; tortoises are to be found, and turtles are easily taken in the ponds by those who understand the way; birds' nests are robbed, and a few dozen of mice easily killed and strung round their waist on their girdle; and, above all, the successive months direct them to different fields as the various fruits and berries of the locality begin to ripen, and they fare sumptuously for the time. Meat is nevertheless the idol they worship, and for this, of which they are in the constant pursuit, they will abandon everything else, and often risk their lives.

I sent parties of Bushmen to traverse the country in different directions in search after elephant spoors, and as they returned unsuccessful, I resolved to go about myself with a small party, until I fell in with some game worth having; consequently, having rolled up a blanket, and taken a spare gun, with a good supply of ammunition, we started eastward for my field of operations the year before. Now, however, after travelling 30 miles over ground that was then covered with spoor, I fell in with nothing, and, making up my mind for a supperless night, directed my followers to cut away the tendrils of a thick grape-vine from beneath a sandal-bush, where, spreading my blanket on a heap of soft green leaves, I was sufficiently sheltered from the threatening dew by the overrunning vines, and some hundred bunches of young grapes that hung overhead. A fire was soon made, and for some time that seemed the only cheer we were to have; but at dusk four young Bushmen lads, who had mysteriously disappeared, returned, one bringing a dish of wild potatoes (*mahubarra*) and wild yams (*morama*); a second, a tortoise and three turtles;* a third, a dish of wild fruit; and the other, who had a small axe with him, brought fourteen young jays,† of which he had robbed the nests in the decayed hollows of old trees. These were all laid down before me, as their chief, and, having made a selection, I allowed them to appropriate the remainder. A stray goose alighting in the dusk on the pond before us, I was fortunate enough to add it to our store.

I continued my hunting expeditions from the camp at the Chenamba hills with indifferent success till the month of March. During much of the time the rain fell in abundance, and thunder-storms were frequent. After these storms, the air was often exceedingly cold, and the Bushmen, like myself, were glad to avail themselves of the warmth of our fire. During the latter part of the period, many of my people suffered from attacks of fever, the prevailing epidemic of this season. Most of the Bushmen were also prostrated. But while I was puzzled for a cure, they made light of their sickness, in

* These turtles burrow underground, like the bull-frogs, during the dry season.

† These birds, before they are fledged, as well as the hen while breeding, are fed by the male bird; the hen never leaves the nest until the brood are fledged. The birds cannot fly well, and if seen on an open field, where an occasional resting-place is not to be found, they are easily run down and eaten by the Bushmen.

which they seemed to suffer most from rheumatism, which they cure by a novel process. One man, more skilful in medicine than the rest, tattooed them over the parts affected, and, having procured the roots of a plant bearing a very pretty flower, something like a tiger-lily, the ashes of this root were rubbed into the wounds, which, in the course of the night, formed pustules, like a blister, discharging inflammatory matter.

In a few days the Bushmen were nearly all recovered, but symptoms of disease continuing in the case of my own men, I administered calomel, with rhubarb, jalap, and quinine, and as they seemed slightly to recover, and flesh was very much required in camp, I started for some vleys about 40 miles east to obtain a supply.

Although we had seen nothing during the day to indicate the neighbourhood of elephants, about 10 o'clock at night the rumbling noise made by those animals was wafted towards us from a distance of more than three miles through the stillness of the forest. This elephant was the first I had ever noticed as having laid down while on spoor; and I felt certain I should fall in with him early, as he seemed very lazy in all his movements, crossing and recrossing his own spoor at intervals, and uprooting the ground in a fearful manner, leaving the marks of his huge tusks, a warning to the too eager hunter, in the soft black and sandy ground at the roots of large mokala and other trees. As I expected, we soon fell in with the creature, and I should have brought him down at a single shot had I been alone; but my followers, while I was quietly stalking in upon him as he stood on an open space, became so excited that they could not keep their tongues from giving vent to sundry ejaculations, when the keen-eared elephant, starting at the first sound, fled in great consternation. However, he soon made a halt, to discover whence the danger threatened, by which time a bullet from my large rifle pierced his shoulder-blade; but still he ambled off with the activity of a steinbok. The dogs, four in number, were immediately slipped, and it being early, cool, and fresh, they were full of spirit, and I knew the elephant would be mine very soon. The noble beast made for the densest forest in the neighbourhood, and furious were the charges he now made at the dogs, and loud and appalling his trumpeting cry, echoing through the stillness of the forest in his career of agony. The blood streamed plentifully from his wound, and I soon overtook him; but the Bushmen, greatly excited, became so clamorous, that the beast heard us and moved on half a mile farther, with only a snap-shot this time after him. Again I followed, and again overtook him; but the Bushmen now became so boisterous that I was obliged to make a demonstration of shooting them. This put an end to their clamour; but the elephant had already observed me, and, dashing his trunk at the dogs, with a shrill utterance of anger, from side to side, to clear them from his path, he backed, in order to obtain an impetus, and charged madly forward; at that moment, levelling my gun at his forehead, while still 120 yards off, I fired, and then ran for my life; but hearing the ground reverberate behind me with his fall in the midst of his mad career, and the Bushmen calling out, "He is dead," I gladly pulled up, and returned to examine my prey, amidst the rapturous cheers of the Bushmen, who could not cease admiring the precision of the shot, yet seemed to regret that the fun was so soon over.

On returning to my camp, upon the 4th of March, finding that my men were not so bad, though again prostrate from relapses brought on by their own neglect, I gave them a plentiful allowance of beef-tea, with wild vegetables and fruits. One of these which I found here was certainly the prince of all wild fruits. They call it *bodododo*. The fruit is of a golden colour, and has a most luscious taste and smell.

It is marked in sections, like those of the pine-apple, and resembles that fruit in colour, but exceeding it in the lusciousness of its scent. It is about the size of a large apple, and though it has a seed to every section, resembling a castor-oil seed, it is the most fleshy fruit I have met with in these parts. It grows on a low shrub, in moist sandy spots, and under the leaves, which point upwards, the fruit pointing downwards. I have often discovered it by the scent. Is this, perhaps, the *Anona squamosa*, known in some parts of India? When green, the bododo is boiled as a vegetable. Another fruit, called *shesha*, now also began to ripen. It is the size of a date, and resembles it when dry, in which state it is stowed away in large quantities for winter consumption. This latter fruit grows on a very low bright-leaved shrub in moist sandy hollows, and is greedily devoured by our dogs, who go out regularly to hunt for it.

Our search for elephants was still continued, but without success. We were now on the edge of the country infested by the tsetse, into which all the animals seemed to have migrated, and whither I dare not take my dogs. We killed three or four snakes, as we do every day when the sun peeps out, and one of my gun-bearers carried a snake in a burning log for several miles, till I pointed it out to him. These reptiles are so abundant at this season, that I felt somewhat nervous when out at night, and the Bushmen in gathering wild fruit are obliged to be very cautious, as they are always found in quest of this, their favourite food.

On my way back to the wagons a small hairy caterpillar happened to come in contact with my hand, to which the hairs from its body so attached that I found it impossible to remove them, although I scrubbed them lustily with sand. A burning heat ensued from the moment of contact, which never ceased for three weeks afterwards, and then only together with the skin in which the hair had lodged. Another kind of caterpillar, which envelopes itself in a cell or hut of grass, and is said, when eaten, to be a deadly poison to cattle, is also plentiful. There are also two other kinds; one encasing itself in a cell made of sand, and another in a cell composed of sticks lashed longitudinally one against the other. These cells, which are lined with the web of the insect, hold the chrysalis.

After the recovery of my people, I had another cause for anxiety in my friend Edwards's long absence. Not having received any tidings from him, and fearing that some mischief had befallen him, I was on the point of starting for Linyanti, to relieve my mind, when on Sunday, the 11th of March, three Bushmen, whose quick steps betokened an errand of importance, came in, bringing a letter on a stick. This letter informed me that Edwards, and all his people, had been struck with fever, and that his oxen were dying of the fly-bite. He had struggled on to within about 30 miles of my encampment. Hurrying off, under the guidance of the Bushmen, I found Edwards lying in the wagon as yellow as gold and as weak as a child, reduced to a skeleton; under the wagon lay the three servants who had accompanied him, groaning with the fever which was devouring them, and calling constantly for water.

On the way we were heated and thirsty with the march, and although fresh elephant spoors were plentiful, we found no water; but on coming to a forest of large dark green and glossy-leaved trees, my guides halted, and having each cut a stick of wood, twisted it in their hands until the bark became loosened, and the wood was extracted, leaving the bark an unbroken tube. With this they ascended the trunks of the trees, aided by rude ladders of forked sticks, and, entering their tubes in the cavities of the decayed trunks, drank their fill by sucking up the water. It puzzled me much at first to find out what they were doing. The trunks of these trees are generally much decayed at the core, though the tree looks healthy enough. The cavities are filled in the summer with rain-water, which lasts nearly the whole of the winter, and I have more than once subsequently quenched my thirst when a novice would certainly have perished: as the sandy soil in which these trees flourish admits of no pools or vleys, nature seems to have supplied the tree itself with vleys, which the earth here refuses, nearly equal to living fountains.

I learnt from Edwards that he had been disappointed, or partially so, in his object at Sekelètu's; that he had struggled hard to get there through the flooded country, and through the fly at night. He had met with a rather unfriendly reception from Sekelètu; and although he had courted his favour by kindness, as well as by presents of rather an extravagant kind, he was obliged to leave without realising his object, the chief prohibiting all assistance being given him, so that he had to find his way back without a guide; and although he was hearty at the time he left Sekelètu's Town, subsequent exposure and hardships brought on the fever, of which he partially recovered only to get a relapse.

For the next two days, and part of the third, we went on cutting our way through the forest, felling, probably, several hundred trees each day. At sunset on the third day, however, we reached Chenamba, and Edwards, who had been growing gradually weaker, now became alarmingly so. He grew perfectly helpless, and unable even to sit up, so that I was obliged to support him in my arms, and convey a little sago broth to his lips with a tea-spoon, while he sometimes fainted in the act of taking this nourishment; and I certainly thought I should soon have to bury him in this wilderness.

CHAPTER XIV.

Dr. Livingstone and Sekelètu—South African Missionaries—A Bushman Revel—Boa-constrictors—At Chapo's—Potato Culture—Hunting on the Botletlie River—Return to Lake Ngami—Transactions with Lechulatèbe—An Unhealthy Season—Plague of Snakes and Mosquitoes—Sufferings from Famine amongst the Native Population—Up the Teougé River—A Visit to Lebèbè—The Bakuba Nation.

By the intelligence which Edwards brought from Linyanti, it appeared that Dr. Livingstone had left the Makololo for the west about the month of November, 1853, and had been seen near the west coast by a party of Mambari traders with whom Edwards fell in at Sekelètu's. The chief denied all knowledge of the packages which had been sent by Mr. Moffat from Moselikatze's, until he learnt from Mr. Edwards the particulars of Mr. Moffat's visit to Moselikatze, and that the goods had actually been sent by Dr. Livingstone's father-in-law. Sekelètu now sent a party of his men to take charge of the packages, and cover them with a hut. Hitherto they had been afraid to approach them, fearing they were merely a bait or trap sent by their enemy, Moselikatze, to lure them to their destruction. All the predictions of his wise doctors were now disregarded.

Mr. Edwards had been very scurvily treated by Sekelètu. Because he could not obtain his merchandise at a ridiculously low figure—about fifty per cent. less than they could be purchased in the colony—the chief declined bargaining, and did everything in his power to intimidate Edwards into giving his goods away. Finding all his endeavours fail, he refused him a guide, and declined lending any sort of assistance; and Edwards was obliged to struggle through the return journey as best he could, subjected to the misfortunes and calamities referred to in the last chapter, and from which he was suffering when I came to his rescue and conducted him to the camp at the Chenamba hills.

Knowing, from past experience, what sort of people Sekelètu and his Makololo were, I had certainly entertained some doubts of the success of the expedition; but we thought that, as no traders from the south had visited them for some time, in consequence of their repeated ill-treatment, they might have seen the error of their ways, and would be glad to renew their connection with the white man, which had been established some years before. It appeared, however, by their treatment of Edwards, that they intended to give up commerce with the south. They pretended that Nake (as they called Dr. Livingstone) intended to bring them traders on cheaper terms from the west coast, whither he had gone to fetch them; expecting also, on his return, a large supply from him of all the good things the white men possess. They would have nothing to do with anyone but Nake—he was their trader—who would give them two barrels of powder, a large heap of lead, and sundry other articles, for a small tusk of ivory, or mend a dozen broken guns for a paltry remuneration; and unless Edwards would condescend to such terms as these, they would wait for him, and cut all acquaintance with traders from the south. However the Makololo may have intended to flatter Dr. Livingstone, I believe these assertions to be a slander on his character and reputation, invented merely to aid Sekelètu and his people in their attempts to make a good bargain.

However favourable may be the opinion I entertain of the South African missionaries, there is one point on which I must take the liberty of offering a remark, applying at least to some few of them: and that is, the readiness and credulity with which they give ear to the reports of the natives in preference to the statements of white men. This has been the cause of some unfriendly feeling between respectable travellers and some of the best-intentioned and most estimable of the missionaries. It arises from the interest they naturally take in what concerns the native population. The bias in favour of the black races, and the estimate the missionary is disposed to make of the native character, are probably influenced by impressions made on the missionary's mind by intercourse with the more eminent, intellectual, and well-conducted members of his flock, forgetting that these are rare individual exceptions.

But to proceed with my narrative. That I had myself escaped the fever, was a blessing for which I could not be sufficiently thankful, especially as there were many important things to be done before we could proceed anywhere. We had almost given up the idea of any other exit from the country than by Walvisch Bay, but felt disposed to wait the turn of events—especially as we hoped to get a letter from our kind friend, Mr. Moffat, informing us that the road to Kuruman and the Free State was safe.

The elephants were now fast leaving this part of the country, as the waters were already drying up; and this induced us to hasten our movements for getting out of the desert to some permanent spring or stream. In consequence, we made every inquiry about a path direct to the junction of the Tamalukan river with the Zouga; but, finding the risk too great, from the scarcity of water, and the bush we should have to pass being reported to be very dense, we determined now on steering south for the Botletlie, and following its course up to the lake.

Before starting, it fell to my lot to make two new axles

for an old wagon, which had already served Mr. Moffat for more than thirty years. In this work, our man Abraham was my only, but a most valuable, assistant. Still, not thoroughly recovered from the fever, he was liable to relapses, which compelled him to desist from labour. A new *trek-touw* and *kaatel* were also to be made; and these cost us several days of most severe labour, never having wielded an adze before. Having completed this task by the 26th of March, we prepared everything for trekking next day. Before parting from our Bushman friends, we witnessed occurrences at the last moment which exhibited some curious traits in the character of this primitive race. The passive and effeminate disposition of the men, of which we have had frequent reason to complain in the course of this narrative, was illustrated in the revel which accompanied the parting feast, when the men allowed themselves to be beaten by the women, who, I am told, are in the constant habit of belabouring their devoted husbands, in order to keep them in proper subjection. On this occasion the men got broken heads at the hands of their gentle partners; one had his nose, another his ear, nearly bitten off. We had presented them with some tobacco and a fly-bitten ox for slaughter on the occasion, and were well pleased to see them all happy, and enjoying, in their way, themselves on this their last night with us. To my horror, however, I was doomed to witness what they call the Porrah, or devil's dance, under the influence of which they worked themselves up to such a pitch of excitement, that they fell to the earth as if shot down with a gun, and writhed in agony, foaming at the mouth, till relieved by letting of blood by the use of needles or other sharp instruments. It was a wonder that some who thus fell were not suffocated with the dust the rest raised as they danced in a circle round the fallen, tramping to the time of one, two, three—one, two, three, &c., and twisting their bodies, arms, or legs simultaneously at different stages of the dance and music. The men carried fans of gnu-tails in their hands, a plume of black ostrich feathers waved on their foreheads, and a moana seed-pod, which rattled at every tramp, encircled their ankles. The women clapped their hands and stamped loudly to keep time with their voices and the tramping of their husbands.

On the afternoon of the next day we left Chenamba, and reaching a dry nullah or valley, which bore indications of having been a flowing stream at some recent date, we were brought to a stop by one of the axles of the lumber-wagon giving way, and the invalids of our party being pitched out some feet behind. We put in the new axle I had shaped, and started again the next day, Abraham driving one wagon and I the other, while for leaders we had two Bamañwato, who had partially recovered. At night we were joined by some of our Bushmen friends, who preferred our company for the present, intending to disperse themselves throughout the southern country, in fear of being fallen upon by the Makololo after we left. We killed for them an ox that seemed likely not to hold up long, as it was suffering from the effects of fly-bite.

In the afternoon of the 28th I started a tiger from its lair, and shortly after, while picking wild-grapes, I was startled by observing that I was stooping over a boa-constrictor coiled up in a large heap. Shouting to Edwards at the wagon to send a gun, I shot the reptile through the back; still, however, raising itself on one half of its body, it darted and snapped at me, its fangs clapping against each other with a rattling noise, till I gave it a mortal blow, when I found it had swallowed a large spring-hare, which, though partially decomposed, was greedily devoured by the Bushmen, as well as the boa itself. This they considered a great delicacy; but it was enough to make one recoil with horror and disgust, from the spectacle afforded by the body of the snake, when divested of its skin, still writhing about and full of life. It was twelve feet long, six inches in diameter, beautifully spotted on the belly with white and black, a bright, glossy chestnut-brown, and black and invisible green on the back. By the accounts given by the Bushmen, these reptiles sometimes, though seldom, attain the length of twenty-five or thirty feet, and will swallow a young buffalo-calf, or even a young giraffe. Being in this country hunted for food, and so easily traced, few indeed escaping the hands of the Bushman, they seldom live to attain more than sixteen or eighteen feet in length. In Natal, where this is not the case, I have seen them above twenty feet long, and capable of swallowing a buffalo-calf, or bushbuck-ram. In the Mariqua, and on the Limpopo, they attain a size and strength sufficient to take in a pallah, and are, in consequence, known in those parts by the name of "Lemetsa pallah" ("Swallow a pallah").

On our return journey, we travelled south by the usual stages till we reached Koobye, near the Ntwetwe salt-pan, so often mentioned in these pages. There it was arranged that, as the neighbourhood was healthy, I should leave Mr. Edwards, who, with some of our people, was still suffering from the effects of the fever, while I hunted for elephants up the banks of the Botletlie as far as Lake Ngami, a distance of about 400 miles, where the rest of our party should join me, making it the starting-point for our journey homewards. Accordingly, on the 10th of May, having bagged three fine springboks early in the morning, we started for Chapo's, but at 11 o'clock our progress was impeded by a fire raging at the exact point in the margin of reeds where we wished to penetrate the marsh of the Bakurutsie. Having halted for some time to dine, we endeavoured to head the fire, which was now raging furiously, driven by a strong easterly wind. We lost our road in the midst of a field of tall reeds, and, fearful of falling into bogs or pitfalls, were obliged to halt until it was found. By that time the fire was within fifty feet behind us, sending its crimson flames fifty feet high, and causing me great anxiety, as there was a considerable quantity of gunpowder in the wagon. With good driving, however, the oxen trotting hard for fully a mile, we escaped the danger, and reached Chapo's village before sunset. This chief immediately sent me a fine goat for slaughter, and some water-melons and sugar-cane for drink. At this season they use these instead of beer, the women being too busy in reaping to brew it. Having expressed a desire to purchase some corn for my journey, I obtained, the next day, about two bushels for a trifle of beads, and was delighted to see a dish of fine potatoes brought to the wagon, the first that had ever been grown in the country, and from the seed which I had introduced the previous year; but hearing that they had not many left, I declined the purchase, and advised them to go on planting, to increase the stock for the benefit of their nation, as well as of future travellers.

Hearing that crocodiles were still numerous, I took my rifle and rode down to the Botletlie river, followed by the chief and a gang of about eighty of his followers, all anxious to see a crocodile slain. Upon arriving at the river, which was here dry, except a pool of about 250 yards long and 150 broad, I was truly horrified at the sight before me, and could not have conceived that such a number of crocodiles could live in so small a body of water. On the opposite bank lay at least 200 crocodiles in a very irregular mass, large, small, and middling sizes, and in some parts the white sandy beach, for a space of several yards, could not be seen at all, so thick did they lie. My purpose in going was to shoot a large one for the sake of the head, and in the centre of a very dense group I noticed one huge old brute, which seemed the father of the lot, to judge by his size and the coarseness of the scales on his back, and his enormous white teeth conspicuous on the shady side of his jaw. Taking a deliberate aim, I fired across the river at this huge monster, aiming behind his eye. A dreadful scramble towards the water followed the report of the gun, and a great splashing and plunging set the whole pool in commotion; but I was delighted to see the old crocodile remain behind. The eyes of the natives were fixed upon him, and I was glad I did not disappoint their expectations. The brute struggled for a long time in vain endeavours to reach the water, and at length he lay on the edge of it. I walked round and put another bullet through his heart, but it took him hours to die. Several others came popping up their heads afterwards, but I did not like to spoil the only water we had to drink. The one I shot was the largest in the troop, measuring seventeen feet in length, and was very bulky and fat. I gave orders to have the head preserved, and the remainder of the carcase was appropriated immediately by a gang of Bushmen, who, less fastidious than their northern neighbours, thought the flesh a great delicacy.

On the 12th, having purchased three or four tusks of ivory, and made Chapo's wives some return for cleaning a quantity of corn for my journey, I hired two Makalakas to assist us to the lake, and left the village at sunrise. Soon after clouds gathered in the west, bringing with them a severely cold wind; thunder pealed in the distance, and the rain, which fell heavily, threatened to deluge us in the middle of the swamp. Hurrying to get out of the reeds, we left a fine old buffalo-bull unmolested in the enjoyment of his pasturage, although we passed within a couple of hundred yards of him. We outspanned on the banks of the Botletlie, where I shot a couple of springboks.

During my short stay at Chapo's I had an opportunity of noticing a great increase in the population, caused chiefly by new additions of Makalakas from Moselikatze's country. These people seem to equal, if not to excel, the Bakurutsie in superstition. Some of them were busy inflicting an experimental sort of punishment, too disgusting to relate, upon a runaway wife; she did not love her husband, and, having run away from him, was brought back by her own family, and punished by them in his presence by flogging, until she was brought to a state of most abject humility and cried piteously for mercy.

On arriving at Chapo's, I found that numbers of his people had been sick, and several had died of the fever. Chapo himself was in good spirits, and with his usual inquisitiveness endeavoured to learn from me how Englishmen taught cattle to be so tractable. He did not believe that they were gradually trained and broken in, but supposed that we gave them some medicine which made them serviceable at once. He also endeavoured to obtain some of the gun medicine from me, assuring me that he had once procured a very excellent supply from some Griquas; and on my assuring him that he had been imposed upon, he did not scruple to tax me, in very round terms, with untruthfulness.

The rain having passed to the southward, we travelled on in the afternoon for about 10 miles and reached Masakanyani, where I shot a large male and female luchee as they darted across the road in front of the wagon. We found that the river was coming down, and, overflowing its banks, was inundating the native corn-fields, which were infested with flocks of geese, ducks, and muscovies.

The natives in these parts, as well as the more northern districts of Linyanti and the Chobé, have either a very prolific crop or no crop at all. This latter happens only when they have an over-abundance of water, in which case their lands remain inundated throughout the season; and when this occurs in two or three consecutive years, a famine is the result.

Following up the course of the Botletlie, and hunting as I

went, large game, especially buffaloes, being plentiful, I one day, near Khamma's Ford and Palami's, lighted on a family of Makobas, who had taken fright at the approach of my wagon, and were hastily conveying the whole of their furniture into a small canoe, intending to steal silently up the river. Having headed them, I detained the party for a few moments, while I inquired the news. They informed me that intelligence had arrived of a commando being on his way from Sekelètu to the lake, and that two marauding parties had also been sent out by Lechulatèbe towards the Makololo, but had returned in consequence of this report—news that was far from being agreeable.

The Makobas I captured seemed to live upon melons only, fish being scarce, as the river was too full for them to catch any. Just here, the melons seemed mostly to be of the bitter sort. It appears strange that one melon-seed should produce both bitter and sweet or sourish melons; which is the fact, though it has puzzled many travellers, who generally believe the melons are two different species. But this is not the case, and it all depends upon the manure it gets. The seeds deposited with the dung of elephants are bitter, those manured by the white rhinoceros sweet. Sometimes they lie mixed on one field; at other times the sweet melons are sought in the grazings and near the haunts of the rhinoceros. The bitter melons are not always eaten, on account of their extreme bitterness, but the seeds being taken out and pounded between two stones, the meal is boiled into a nice pottage, or eaten raw.

We passed on our way the site of Samaganga's old village. Our dogs bayed several buffaloes along the river-banks, but we killed none. One monkey the dogs caught, and tore to pieces. Troops of koodoos and great flocks of guinea-fowls crossed our path all day, and one afternoon we fell in with a party of famished Bushmen, moving skeletons, greedily hacking and cutting up a gnu which had died of hunger and mange, hardly waiting while the animal was even partially cooked. During the succeeding day's journey I gathered the seeds of a medlar, which grew on the banks of the river. It is called by the natives *melo*,* and differs little or nothing, excepting in size, from the ordinary medlar of our gardens.

We afterwards fell in with a party of Lechulatèbe's men, bound homeward from Sekomi's. Asking them as they passed where they would sleep that night, they replied by asking me, "Do I know? Ask my feet, or ask the sun." This is characteristic of the race: they will either answer one question by asking another, or tell you to ask somebody else; or if a thing is doubtful, or beyond their power to know, they say, "God will know."

At Makato's we met others coming from the town in pursuit of a man with his wife and child, who had fled. They were now safely at Chapo's, but some of Lechulatèbe's men travelling in company with my wagon seized their clubs and spears, and were going back to murder the Bushmen we passed the day before for allowing this wretched family to escape. I told them if they did such a thing without the express orders of their chief, I would try to prevent them by force, if necessary; and, after much noise and disputation, they gave up the point, but not before I had thrown in a few luchee skins and pieces of native tobacco. The unfortunate wretch they wished to kill had a narrow escape of being made the victim of native superstition, malice, or jealousy. Lechulatèbe, who had for some time had a sore on his heel, made by the chafing of a bad shoe, was persuaded by his prophets or dice-throwers that his uncle Makasana was killing him; in truth, he would have put his uncle to death for many other reasons, long

---

* "Melo," or "Mashulesa Pelu," a wild medlar. The natives will not eat this fruit if they intend going on a journey, believing it will surely turn out disastrous.

before this, but that Magalakoè dissuaded him. Lechulatèbe therefore pitched upon some other unfortunate individual, a follower of Makasana's, against whom he had a spite, and was going to kill him, as the wizard who had harmed him; but, having received notice of the danger he was in, the man fled.

To oblige the natives, I lay by the water several nights. Buffaloes and rhinoceroses approached within a certain distance, but kept me only in a wearisome state of excitement and suspense. A new species of jackal, as yet undescribed, and called *shuatlé*, or "barking jackal," afforded me some amusement, from his appearing to be very much entertained with the beauty of his brush, which he elevated while sitting on his haunches, and, barking laughingly, jumped about in a sportive manner, and made several feints to catch it in his mouth, turning round and round after it.

We now gradually fell in with more and more of the Makobas. Though at other times fine, stout, and athletic men, famine had reduced them to mere skeletons. All their food is obtained from the river, which, being overflown, was too deep for their nets; nor could they get at the roots of the tsetla (*Juncas serratus*), which at other times formed their principal diet. Beads were now quite at a discount; but anything they possessed was obtainable for a pound of flesh or grain.

I sent a man to give notice to Lechulatèbe of my approach, and slept above the junction of the Tamalukan with the Zouga. Here we were visited by a large black dog, that came to share our fire, and lay alongside of me. Suffering greatly from the soreness of my eyes, and being hence deprived of rest, I sat all night by the fire, or walked about nearly blinded by pain. Leaving my veldstoelje* for a moment, to obtain another dose of alum-water from the fore-chest of the wagon to bathe my eyes, I heard a howl

* Camp-stool.

behind me, and, looking round, dimly beheld a cloud of dust mingling with the flames of the fire, my seat upset, and the black stranger missing. The men, waking up, insisted that a tiger had carried off the dog. I could hardly believe it possible, for I had only just turned my back, and moved only four yards from the fire; their surmise, however, proved to be correct.

On the 31st of May I reached the site of Lechulatèbe's Town; but, to my surprise, the tribe had removed to the north side of the lake, above the junction of the Teougé river, whither I sent a messenger to acquaint the chief with my arrival. Shortly afterwards I received the intelligence that four wagons, with white men and a boat, had arrived from the west coast, and, being anxious to know who the strangers were, I started on horseback, and, reaching the town in eight hours, met with a hearty welcome from Lechulatèbe. I was obliged to leave my horse on the eastern bank of the river for the night, as the crocodiles were reported to be very vicious hereabouts. Indeed, it was with great difficulty that I got the Makobas to take me across, and I only succeeded after an hour's wandering up and down the banks in the dark. At last I came to a village where they had a large boat, and, by threatening to discharge my rifle among their huts, I got them to come over for me, the boats being all secured on the western bank for fear of enemies. The white men from the west proved to be Mr. Green and Dr. Wahlberg, of both which gentlemen I shall have hereafter to record some sad and intensely painful tidings.

My wagon having arrived, dealings were opened with Lechulatèbe, who seemed desirous of purchasing one of my horses, to which he had taken a fancy, but we had to wrangle over the bargain for several days. Fifty tusks were brought to the wagon regularly every morning, and as regularly carried back in the evening, without our coming to terms, and at last any attempt at dealing, either for the horse or for other articles, was dropped for a time. In the end, however, having found out the folly of the course he had hitherto pursued in this matter, Lechulatèbe sent to inquire whether he might come to take coffee at my wagon, when, using the advantage thus afforded me, I pressed him for boats with which to fetch up my friend Edwards. A bargain was accordingly made, and I despatched several canoes to bring up my ivory, with a letter to Edwards, informing him of my intention to proceed westward.

I visited the lake again the following year, at the commencement of the unhealthy season, which begins about the month of November. Lechulatèbe was then awfully frightened at the many deaths which were taking place daily from the fever, and was seldom seen outside of his hut or in the town, where, he said, "death was stalking about." All his doctors were in constant attendance, sprinkling decoctions of herbs about the doorways to cleanse them from the "taint of death." All those in whose family anyone had died were required to be well cleansed, and were only admitted into the company of others after a certain number of days, when the taint of death was said to be gone.* The chief's own women and children, in particular, received frequent ablutions, and underwent other ceremonies, before they were permitted to show themselves. Day after day, and night after night, the wailings increased. Nothing can be more startling and doleful than the midnight wail of several hundred women, and we were often disturbed by them during the few intervals of repose that we managed to steal from the tortures inflicted by the myriads of mosquitoes which visited us nightly.

* The chief's sanitary regulations might do honour to communities in the enjoyment of a much higher state of civilisation.—ED. [B.]

Makasana, the chief's uncle, who had been suffering for years, died while I was giving him the best medical aid I could, so that, according to native ideas, I had killed him. As he was generally disliked by the chief and his followers, nothing serious resulted to me from this occurrence. Makasana was always looked upon as an old wizard, and Lechulatèbe exhibited not the least sign of regret, but much of rejoicing, at the event. One morning, after having given my patient his medicine, he walked homeward, but soon lay down from exhaustion. Lechulatèbe passing exultingly remarked, "That wizard's gone; he'll never come back again." He rallied, however, sufficiently to enable him to reach his home. Shortly after, a general wailing commenced, and it was thought that Makasana was really dead; but while they were tying up the body in the posture proper for interment, he recovered from the swoon into which he had fallen, and resisted the premature office. However, a day or two afterwards he actually died, to the great joy of Lechulatèbe.*

In the midst of several holidays caused by deaths, a festival was held, called "The Feast of the Question;" that is, the asking a parent for his daughter in marriage. The father of the girl, on giving his consent, slaughters cattle and sheep according to his means, and entertains all his friends and relations. Neither he nor the bridegroom partakes of the food destined for these occasions. Processions of men may be seen moving about with pots and dishes to the abodes of the different branches of the family, feasting to their

---

* When Mr. Edwards last visited the lake, Lechulatèbe's favourite slave, and almost constant attendant, had been killed by an accident in the field. The women of the place, on hearing the news, set up their doleful wail, which, recalling to the chief's mind the image of death, so irritated him, that he vented gross anathemas upon them, driving them to the field where she had died, and cursing the dead besides. His cousin Thalè ventured to remark, "The people are afflicted," when, looking contemptuously on the speaker, he bade him take the mourners to his own house.

hearts' content. Processions of women likewise do the same, after which they strike up a dance, and continue singing and dancing for several days after. The most interesting ceremony of this kind that I have ever seen was upon one occasion when Lechulatèbe bestowed upon his own brother-in-law, Thabè, his daughter by Thabè's own sister!

During the month of November, we were annoyed with all kinds of vermin, snakes, scorpions, spiders, centipedes, mosquitoes, sand-flies, &c. One evening, during a severe and very windy thunder-storm, a large snake crept down the overhanging ranches of a tree, and, dropping into our wagon, drove us out of it, and in the confusion that prevailed we groped about for an hour in the dark, fearing to return again to our beds. We spent a most miserable night in the neighbouring hut, which we had constructed as a defence against the mosquitoes, lying down without a blanket, which we dared not fetch for fear of bringing the snake with it.

Snakes are here very numerous. They came constantly after the mice, which had built their nests in our wagon, and all night long drummed with their little feet on the top of it. One day while out with the wagon I marched ahead of the oxen to pick a road, while Edwards sat on it to guide it safely through the trees and fallen logs. On looking round to see how the wagon got through a difficult pass, I felt myself treading on something soft and slippery, and, looking down, found an enormous black snake, eight feet long, extending its coils round my boots and striking at them with its fangs. I leapt off the reptile; but my foot slipped, and, in trying to regain my equilibrium, I jumped up two or three times, but always alighted within his coils again, and, as if bound by some spell, could not get away. At length, making a desperate effort, and uttering an involuntary shriek, I made a tremendous leap, having probably escaped being bitten only by the constant motion of my body.

The reptile glided immediately into a hole, and disappeared before Edwards could come to my aid with his gun. Still nervous from the effects of the fright, I soon afterwards stumbled over an egoana (*Varanus niloticus*), which, opening its jaws and hissing loudly before me, nearly frightened me as much as the snake had done. On another occasion, while sitting at breakfast, a snake fell from the tree over our heads into our fire-place, and our servants have frequently found them under our blankets.

At this period the natives had gathered very bad crops, owing to the long-continued drought, and partly to losses of grain by clouds of birds the preceding year. Consequently the Batoana themselves were in a state bordering on starvation, not to speak of the Makobas and poor people, some of whom were dying; and many children taken in battle, or otherwise robbed from their parents, were perishing before our eyes. We saved the lives of some, but there were others beyond our power of assisting, who died of hunger, and the few we saved were preserved with the greatest difficulty, and with much quarrelling with the native women, who were so jealous of our attentions to these poor creatures that they frequently attempted to throw them into the river, and always watched an opportunity of robbing them of the pieces of hide, flesh, and bones we gave them. One poor little creature, whose shockingly-emaciated condition nothing could exceed, we learnt with regret had been burnt to death by some old woman the day after we left, for having stolen the flesh out of their pots. So reduced were the natives that we were obliged to send 350 miles for corn for our servants—a distance, however, which in this country is thought trifling. In the dearth of other food many of the natives busily employed themselves in fishing on the lake.

Among other annoyances with which we were constantly pestered at the lake, was a kind of fly, the size of the common house-fly, but very thick about the abdomen, and rather white. These bit us very severely, and the ears of our dogs were quite excoriated by them. But, above all, nothing can be so painful to the traveller, at this season, as the tortures he endures here from the mosquitoes. Their numbers are so overwhelming that there is no escape, and remedies that at other times, and in other places not so much infested, would be of some relief, are here of no avail: you are obliged to succumb and be bitten into a fever, and rise next morning with your face and forehead quite disfigured. Often have we locked ourselves into a small hut and steamed ourselves in the smoke of cattle-dung, almost to suffocation, without relief; some will make their entrance through crannies, and contrive to escape the fumes. We have often been driven out of our wagons, and obliged to run about in the open air to get rid of them, and urged back again to undergo the suffering. At the present season the mosquitoes visited the lake in such clouds that they nearly obscured the sun from view before its setting. The sufferings from them are enough to bring on fever without the aid of malaria or anything else, and the loud buzzing they make at night can only be compared to the croaking of a multitude of frogs in the distance, for which we at first mistook it. Nor is man the only sufferer from these insects. Our cattle were in a most deplorable condition. Cattle, horses, and sheep fled in terror into the desert, lowing and bellowing as they went, and often were not recovered till late the next day, having been guided by a natural instinct to take to the dry country, several miles from the water-side. A gust of wind or a cold air is the only relief we had from the terrible plague.

After I had finished my transactions with Lechulatèbe, I had leisure to enjoy the company of Mr. Green, and had many a pleasant sail in his boat on the lake, which, I was sorry to find, was becoming alarmingly shallow, not containing more than three and a half feet of water on the average. We also made up a little exploring party northward towards Lebèbè. For several days we continued, with most laborious exertion, the winding stream of the Teougé river, in Green's large boat; though a light canoe, pushed along by poles, would have answered better. We shot several quaggas, pallahs, and tsèsèbies—here called kahubs; killed or wounded two sea-cows, which we lost; and wounded an elephant and several rhinoceroses, most of which, as we afterwards learnt, had died. My object was chiefly to search for the nakoñ and poku, both of which I heard might be found up the Teougé, or Serpentine river. Professor Wahlberg, a thoroughly indefatigable sportsman and pedestrian, had meanwhile gone overland to try experiments with the tsetse fly, and to search for new specimens of natural history. On our way up we occasionally walked on the east bank, as being less inundated, in search of the rhinoceros, but we were told there were none in any direction northward or eastward, excepting a few black ones more south, and near the lake. Buffaloes were plentiful in the reeds and marshes, but there we could not get at them; and as we were obliged to trace the river with the utmost caution, for fear of losing its winding course in the endless maze of reeds and bulrushes, we dared not venture far from it, and, although we killed abundance of game, we were often half-starved, owing to our own prodigality or mismanagement, and were obliged to buy corn and beans, as well as goats and milk, from the natives as we went along.

As my own short sojourn on the banks of the Teougé, with the result of the few inquiries I was personally enabled to make, would furnish but very scanty information respecting the people dwelling on the upper waters of that stream,

I avail myself of an account published by Mr. Green* of an expedition, in which he was accompanied by Professor Wahlberg, to the town of Lebèbè, chief of the Baveko nation, lying some 400 or 500 miles north of the lake, in about lat. 18°, long. 21° E. I shall confine the extracts from this paper to what occurred during the visit to Lebèbè's Town, omitting Mr. Green's interesting description of the scenery of the Teougé; his account of the vast quantities of game, especially elephants, they fell in with, as well as the details of their tedious wanderings on the long journey, during which they endured sufferings and privations of the most painful description—deserted by their hired followers in the midst of their difficulties, and having to carry their own baggage, guns, and provisions, bending under the weight of which they staggered forward, now overcome with exhaustion and thirst, and then falling prostrate with fever. Suffice it to say, at length they reached the town of Lebèbè (god of the waters), the chief of the Baveko. Their disappointment here was only equalled by their previous sufferings. Suspected of the most nefarious designs on the chief himself, the travellers were refused admission to his presence; they were rudely treated, efforts made to detain them in the country, and at last compelled to return to the lake without assistance, and without accomplishing the main objects of their journey. However, Mr. Green's description of these people is the only account extant, their tribe never having been visited before by any white men from the south.

MR. GREEN'S ACCOUNT OF HIS VISIT TO LEBÈBÈ'S TOWN.

"We arrived at Lebèbè's Town by moonlight on the 22nd of October [1856], and were soon surrounded by people from every

* This account was published in the "Eastern Province Monthly Magazine (Cape Colony)."

quarter. To obtain a good view of us a large fire was lighted, which, fed with a constant supply of dry reeds, enabled them to satisfy their eager curiosity. Our exterior could not have given them much idea of the habiliments worn by Europeans, ours consisting merely of a shirt and trousers, and it would have been difficult for a superior judge to have decided the original colour of these garments; indeed I must admit that, could we have been seen by any of our own countrymen, we should have been pronounced to be two most disreputable-looking ruffians. Having despatched a present which we had provided for the occasion, with a message to the chief explaining the nature of our visit, he sent us in return a dish of ponka, one of earth-beans, and another of Kaffir corn, with a message that he had no ivory, and therefore could have no dealings with us unless we would purchase slaves. This we of course declined. The chief added that if we wished to remain in his dominions, he would be very glad. We could employ our time in shooting elephants, and keep him supplied with meat, he being anxious to partake of the flesh of animals, which we were to have the honour of shooting. We felt exceedingly grateful for his kind intentions, and informed him that our anxiety to return to our own country alone prevented our complying with his wishes. An account of this unexplored portion of Africa would no doubt be a subject of great interest to the general reader; but as my journey was a short and hurried one, my stock of information, I am sorry to say, is on an exceedingly limited scale.

"The country of the Baveko, or, more properly speaking, the Bakuba, appears to be thickly populated, and, according to their account, is much more so beyond and to the westward of them. They spoke of a powerful tribe to the westward, stating them to be very far distant, and exceedingly rich in copper and iron, which must be, I think, the Ovambo, so called by the Damaras. They say that they have never visited the country of this nation, but have heard much concerning it and its people. I should infer their information to have been gathered from the Mambari traders, natives employed by the Portuguese, who frequently visit Lebèbè from the west coast. A party had taken their departure only the day before our arrival. They had purchased slaves and ivory to a considerable extent; which accounted for Lebèbè having none of the latter commodity on hand when we arrived. It is my impression that the Mambari traders are accompanied by the people of the Ovambo tribe on their inland commercial expeditions, from the fact of one of Mr. Wahlberg's servants, a Damara, conversing in the Ovambo language whilst at Lebèbè's Town, and, as he believed, with people of Lebèbè's tribe. From this circumstance, Mr. Wahlberg inferred that the Bakuba were really the race of Ovambo. He surprised us much by informing us that this was the case, saying, when we overtook him, 'Do you know that I have been amongst the Ovambo?'—'Impossible!' we replied. 'Oh, no! there is no doubt about it,' was his rejoinder: 'they recognised the name of Nangoro, who, according to Messrs. Galton and Andersson's account, is the paramount chief of that country.' Had this been the case, either Mr. Wilson or Dr. Livingstone would have learnt the fact long ago from natives at the lake; but I attribute the mistake to Mr. Wahlberg's being unacquainted with the Sechuana language, as well as his only possessing an imperfect knowledge of the Damara dialect. We made strict inquiries respecting the Ovambo, it being my impression that they would surely be known by the Bakuba. But the name Ovambo was not known or understood by them; they term all who come from the great water to the westward—meaning the sea—'Mambari.' Mr. Wahlberg had no knowledge of the fact that any Mambari were located at Lebèbè's, although they were there at the same time as ourselves, and had been for months previous.

"Although I have generally spoken of Lebèbè's tribe as Baveko in alluding to them, such is not their appropriate name; they state that they are the original and highest of the Bakuba nation, not Bakoba. We took a deal of trouble in having this properly explained to us, and inquired if they then acknowledged themselves as slaves? But they laughed at the idea of such an epithet being applied to them, saying, 'Isn't Lebèbè a great chief? How can we be slaves? He is the father of us, the Bakuba race.' We then interrogated them as to whether they were of the same tribe as the people called Bakoba, in Lechulatèbe's territory, and their answer was, 'We know the Bayèyè (men) you speak of; we are the head of that nation.' This answer clearly indicates that Bakuba is a word of their own language, distinguishing the name of the tribe, and not a Bechuana word, or appellation applied to them by the Bechuanas. According to my idea the word 'Bayèyè' simply signifies men, and is not intended to apply to a nation. For instance, when asked what they call themselves, their invariable answer is, 'Are we not Bayèyè (men)?' 'But have you no other name as a nation, such as Lechulatèbe's people?'—'They are Bechuanas; our nation is the Bakuba.' Lebèbè's tribe have quite a different dialect of the Bakuba language to that of their kindred, being near the lake, which at first inclined us to believe they were of a different race.

"Their manners and customs also widely differ from the other Bakuba. Their mode of telling news is very singular; a general clapping of hands first takes place, which is sometimes continued through the whole discourse. The conversation goes on with short interrupted sentences of two, three, and four words each, it being generally repeated by the hearer; if not, other words are used in place. The news is always related in a most solemn manner, and, when terminated, there is another clapping of hands all round. The chief personage, who has heard the news from his friend, has now to recite his in return, and this formal practice takes place when and wherever they meet. They must be well acquainted (allowing for a little difference of time) with every incident that occurs in their country, as parties are continually travelling either by land or water. These people are most inveterate smokers; their greatest pride seems to be centred in their pipes, the bowls of some of which are very neatly carved, most commonly with a figure representing the head of a man. The stem varies in length, some being at least four feet, and usually composed of a portion of gun-barrel, with an additional iron tube of their own manufacture. Strolling about with these quaint pipes they certainly look singularly ridiculous, though as proud and seemingly as contented as if the whole universe belonged to them. My friend Mr. Wahlberg remarked that they reminded him of so many Germans, with these long pipes of theirs. It was with extreme difficulty I succeeded in purchasing two, and these were poor specimens compared to others I had seen. Indeed, some of the possessors would far prefer parting with their wives; and, moreover, they actually refused to sell a pipe for a quantity of beads which would purchase the largest elephant's tusk they had ever seen.

"Lebèbè's market is well stocked with beads of every description. We had a great variety with us, some of them esteemed of high value among the Bechuanas, but we found them commonly worn by the people of this tribe. They have a very fair supply of arms, most of which are of Portuguese manufacture, which for service are far inferior to our common musket. Powder they appear to be well supplied with—so much so that the Batawana from the lake receive constant supplies from that quarter. Indeed, I may mention here that a party of the latter tribe had been on a visit to Lebèbè's territory with the same intention. In lieu of lead they use iron bullets, which are beaten roughly into shape, and they do not seem to care if the bullets are several times smaller than the bore of the guns. Copper seems to abound in their territory, unless it be obtained from some other neighbouring tribe; in fact, I have little doubt but that much of that metal is procured from the Ovambo. Large solid rings are manufactured from this malleable article, with which they, the females in particular, decorate their arms and ankles, some wearing from ten to twelve of them. An observation applied by the chief at the lake regarding the quantity of beads worn by the females of his tribe, and mentioned by Mr. Andersson in his interesting work, 'Lake Ngami'—viz. 'The women,' who chiefly wear beads, 'grunt under their burdens like pigs'—might certainly be applied to the women of Lebèbè's tribe, substituting their ornaments in copper for those of beads.

"The men of this tribe show a rude but ingenious taste for carving; most of their kerries, or knob-sticks, and more particularly their pipes, being ornamented with figures representing heads, and other various species of game. Many of their wooden water-vessels, &c., are neatly executed, and have some device or other engraved upon them.

" On the morning following our arrival, we were awakened, at an early hour, by a hubbub of voices proceeding from a crowd of inquisitive strangers. Mr. Wilson suggested that we should conceal ourselves, and establish a rule that each individual should contribute something of value for inspecting us. We inquired if we were the first white men they had seen, when, after some hesitation, they explained that a white man had visited their tribe some time ago, but that he was not so pale as we. They further stated, in answer to our question, that the person alluded to had travelled with a mule, and proceeded to the north. I concluded that they referred to a Portuguese from the west coast, in all probability en route to the Zambesi river.

"Upon asking for Lebèbè, and expressing surprise at his not honouring us with the customary visit, the interpreter answered in rather a dignified tone, ' Is he not then a great chief? is he not God?' This silenced any further inquiry on our parts on the subject. At length one of the men whom, upon our arrival, they had tried to impose upon us as being Lebèbè, invited us to cross over to his town, to which a boat was waiting to convey us, it being situated on an island, like most of the other villages, and informed us that we should be allowed the privilege of being admitted to the illustrious presence of the great chief. Being afraid to trust in the honesty of these savages, by leaving our guns and ammunition behind, we were preparing to buckle on our bandoliers, which I must confess looked rather like a hostile way of paying a visit. As such they evidently considered it, as the interpreter immediately stopped our further proceedings with an excuse that he must first obtain the chief's consent. He shortly reappeared to inform us that we could not be admitted into his presence. Probably they imagined it was our intention to go over and inflict some injury on the chief, bewitching, at least, if not despatching him. It is strange that Lebèbè never visited Mr. Wahlberg, nor was the latter allowed to pay his town a visit, although he assured me that he frequently expressed a wish to be permitted to pay his respects to the sovereign.

"I have not, however, the slightest doubt but that the chief came to see us incognito, as we were struck with the arrival of an elderly personage, such as Lebèbè is described to be, creating a great stir amongst the people round us, who all made room for him and treated him with an unusual degree of respect. One of his subjects followed closely in his footsteps, carrying a battle-axe over his shoulder, and would not be seated until the former had shown the example. This certainly was a person superior in appearance to those of any of his nation who surrounded him. Of a tall and robust frame, with a shrewd expression of countenance, and hair and beard longer than is commonly seen in the natives of this country, the whiteness of which added to his patriarchal appearance, his mien was that of a thorough semi-barbarian chief. We took good care to explain to him the assortment of merchandise we had brought so far, and the difficulties attached to its transport, and which we were obliged, through the treachery of his own subjects, to leave four days in the rear. He expressed much anxiety concerning the absence of the goods, and repeatedly told us that we should not have come without them. Some of his people seemed to doubt whether our statement regarding the riches we had brought for the benefit of the chief was not false, and, indeed, two or three of them had the audacity to tell us so, remarking to their friends that we were only trying to impose upon them.

"These people tormented us exceedingly with entreaties to hunt and shoot their buffaloes or hippopotami, but our supply of ammunition was nearly exhausted, the few bullets which were left being required to supply food on our journey back: we were compelled to tell them that the guns used for that species of game were left with our merchandise, which, in fact, was the case. A message to this effect must have been sent to Lebèbè, as, shortly afterwards, the interpreter came over to inform us that it was the intention of the chief to furnish us with an escort to return, and assist in transporting our effects to his town; at the same time we were distinctly to understand that we should be expected to kill numbers of elephants for him on our return. I really believe that this was merely a scheme to entice us to bring our property up, and detain us hunting for their benefit. This they imagined could easily be effected by refusing to render us the same assistance in taking the things back; but, rather than have remained in their country upon such terms, we would have destroyed what property we possessed. As it was, however, we had discovered that the market was empty of ivory, and the unhealthy season rapidly approaching.

"This, combined with the difficulty of obtaining assistance, obliged us to abandon the idea of further exploring this country, much as we wished to have done this; but, as the escort was voluntarily furnished us, we did not think proper to inform the chief of our determination to proceed back to the lake without delay. The Griqua party that reached Lebèlè's in the year 1852 were detained under some pretext beyond the healthy season, and during their stay were employed in keeping the whole town supplied with meat, the fruits of their hunting, much to the satisfaction of the chief and his people; but the Griquas on departing* left ten of their number behind, it being matter admitting of some doubt, as I consider, whether these unfortunate Griquas died a natural death by the epidemic fever or perished by poison. It is evident that, after their ammunition was expended, the Bakuba would only be too glad to get rid of them.

"The Bakuba nation admit that their country is more subject to the fatal epidemic, prevalent at certain seasons, than the vicinity of the lake and the innumerable rivers and streams that water these regions; still, at all risks, they would endeavour, if possible, to detain a good hunter as long as his powder and lead lasted, and so long he would be considered their friend. They are very indifferent hunters themselves, although their country appeared to abound in a great variety of game, especially buffaloes. To find these grazing in sight of their villages, as Mr. Wahlberg informed me was

* Mr. Green states that the remainder of the Griquas, after losing all their oxen by the fly, were obliged to abandon their wagons, which were destroyed by the natives.

the case during his visit, appears at first sight to a stranger as if he saw herds of domesticated cattle. Elephants and rhinoceroses were said to be numerous about a day's journey from the town. I saw traces of the former within an hour's walk, where they had passed through the cultivated lands on their way to water. This plainly indicates that they can seldom be disturbed. For a sportsman desirous of hunting the larger game, and who can undergo the hardships of following it on foot, to which he must necessarily submit, there is no portion of South Africa I yet know equal to a considerable extent of the country on the Donka river as a hunting ground, more particularly for the noble sport of elephant-shooting. But it must be followed during the dry or winter season, commencing in the month of June and ending in October, which affords the hunter a period of five months for his sport. Should it prove, as has been asserted by some South African travellers, that the bite of the tsetse fly has no effect on mules, they would become an invaluable acquisition to hunters or traders selecting this route."

CHAPTER XV.

Preparations for Journey westward — Professor Wahlberg — Departure from the Lake—The Western Desert—Increasing Scarcity of Water and Game—Elephant's Kloof—The Damaras—The Namaqua Chief, Lamert —The Nosop River—Jonker Africaner— Mission Station of Barmen— Otjimbengue—The Knisip River—Missionary Life in South Africa— Walvisch Bay—The Voyage to Cape Town.

I HAD long entertained the idea of travelling westward, and of our making our exit by the western coast of the continent, instead of returning on our steps by the road we had come from the south-east. If this plan were practicable, we should escape all danger from the Griquas and renegade Bechuanas, who, we had been given to understand, would endeavour to intercept and plunder our valuable convoy as it approached Kuruman or Griqua Town in our progress to the colony.

Besides this, the plan presented the great advantage of affording an opportunity for exploring a new line of country, of acquiring information respecting its pastures and resources, and of forming some acquaintance with the populous tribes which overspread the vast region extending from the colonial frontier far to the northward, beyond the tropic, and from the great Kalahari desert to the shores of the Atlantic Ocean. Above all, it would further the design, so dear to my heart, of assisting to open fresh channels of communication for commercial purposes with the south-central regions of the continent, for which the harbour of Walvisch Bay, frequented from Cape Town, and at which I now proposed to embark for that place, offered a convenient point of departure. Mr. Edwards had learnt at Linyanti that Dr. Livingstone, with the indefatigable zeal he has devoted to the opening up of these central regions of South Africa to the civilizing influences of missionary and commercial enterprise, was engaged in a similar undertaking, having for its object to trace a route to the western coast. But its proposed terminus was understood to be the Portuguese settlement of Loanda, while as the doctor's course lay in the path of the slave-traders, and was, therefore, exposed to the demoralization produced by that nefarious traffic, such a channel of commercial communication with the interior, even if practicable, appeared to me less desirable than the route to Walvisch Bay; besides the fact of the latter being much shorter and more direct for the Lake country and Linyanti, the centres from which the existing trade may be best developed. In choosing this line of road for my own journey, I confess, also, that I promised myself the gratification of being able to accomplish the ambitious design of crossing the continent from the eastern to the western coast, which I had long cherished. There were difficulties to be contended with, but those I hoped to overcome in the same resolute and persevering spirit which had hitherto supported me in so many dangers, privations, and disappointments. The road, I learnt, was good, water generally not deficient, and the distance to be traversed between the lake to Walvisch Bay not exceeding from 700 to 800 miles. It had been represented that the principal obstacle we should have to encounter would be some trouble with the Namaquas in passing through their country; but apprehensions of that sort failed of deterring me from the present enterprise, after having successfully surmounted the serious hindrances thrown in our way by the jealousy of the Boers and the cupidity of native chiefs in the former stages of the expedition.

The desert now before us forms the northern extremity of

the great Kalahari desert, the eastern side of which we had often skirted in our journeyings to and fro between Kuruman and the Lake Ngami. Even at this, its narrowest part, the desert presents to the traveller the formidable obstacle of some two degrees of longitude, during the transit of which he has mainly to depend upon certain vleys, or pools, and springs, the supplies of water at which are very uncertain, and, as we had the misfortune to find, frequently fail altogether. Koobye is the first regular station from the lake, which it took us from three to four days to reach, and then we had a journey of 60 miles before us, without any certainty of finding water on the road.

It is unnecessary to task the reader's patience by describing in detail the incidents belonging to our route from Lake Ngami to Walvisch Bay; more especially as, in a subsequent visit to the interior, the record of which is given in the succeeding chapters of my narrative, I went over the same ground, and with fuller means of observation at my disposal. I omit, therefore, all but the more important circumstances which marked this line of travel on the present occasion.

On the 23rd of July, 1855, Edwards having arrived from Chapo's, we commenced seriously to make preparations for our long journey to the west coast. By an arrangement with Mr. Green, we engaged three of his Damara servants, who were anxious to return home, and also a Griqua and a Hottentot in Mr. Wahlberg's service, who had accompanied him in the capacity of guides, though they proved to be spies from Lamert, chief of the Namaqua tribe, residing at Twass, about half way to the coast. These men were to act as our guides over a great extent of country only once traversed by wagons, the spoors of which were probably either quite obliterated, or likely to prove very difficult to follow. By four of our attendants, who preferred returning to their own country to accompanying us farther, we sent letters to

Mr. Moffat, acquainting him with our plans. On taking leave of Mr. Wahlberg, I gave him a rough map of the previously unknown country to the eastward through which I had travelled and hunted; but I could not help expressing to him my fears for his safety—fears which events too sadly justified.

Taking a friendly farewell of the chief and his people, we left Mr. Green's encampment at the lake on the afternoon of the 1st of August, our two wagons heavily laden with the hard-earned spoils of a long and tedious expedition. Mr. Green accompanied us a short distance, and we passed a final evening of convivial enjoyment, alloyed by the shade of melancholy which always attaches to leave-takings, in the present case rendered deeper than usual by surrounding circumstances—talking over past incidents and discussing probabilities of the future. Proceeding, finally, by way of Koobye, Thounce, and Ghanze, sometimes obtaining only, with much difficulty, an insufficient supply of water,* and on the look-out, not always successfully, for game, we struck, on the 3rd of September, the bed of an old river, Otjembendi, where we obtained a supply of water for our cattle. A fine salt-lick, which we found close by, was visited nightly by troops of giraffes and elands. At Gunegga, the stage immediately preceding, we had fallen in with several Bushmen.

The inhabitants of those parts of the country through which we had now been travelling are a much finer race of Bushmen than we generally met with. Freedom, and the enjoyment of their own game for food and skins for clothing, are, of course, the main cause. They acknowledge no chief, and being in the habit of defending themselves against oppressors and intruders, either from the Lake or the Namaqua regions,

---

* In all the larger pools of stagnant water we find the *Dytiscus*, or water-beetle. They do not look much unlike a small water-tortoise; they are more than an inch long, of a dark olive colour, margined with yellow. They fly about at night, are very active and strong, and their larvæ are of the size and general appearance of a shrimp.

their minds are free from the constant apprehension of human plunderers, and the life they lead is comparatively a fearless one. The population is numerous, and they are more attached to each other than in other parts, and in former times have often combined to resist the attacks of marauding parties sent out by the Batawana and other tribes.

On the 4th of September, eight hours travelling in the bed of the winding and often very sandy river brought us to some *sucking-waters* of the Bushmen. There we quenched our thirst, and dug out several lerushes, which, growing abundantly against the sides of the valley, is the Bushmen's principal article of food, serving for both meat and drink.* We travelled three hours farther, the wagon-wheels sinking twelve inches into the sand; we also passed two small wells. Next day we travelled twelve hours in the same winding valley, passing another little well, and several large ones, dug in former days by the Damaras for the use of their cattle, before they were robbed of them by the Namaquas. At that time

* Besides the lerush, which I have just described as being an article of food as well as drink for man in these wastes, I found here another root, which, though not so pleasant to the taste as the last, is still worth knowing. It is rather of a flat form, with a concavity on the one side, and perfectly convex on the other. It tastes like a raw potato, but is very juicy, and is generally found on the sandy plains, entwined in the middle of a moretla bush, the root scarcely two inches under the sand. The roots of some kinds of trees are also chewed by the natives, in case of extreme thirst. Everyone travelling in this country should make himself acquainted with these as soon as possible, as well as with all the edible roots. This knowledge will often prove of great value, and may be the means, perhaps, of saving life. Another plan, though not a very delectable or delicate one, I can also recommend in cases of extreme thirst. Everybody, of course, carries a gun in this country, and is almost sure to fall in with a steinbok or duiker, at least, on the most parched soil. Towards evening, or in the early morning, they are very tame, and may be stalked closely, and killed, even by an indifferent shot, if cautious. If you kill one of these animals, skin it, make a hole in the earth, lay the skin over it, press it into the hollow, forming a basin; then take the stomach of the animal, which is generally full of juicy roots, barely masticated, squeeze these out well, and you will at least be able to save life.

VOL. I.   Y

this country belonged to the former. The next morning, two hours and a half brought us to the Riet Fountain, or Thanobis, where we arrived nearly famished, and were compelled to kill one of our oxen.

Next day, the 14th of September, passing over several ridges thickly strewn with large fragments of white quartz, we reached Elephant's Kloof, or valley. This may be considered the eastern boundary of the chief Amraal's territory, but this eastern portion of it is at present governed by his son Lamert. Here we found a pretty fountain in a barren rocky valley covered with aloes, euphorbias, and several kinds of cacti. The variety of low thorn bushes, densely clad with yellow downy blossoms, looked quite refreshing, and emitted odours quite exhilarating. This is the only real fountain I had seen since leaving the Bamañwato country; and even this, though a running spring, seems to be fast failing. Although there is an abundance of water at present, it becomes absorbed in the sand; nor do I hear of its making its escape anywhere farther south. At one time of the year, however, I believe, it still has its full flow. Game of all kinds have been in the habit of visiting this fountain in great numbers, and offering, like Ghanze, every advantage to poachers. The destruction of game has been great, as the bleached bones of elephants, rhinoceroses, and other animals plainly indicate. Rhinoceroses and giraffes still frequent this fountain, and occasionally a few stray elephants. Lions are regular visitors.

At Elephant's Kloof we met with a party of Damaras—poor, lean, emaciated, shrivelled, and scabby creatures, equalling in poverty some of the most wretched Bushmen I had yet seen, although they had once been the possessors of immense flocks and herds, and the owners of the soil on which they now grubbed for roots in order to obtain a scanty subsistence. Fine, tall, and noble-looking, even in their poverty, they give one an idea that they must be a spiritless race to brook their thraldom under a tribe of the most besotted, dissipated, and debauched wretches on the face of the earth, and those, too, of such small numbers. The Damaras must either be a very cowardly or enervated people, or they must have lived, like the Makobas on the Botletlie, disunited or in small clans, ever to allow themselves to become subjected to so despicable a race as the Namaqua Hottentots, for whom every other black tribe in South Africa has a most consummate contempt. I could not help pitying these people in their abject condition, and wished that I could speak their language, that I might offer a word of sympathy and consolation to them. As there is no law amongst the Namaquas for punishing ill-usage to slaves, any man possessing a Damara may kill him at pleasure. One of my guides from the lake, who was a Namaqua, had a Damara slave with him, and I verily believe that nothing but the inconvenience of being without a servant prevented him, on many occasions, from shooting the poor fellow. And now, being so near home, he did attempt to take the man's life, by throwing large stones at his head; but, as I told him that for a murder committed in my camp he would dangle on the first tree we found, he thought it most prudent to smother his wrath.

At intervals, wherever the ground is suited for the purpose, there are large wells dug by the Damaras to water their cattle, and as the land is generally of a sandy nature, they have wooden troughs into which they diligently scoop up the water as fast as it percolates through the sand. Large herds of Namaqua cattle were grazing in every direction, herded by strapping Namaquas, seldom less than six feet in height, but often nearer seven. We were visited by some of these, who, being privileged to use a portion of the milk and butter, looked sleek and shiny, and showed to better advantage than their friends of Elephant's Kloof. We also observed that in their dress they very much resembled the Maslinnas and other tribes near the east coast; and, besides the apron of sheep-skin which they wear round their loins, they are always girded with coils of cords plaited of thin strips of leather round their waist. This appendage seems very cumbersome and obstructive to agility, but they turn it to good account by carrying half a dozen large-headed bludgeons, or kerries, inserted between it and their bodies, ready to be unsheathed in an instant. They wore sandals, resembling those used by the ancients; and their hair, which they allow to grow long, is twisted into cords, profusely greased, and often slightly parted on the forehead, which is, moreover, generally decorated with a round striped shell. They have no shields, but large broad-bladed spears, all iron, ornamented with the tail of an ox appended about the middle of the shaft; but this spear, which is constantly polished and glittering, seems to be more ornamental than useful, or they would long ere this have turned them upon their cruel and unsparing oppressors.

The Damara women, besides being decorated with several strings of heavy and cumbersome iron beads round their legs, wore also a few common Portuguese beads, which they informed us came from Nangoro's, a chief of a large nation considerably to the north-west, near a river, Cunené, flowing from the country of the Baveko. They say there are many rivers there running westward, and that the tsetse is unknown. One remarkable feature about the Damara women is, that we never saw their heads uncovered. Their head-dress consists of a curtained hood or bonnet of skin resembling a helmet, ornamented with a string of cowries, and having three leathern horns erect at the back, besides a heavy appendage of rude iron beads (most prized), probably 10 to 20 lbs. in weight, streaming down behind their backs.* A Damara

* These back chains are commonly appendages to the ornaments worn by the Kaffir women far eastward.—ED. [B.]

mother takes great pains with her offspring, greasing it daily, and stretching and straightening every limb morning and evening; and although the women are fat, and the men generally thin about the waist and long-backed, they are in stature the finest race of people I have ever seen in Africa. Heart and courage only are wanting. In cleanliness they are very deficient, constantly smearing themselves with grease and ochre, and they smell most disagreeably: from a superstitious idea, they never wash their milk vessels or other utensils, for fear of the milk cows drying up, or their cattle dying. The Namaquas seemed very distrustful of the Damaras, making them drink first of the milk they brought. The Damaras do not practise agriculture of any description. They are the only African people that I am acquainted with who do not take snuff, but they are fast learning to smoke from the Hottentots, and tobacco has produced the same violent effects upon them as upon the Bushmen. The Damaras live in round huts like Zulu kraals, plastered thickly with cow-dung. Their villages are generally enclosed with a thorn fence, as a protection against lions. Their language is, like the other languages of all the black tribes of South Africa, a dialect of the Kaffir, but more removed than that of the Bechuanas.*

Having had an opportunity of hearing a party of Damaras from the west coast converse in the presence of the Bayèyè, or Bakoba, on the Teougé river, I could not help noticing the great affinity between their respective languages. Indeed, most of the Damara words bear a great resemblance to, at least, if they are not quite the same as, the Bayèyè. All their domestic animals, as well as the wild ones, known to

* To persons taking an interest in the languages of South Africa, I would recommend the Damara Grammar and Vocabulary, published by the Rev. Hugo Hahn, a very intelligent member of the Rhine Missionary Society, labouring in Damara Land.

each other, as also implements and utensils, such as pots, ladles, &c., go by quite the same name. I also found that the Mochuèri tree worshipped by the Damaras, who call it Umborambunga, as the source from which they sprang, is also superstitiously reverenced or *danced to*, among many other things, by the Bayèyè.* I also ascertained from some of the oldest of the natives about the lake, that they have a recollection of having heard that the Damaras came originally from the east; from whence, having been driven by some more powerful tribe, they found their way to the lake. From the lake, some steered directly westward, while others, having gone northwards to Lebèbè's present territory, remained there for a number of years, and then migrated southwards and westwards. From subsequent information, I have traced their original country to the south of the Zambesi, that country being inhabited by tribes of Banabea and Batoka; and having from personal observation and comparison found the language of the Batoka to be as like the Damara as the Bayèyè is, and having, moreover, found a similarity in their religious rites and their customs, such as chipping out the two front teeth, &c.. I feel convinced that the Batoka country, on the south side of the Zambesi, is that from which they originally came. In some parts of their dress they also resemble that people. One thing, however, in which they greatly differ is, that the Damaras practise no agriculture, and for this difference I can only account on the supposition that their flight must have been so precipitate that they carried no seeds with them; or, perhaps, that famine had compelled them to devour every grain of corn; or that, having been harassed by their neighbours in their newly-found country, they preferred depending upon the supply of game, with the other resources of a pastoral or nomadic

* The Bayèyè say that the Umborambunga or Mochuèri was given them as a medicine at the creation of the world.

life, to the risk of tilling the soil and depending upon its uncertain fruits. In this manner very likely also they lost the seeds of tobacco; and this only can account for this isolated nation, being the only one amongst so many in South Africa, not having known the use of tobacco until they learnt it from their conquerors, the Hottentots. Damara tradition would also seem to corroborate the fact of their having come from the east about seventy or eighty years ago.

On the 19th of September we changed our course, and steered in a northerly direction through a pass in a quartz range, and, coming to a beautiful patch of green camel-thorns, we bivouacked for the night. Next day we again steered westward, making direct for Twass, the usual residence of the chief Lamert, but from which he had lately removed in a northerly direction. Our course varied for six hours from west to south-west, crossing several spruits. The next day we travelled with great toil, as on the preceding one, along a heavy, sandy road, and then sent the oxen to a fountain in search of water.

While waiting for the return of the oxen, about twenty men mounted and armed issued from the grey kroutze bushes north of our encampment, and, having tethered their horses, came to the wagons, and Lamert, the Namaqua chief, with a very ill-looking guard of honour, presented themselves to us: he approached us with a most profound and respectful bow, and we were well pleased to shake hands with him, a civility he evidently did not expect. Having seated him on a vatze or water-cask, we scrutinised the group before us, the majority of whom looked more like plunderers and assassins than anything else, and if the physiognomy of man is indeed the index to the soul, nearly every one of these might have been condemned to the gallows by that alone. One or two young men there were of a milder expression of countenance, appearing intelligent and well dressed; these were the chief's sons; but the remainder were men of short stature, pale copper-colour, withered frames, and dilapidated constitutions. They wore leather trousers. Their hats, of a coarse colonial manufacture, with broad brims and low crowns, they seemed to be very proud of, and had them decorated with broad scarlet ribbons and huge tassels. Some wore trousers and coat without a shirt, or exhibited other strange costumes, among which were old cast-off long-tailed coats and black beaver hats. The chief, however, was an exception. He wore a snuff-brown cloth coat, the usual hat, with a gaudy ribbon, a coarse white shirt, a tweed waistcoat, moleskin trousers, and veldt shoes. Lamert has a pleasing but idiotic expression of countenance. He was the only one of the party who pretended to speak Dutch, but this was policy. Having come to purchase gunpowder, I anticipated his design by making him a present of what we could spare, in order to avoid trading with a troublesome people, although I have no doubt that we could have bought many a fine large ox at the rate of 2 lbs. of gunpowder each, or fifty leaden bullets.

Lamert Lamert is the son of Amraal Lamert, the paramount chief of the Namaqua tribe, who, being now an old man, and more inclined for rest and devotion, being a good Christian, allows his son to rule this, the largest portion of the tribe, finding it difficult to prevent them from making frequent and destructive raids upon the Damara tribes possessing cattle. Although Lamert is allowed to be a very moderate man, he is accused of sanctioning these expeditions and taking part in them; and in order to carry on these operations without fear of paternal reproach, he keeps himself thus far apart from his father, who was living with a missionary at Wesleyvale, 50 miles to the south-west.

Amraal, who is upwards of seventy, but remarkably hale for his age, was formerly a slave at Cape Town, in the time of the Dutch East India Company, and fought for them against the British forces at the Cape and at Blueberg. After passing through a variety of adventures, he is now the chief of a numerous people, called *Gam Naka* (Lion's Hair), consisting of 1000 or 1200 souls, besides slaves.

The Namaquas practise no agriculture; they have not patience enough for that; they are men of impulse, prodigal in the extreme, and of idle and dissolute habits. They dwell in mud huts, like the Korannas, and live on milk, meat, and tobacco, which latter they may be said literally to imbibe with their mother's milk; so young do they learn to use the pipe. The Namaquas are exceedingly fond of intoxicating liquors, and have most ingeniously learnt to distil spirit from a wild berry (morètla: *grewia flava*), with the aid of a gun-barrel, an iron pot, and an inverted kettle. They are very excitable, and when in a state of intoxication are dangerous and desperate; but in their sober senses I have always considered them as arrant cowards. Indeed, they are well aware of this, for when they contemplate mischief they invariably resort to intoxicating liquors to get their courage up: it is almost a pity that nature has so bountifully placed within their reach ample materials for gratifying their propensity. One of the principal ingredients in beer-making is a kind of barm called *moer*, which, I believe, is made of the root of a gourd resembling the cucumber, which grows all over the desert. This barm, put into a pail of water with a bottle of honey, ferments it in four hours on a warm day, and produces a pleasant drink, like ginger beer, but intoxicating if taken in any quantity. Upon Europeans this drink has scarcely any effect, but the excitable Hottentots are easily prostrated by it. They also make a kind of cider by the same process of the morètla berry, as well as the mogoana.

The Hottentots practise polygamy, but the missionaries informed me that where this is the case very few children are born. One of the tribes tried the experiment with the view of increasing their numbers and power. The men took each as many wives as they could support, but in a few years the result proved no increase, but a decided decrease, whereas among those who lived at missionary stations and confined themselves to one wife there was a considerable increase of population.

As the Namaquas practise no agriculture, they have no employment for the servile Damaras but as cattle herds, and they evince fear of this tribe some day running off with their ill-gotten herds, or seeking revenge on themselves. Under this apprehension they often butcher whole villages of these wretched people, who are scarcely ever known to resist. One starving party of Damaras having lately fled, and taken some cattle with them, the neighbouring Namaquas resolved, when the winter came, to go round the country, and kill all the Damaras for whom they had no employment, especially those we met at Elephant's Kloof, who, being half-starved, they feared would run off with some of their cattle.

Having had a piece of iron welded by the chief Lamert, who is a good blacksmith, we started on our journey, travelling four hours; and the next morning reaching the deserted village of Twass, we found pallahs, springboks, and steinboks grazing amongst the deserted habitations of the Namaquas. Some of these, and a few guinea-fowls, we shot.

We were surprised at the number of bones of different animals that lay bleaching about the fountains all the way from the lake. Skulls of every kind of animal, from the elephant to the antelope, lay scattered about in such numbers as to show plainly that, not merely destruction, but great waste, must have been committed.

After leaving Twass, two short treks brought us to Elephant's Fountain, where we found the ruins of a missionary house and chapel, with a Damara village. The valley of Elephant's Fountain is exceedingly pretty, and was much enlivened towards evening by a beautiful sunset tinging the crowns of the mimosas with a bright golden hue. The place is famous for the growth of good tobacco, the culture of which is monopolized by one or two half-civilized Damaras. There are here several fountains, and one which bubbles out within a few feet of the missionary's house used to be visited nightly by troops of elephants when the missionaries lived there; but now lions, zebras, and antelopes only frequent it.*

On the 29th of September we bargained with a guide to take us a short cut to the Noosop river, a distance of 30 miles, which had not yet been travelled, for the paltry remuneration of ten bullets. Springboks, hartebeests, quaggas, and ostriches presented themselves constantly to our view, and there were many indications of the presence of lions and giraffes occurring here.

Two days later, a few hours brought us to a Namaqua village at Kalk Fontein, or Wittvley, inhabited by notoriously the most lawless set of vagabonds in this country. They gave us a great deal of trouble, importuning us for gunpowder and lead, of which we had none to spare. On making our preparations to depart, one of these fellows, in a very authoritative manner, and with violent gestures and threats, forbade our leaving until their chief had seen us. Being neither inclined to delay our journey for the chief's convenience, nor to acknowledge the authority, nor be intimidated by fear of these ruffians, we coolly went on with our preparations for starting, anxious enough to place miles between us, as the gang were fast getting up their courage by drinking honey-beer, to make matters still worse.

* In a subsequent journey I found here the Rev. Mr. Eckard, and received from him such kindness and warm hospitality that, although our acquaintance lasted only a day, years will never efface it from my memory.

We were glad to leave these detestable people behind us, and in the course of five hours struck the northern bank of the Noosop river, the banks of which were shaded with an unbroken line of mimosas, indicating for a great distance the windings of the stream. This periodical river, which is the largest I have seen in this part of the country, has its source in a mountain-range a little to the north of Jonker Africaner's Town in Damara Land. From this point, at which we struck it, the river flows south-east, till, joined by the waters of Elephant's Fountain, it makes a sweep to the south, when, having passed the missionary station at Wesleyvale, it flows, I believe, into the Great Fish river. The Noosop is crossed at intervals by ridges of quartz, hornblende, and schist, &c., succeeded occasionally by plains covered with calcareous rock, intersected by hills of what is commonly called puddingstone, and other conglomerates, which appear to have been at some period in a state of fusion. These hills were strongly impregnated with copper, and one stratum of schist crossing our path, and bearing nearly east and west, was strewn on the surface with lumps of copper ore, some samples of which we brought away with us.

Although the Noosop is so large a river, no water could be found at this season in its channel, except at intervals, and by digging shallow wells, into which the water quickly percolated, sparkling with innumerable particles of mica.

Two days afterwards, travelling over a wide plain, we struck the Noosop river again. Some few hartebeests, wildebeests, ostriches, and springboks, occasionally enlivened the monotonous scenery, and shooting a few of the latter, or catching a young one, caused some diversion to men and dogs.

In the morning of the 6th, after a trek of an hour and a half, we stopped at a Berg-Damara village. The people seem to kill abundance of game, and, like the Makobas of the lake, hang up on a tree the hoofs, horns, head, and gall, to propitiate their deity for future luck. At night the lions roared around us. On again crossing the river, and another stream joining it from the south, we saw for the first time the Wilde-paard (*Equus montanus*), which generally inhabits elevated or mountainous regions. The only game we bagged were springboks, steinboks, and pauws.

We now travelled, in a westerly direction, over grounds sparkling with mica, and also containing ironstone, quartz, granite, slate, and sandstone, with mountains of mica-schist and talc glowing in the sun. In the afternoon our course turned to the sun for about four hours, and the next morning, after trekking about three hours in the same direction, we struck into the main road on the height of Jonker's mountain. Here we were about 5000 feet above the level of the sea, but some of the mountain peaks must be fully a thousand feet higher.

In the afternoon of the 10th of October we descended the steep mountain-range skirting the abode of the chief, Jonker Africaner—at one moment ascending almost perpendicular mountain-tops, and then descending precipitous declivities with terrible velocity, our wagons well secured with drags, while the voices of the drivers and the loud crack of their whips re-echoed from the crags and caverns. After five hours we found ourselves in a deep valley, and in a warm, close, and apparently unhealthy region.

Next morning, an hour's trekking brought us opposite to Jonker's Town, having just before passed the ruins of a mission house at Eikhams, a lovely spot, lying on the slope of a hill clothed with mimosa trees, down which pours a clear stream from a hot mineral fountain. Here we met two singular individuals, slaves who had been recaptured by the English from the Portuguese, and set free at Cape Town. They informed me that they came originally from Lake Maravi, near the east coast. Having adopted the dress of the Damaras, and married into Damara families, they could not be distinguished from that tribe. They spoke their language fluently, and as they still remembered their own, I found that many words in their language were common to both.

A few hours' journey down the valley, on the bank of the river, brought us to an old town, which was enclosed with stone walls as a defence against the Rooi-volk, a tribe of Hottentots living southwards, who had been recently attacked by Jonker's people, but with small success.

We now passed through beautiful and varied scenery of hills and rocks, with mimosa groves, clear pools of water, and running streams—the latter finally losing themselves in the pebbly bed of the Schwagoup river. An accident to the pole of one of our wagons brought us to a stand in the bed of the Schwagoup, where I shot a beautiful salamander on a ledge of rocks. It measured ten inches. The head was a bright orange tinged with purple, the body glossy indigo, the tail coral with a transverse band of yellow, one inch broad at the base.

On the 17th of October we reached the mission station of Barmen, the residence of the Rev. Hugo Hahn, who was now absent, but whose acquaintance I had afterwards the pleasure of making. The country about here abounds in copper ore, indications of which were found on the surface near this station. I observed the ground near Barmen to be composed in parts of angular pebbles of iron pyrites. The hills are composed of schist and slaty sandstone, greatly impregnated with mica. The country abounds in hot mineral springs.

We saw at Barmen two females, who were real objects of pity and compassion, having had their feet cut off by the ruthless Namaquas for the sake of some brass and iron rings they wore. We were told of another woman who had her feet cut off for trying to run away from her thraldom, after being made a captive by the Hottentots.

Crossing the river several times in a south-west direction, we came in an hour or two to another hot spring called Otjikango Katiete, bubbling up between large granite boulders. Many hawks and other birds, besides dogs and other animals, had evidently scalded themselves to death some time before, their remains still lying scattered amongst the rushes. Quartz and limestone abound, and baboons perched everywhere on the distant heights.

Otjimbengue, which we reached in two days, is the residence of a missionary, the abode of many Damaras, and a depôt of a mining company. It has become a place of considerable importance, and is in fact considered the capital of Damara Land—now by conquest territory of Jonker Africaner. The Kuisip river, flowing into Walvisch Bay, forms the southern boundary of Damara Land, and the northern boundary of Great Namaqua Land. Many Damara villages lie interspersed among those of the Namaqua amid the hills and valleys of this country, but all acknowledge the supremacy of Jonker, the Namaqua chief. Indeed, the nationality of the Damaras is completely lost; their language is fast becoming corrupted by that of the Namaquas, and the same may be said of their manners. According to the account of the missionaries, they are a cowardly race, and never made the slightest resistance when attacked by the Namaquas. There are many hundreds, perhaps thousands, of Damaras throughout the land, living on roots and berries, and exceeding probably ten or twelve fold the number of Hottentots, and if they only knew their strength, and had the courage, they might sweep every Namaqua out of the country. But there seems to be no unity—no combination whatsoever among them. This may be accounted for by their having chiefly lived in small clans, each jealous of its neighbour, and plundering one another of their cattle. The missionaries describe them as being even now so faithless to each other that they would prefer joining the Hottentots in a raid against their own people to uniting with them against their common enemy.

Many Damaras are living at the missionary station, both for protection against the Hottentots and for instruction. Besides his religious teaching, Mr. Rath encourages them to make gardens and grow crops for their subsistence, instead of depending on the few cattle they can rear, and what they can pick up in the wilds. Many of them were now working for the mining companies, the only wages being their daily food, at which rate they were glad to find employment. They are an industrious people—wonderfully so, considering their past life of idleness. Those under Mr. Rath and Mr. Hörnemann's tuition were taking rapidly to garden making, and patiently watering for the long-coming crops. We had many servants to work for us for one meal of flesh per day, and many who wished to go with us to Cape Town, and become our slaves for life. All they asked was food. We might have carried off hundreds of such slaves without any opposition from the Hottentots, who, indeed, would only be too glad to see some of them depart.

Until very recently the Hottentots asserted no right over this country. They had subdued the Damaras for the sake of their cattle, and such lands as they required for their pasturage. This was all they required; and as they never grazed their cattle beyond certain bounds, the territory they acquired was very limited. From the coast, 170 miles inland, and to a considerable distance northwards, the country was inhabited by Berg-Damaras, whose habits, like the Bushmen of Natal, were not well calculated to encourage cattle grazing in their neighbourhood. Recently, however, the people belonging to the mining company here, probably without the knowledge or sanction of the shareholders and chief manager, who reside at the Cape of Good Hope, have led Jonker to believe that this country belongs to him, even as far as the coast, and that it was optional with him whether missionaries were to be tolerated at Otjimbengue. In fact, he was prevailed upon to write a letter to the missionary, in which that gentleman was given to understand that he was there merely on sufferance, and was subservient to him. At the same time the then manager of the Walvisch Bay Mining Company, in Damara Land, fearing that perhaps it was our intention to form a rival company on returning to the Cape, induced the chief to write us an insolent letter, ordering us out of his country at once, and forbidding us ever to come into it again. Such conduct from Jonker, without any cause whatever on our part, rather surprised us; but as it was impossible for us to leave the country till a vessel came to Walvisch Bay, from which we were still 120 miles distant, I paid no attention to the letter, and consulted my own convenience as to the time of our departure.

Shortly afterwards, having an interview with Jonker, he honestly confessed that the manager had made him drunk, and got him to sign a letter with the contents of which he was unacquainted. Indeed, he strongly pressed me to come back to his country, saying that he would throw the whole of it open to me, and that I might even establish a mining company in any part of his territory. On my again passing through that country, I received much kindness and civility from him; but the dislike generally entertained towards the missionaries by the whites in the colony showed itself here in even a stronger degree.

Now, as an impartial observer, and having seen a good deal of missionary life and labour, I may be allowed to take this opportunity of offering a few remarks on the origin of this dislike. It is often foolishly asserted that the missionaries in this country are all traders. This is incorrect. That they trade occasionally is true; but that they trade for produce which they can sell again or profit by, is not the case. The lot of a missionary in Africa is a hard one; his life is one of trial and self-denial. Deprived, often for months together, of the common necessaries of life, cut off from society, from friends and relations, with the prospect of never seeing them more—it is cruel that they should be looked upon with suspicion by those whom they have come to benefit, and be despised and slandered by their own countrymen. That a missionary trades in this country is only because he is compelled to do so to obtain the supplies necessary for the wants of his family. You never hear of missionaries exporting cattle, ivory, or any other commodity. They trade for cattle with merchandise, because money is neither known nor esteemed. Could the missionary send to the butcher and the baker every day, and buy his few pounds of meat or bread, he would not be obliged to purchase the cattle from the natives. He is compelled to keep a small number of cattle, and slaughter the increase; and if his wife wants a little milk for her young children, they must have several cows to furnish the supply, as a dozen Damara cows give scarcely as much milk as one European: if God blesses this little flock with a healthy increase, they are pointed out by the jealous and selfish white man as the profits of trade. I have seen a great deal of missionary life, and have every reason to sympathise with them. Their labours are difficult, their trials many, their earthly reward a bare subsistence. I believe that the real causes of dislike to the missionaries in South Africa are the avarice of trade, and jealousy of the influence they possess, and the check they are upon those who would like to exercise an arbitrary and unjust authority over the natives. I could say a great deal more on this subject; but the missionaries are a class of men, generally speaking, so irreproachable, that the scandals of the unprincipled cannot affect them with well-thinking men; nor do their characters require any further defence by me.

The country here is in all directions well adapted for cattle-farming; sheep and goats also thrive everywhere. There are certainly not many running streams, excepting the hot mineral springs; but the periodical rivers everywhere produce water, even by scratching with the hand a depth of six or twelve inches in the gravel, and at intervals the sandy beds of the periodical sand-rivers send out running streams, which leave large pools of water. In the summer season, after rains have fallen 200 miles east, overwhelming floods come roaring down so suddenly that they often carry off some unfortunate family while sleeping on the river-bank.

For fine-woolled sheep, the country seems particularly adapted; grain is cultivated in the gravelly beds of the rivers, in order to save the trouble of irrigation, and the Damaras have already reaped some fine crops. Vegetables of every kind grow most luxuriantly on the low, moist banks of alluvial soil, which, extending as far as the sea-coast, might be converted into gardens, and cultivated with but little or no water. Gourds, such as calabashes, cucumbers, &c., are the only vegetables I noticed that required a little water; everything else was left to chance.

Besides copper ore, ivory will shortly be an article of export from Walvisch Bay, as the natives are becoming alive to the value of that article; and there are other valuable products of the country for which trade might be established. Sweet gum and gum arabic are abundant, and could be collected in great quantities if they had a market for it; hides and skins also will always form a principal export.

The country to the north is inhabited by some independent clans of Damaras still possessing cattle, and living in mountain fastnesses, where the Namaquas have either met with repulses or are afraid to venture an attack. Still farther north we met the Ovambo, a powerful people of agricultural habits. This tribe was first visited by Messrs. Galton and

Andersson in 1857, and the last-named gentleman has since published a well-known and intensely interesting volume, under the title of "The Okovango River," in which the most reliable information respecting them, and the country they occupy, is to be found.

It is astonishing what a similarity there is in the manners and practices of the human family throughout the world. Even here, the two different classes of Damaras practise rites in common with the New Zealanders, such as that of chipping out the front teeth and cutting off the smaller finger. In their mode of expressing themselves, they are very like the Bechuanas. Thus, when our man Jack was called first to one spot and then to another, he exclaimed, "Is Jack a river?" meaning, can I run here and there at the same time.

The different clans (or "aendas") of the Damaras worship each a different deity, under the common appellation "Omukuru" (Old or Great one). These superstitious rites descend from one generation to another, not through the male but through the female line. The Bechuana clans, in like manner, worship the crocodile, various fishes, the lion, or other objects. If they do not directly acknowledge in words the belief in a future state, many of their customs, such as their propitiatory sacrifices to the dead, show plainly the belief in the continued existence of a spiritual part of man's nature.

The Damaras are said to wail over their dead years after, returning to their graves and performing certain ceremonies, and making sacrifices to propitiate them. They sing a doleful dirge when they see the cattle that once was theirs, and weep over their losses. Their notions of death and a future state must be peculiar, for it is said of them that they die happy if they know there is any chance of a wolf or jackal benefiting by their body after death.

They say, also, that when men and animals first came out

of the tree (Omborombonga) all was darkness, and that a fire having been made by a Damara, it caused them all to flee, saving dogs and other domesticated animals; and thus they account for the wild animals being still unsubdued, and terrified at the sight of men and of fire.

The Damaras believe in witchcraft. Their doctors, or wizards, are called "Omundu-onganga." These people impose upon their credulous neighbours in the same manner, and with the same facilities, as the *ngaka* or *balooi* of the Bechuanas. The Damaras hold many superstitious ideas common to the Bechuanas, Bushmen, and other tribes of South Africa, such as forbidden meats, which they dare not eat for fear of misfortunes or bad luck attending them.

Polygamy is everywhere common amongst the Damaras, some having as many as ten or twelve, and even twenty wives—as many as they can afford to purchase and maintain. Like the Bechuanas, too, the girls are disposed of to husbands while still babies. The price of a wife varied formerly from two to ten head of cattle, but now, in their poverty, the parents are often glad to take one cow for a daughter. As with the other tribes of South Africa, the eldest son of the principal or first wife inherits his father's property.

Besides chipping the teeth, the Damaras practise circumcision, but they do not seem to attach the same importance to it as the Bechuanas; at least, the ceremony is not attended with the same pomp and circumstance. The first fruits of the fields are also, as with the Bechuanas, first tasted by the chiefs, or the heads of clans. They bury their dead in the same sitting posture as the Bechuanas. Though the Damaras are, generally speaking, great gluttons, they would not think of eating in the presence of any of their tribe without sharing their meal with all comers, for fear of being visited by a curse from their "Omu-kuru," and becoming impoverished.

The great majority of Damaras live entirely on roots, their

cattle, of which they formerly possessed great numbers, having all passed into the hands of the Namaquas, who are so prodigal that one may expect to see them, after exhausting their store, reduced to the necessities of their former habits of existence.

The Damaras, with a fine figure and well-chiseled features, have, on the whole, a care-worn cast of countenance, depicting grief or melancholy. The females in youth are very good-looking, but soon after maturity they begin to wither. Whether from intercourse with the Hottentots, or from whatever other cause, they are certainly not a very chaste or moral people, and are the greatest gluttons, without exception, that I have ever met with. Their weapons are a large-bladed spear, beautifully polished and bright. This scarcely admitting of a firm grasp, they fix the skin of an ox-tail, with the brush at the handle, both for convenience and ornament. Besides their spear, they carry a number of clubs, or kerrie-sticks, strung in their girdle, and a bow five and a half feet in length, with poisoned arrows three feet long.

The Berg-Damaras (Obalorotwa, or Ghou-Damaup, signifying Locust-eaters) are a fine, well-built race, of shorter stature, but more robust, than the Obahèrrèru, or Cattle-Damaras. Though differing vastly in their physical character from the Namaquas, they speak the same language. They practise the rite of circumcision by cutting off the little finger of each hand. Like the Bayèyè, they make their offering to the Deity on a tree in the middle of their village, by hanging the horns and hoofs of all the game they kill on it, as well as all the reptiles, such as lizards and chameleons, which they do not eat. They kill numbers of animals by infecting pools of water with the juice of a poisonous euphorbia, and another milky bush. They also kill the white rhinoceros with the same drug, although the black one eats greedily of the same bush with perfect impunity.

Before we left Otjimbengue, Mr. Rath, who took a great interest in whatever tended to improve the condition of the Damaras, called my attention to a silky substance contained in certain bulbs, and begged me to take some of them to the Cape, and see whether anything could be done with them. I also brought some of the sweet gum, of which large quantities could be gathered, but I could find no one in Cape Town disposed to interest himself in the matter, and the thing fell to the ground. If these bulbs could be turned to account by manufacturers, many tons of them can be gathered by the Damaras.

Having settled our business, and sold our wagons and oxen to the mining company, under the condition of our usage of them as far as Walvisch Bay. we left Otjimbengue in company with a train of seven or eight wagons, laden each with about 1200 lbs. of copper ore. The company's oxen galloped off, and soon left us far behind; but our drivers, shrewdly observing, "Wait a bit till we get to the heavy road," walked steadily alongside of our teams, each dragging more than 4500 lbs. weight. Next morning we crossed the bed of the Tsoubis river, near its junction with the Schwagoup, passing some granitic peaks of very fantastic shapes. Here we found only a small muddy pool and a deep well, at which it was very difficult to get the water; but anyone who has travelled much in this country is not easily daunted by difficulties of this kind. Being at a loss for a cup or vessel to dip up a supply, I tied a small tapering bunch of grass together, and, having soaked this well, held it over my mouth, and the water streamed from it faster than I could swallow it. A small bunch, well constructed, will absorb, like a sponge, about a tumbler-full of water at every dip.

Leaving Tsoubis, and travelling all night by a very circuitous and stony road, through ruts worn in the granite rock, we reached Onanies next morning. Thence, after resting the cattle a day, we crossed a wide plain, covered with vast herds of springboks, and passing between some conical hills reached next day a dry nullah called Tingas, in the bed of which were some pools of brackish water. From this last place we had about thirteen hours to travel without either water or grass. The neighbourhood is a notorious haunt of lions, the numerous recesses in the adjacent rocks favouring their concealment.

Starting in the afternoon of the following day, and travelling all night, we next morning reached Oesip gorge, at which place we rested a day, sending the oxen towater at a spot seven miles distant. The barren plain in which we now were exhibited scarcely any trace of vegetation, owing probably to the almost total absence of rain, which scarcely ever falls in this part of the country—perhaps only once in ten years.

We had still to trek across the plain of the Narriep desert, than which nothing can be more dreary. The oxen dragged slowly along through the sand, halting only for half an hour while we made a kettle of coffee, under shelter of a hill composed of a huge mass of granite. Solid masses of this kind are numerously met with in this region. After travelling fifteen hours without unyoking the oxen, we reached an outspan, appropriately termed Sand Fontein. The only water we obtained was through percolation into the hollows of the sand-hills, which was brackish in quality. We had to send the oxen to a somewhat better watering-place, of like description, three miles off, where they obtained also a few reeds, or hard kind of grass—poor provender for hard-worked cattle.

Amongst the white sand-hills here are found many narras, a kind of gourd of a pale green colour, and pleasant, though mawkish, to the taste. It grows on a prickly creeping shrub which forms the base of these shifting sand-hills, as against these bushes the sand accumulates as it is deposited by the wind, until it forms, with the growth of the bush, a large sand-hill. The natives collect the seed of these gourds, and eat them. They seem very nutritious, tasting like almonds, but are rather rich and oily. The pulp is converted into a jam, and, being spread in its liquid state on the level sand to dry, it is packed in sheets, resembling card-board, for the winter.

This part of the country belongs to the Beach Hottentots, a small tribe living at Sheppmansdorp, a village 18 miles south of Walvisch Bay. These people subsist chiefly on the narra, with fish, which they impale on the horn of a gemsbok, or with an assagai; they possess a few cattle and guns. The ground is equally divided amongst this tribe, each family having its portion of narra-bush, which grows only in this locality. The narra is eaten by every animal and almost every kind of bird, particularly the ostrich, and were it not for this gift of Providence no human creatures could exist here.

Only jackals, hyenas, and reptiles, with a few ostriches,* inhabit the dense dabby-bushes in the bed of the Kuisip river, which falls into Walvisch Bay, south of the Schwagoup.

An hour of toilsome trekking in the bed of the Kuisip river now brought us, through heavy sand-hills and dense dabby-bushes, to the verge of an open plain, two miles in extent, where we came in sight of the only habitation at Walvisch Bay—a small wooden house near the beach, surmounted by a flagstaff, and planted on a mound artificially thrown up to raise the site of the house, as the plain is inun-

* His Excellency Sir George Grey recently communicated to me the result of an experiment which was attended with great success. The white feathers, about forty in number, of the ostriches kept in the gardens at Government House, having been plucked while the birds were under the influence of chloroform, were replaced in a few months by others of equal beauty. The success of the experiment so humanely conducted is not without its commercial importance, and may perhaps lead to the domestication of these valuable and magnificent birds, whose feathers are worth in the Cape Town market £10 the ton weight.

dated by the sea at every change of the moon, as well as, at much rarer intervals, by the river. The first thing that strikes the observer here is an extraordinary mirage, which contorts everything visible into the most fantastic shapes. Birds skimming the surface of a distant inlet or lagoon assume the appearance of large ships in full sail, and any little black spot in the distance seems as large as a man. Observing a flock of flamingoes near an old boat, I regretted that the intervening water prevented my going after them:

DISTANT VIEW OF WALVISCH BAY.

but it being pointed out to me that this was merely the effect of the mirage, I started after the birds and soon killed a couple of them.

Besides large flocks of flamingoes, pelicans and cormorants of various descriptions abounded. The former pursue their prey nightly in the shallow lagoons, driving the fish, by flapping their wings on the water, near the shore, and catching them while floundering.

During our stay at Walvisch Bay we saw, nearly every day, two or three whales playing about in its waters. They abound on this coast. A captain of my acquaintance has taken, in one month, twenty-two fish, yielding 200 barrels of oil. A general fishing establishment might perhaps be successfully formed here, the only drawback being the distance from the sea, which is three miles. Seals are occasionally killed, and several kinds of sharks, yielding large quantities of oil; and the oil of one kind, the black shark, which I tested, proved to be as good, if not better, than cod-liver oil. It would be a great boon and benefit to the poor, as I was told by a Cape Town medical man, if brought into general use. Mr. Hahn cured a Damara in the last stage of consumption with the oil of the liver of this fish. A shark peculiar to this coast is the bone-shark, or man-eating, which attains the length of thirty feet, and yields a large quantity of oil. Captain Bruce killed a black shark measuring 22 ft. 6 in., which yielded eighty gallons of oil from the liver.

Porpoises are also very plentiful here, coming in large shoals to fish so near to the shore that they can be harpooned from it. Indeed, while in chase of small fish—on which they prey—they often get out of their element altogether, and are seen floundering on the beach. We frequently attempted to harpoon them, but they always ran off with the harpoon and line. The smaller kind of sharks we speared all day only a few feet from the edge of the water; catching also, in the same way, a number of rays of various descriptions, amongst which the most common was one having a venomous sting on its tail, with inverted barbs. Soles also are very common here.

While here we had frequent opportunities of seeing the Beach Hottentots, a filthy race, who are generally employed by the white men to run on errands or carry letters. Here we made the acquaintance of Mr. Bam, the missionary stationed at Sheppmansdorp, and his family—very kind and hospitable people, living a lonely life amongst these utter savages. A month or two later, during a fever which carried off a number of his small flock, this worthy man died in my arms. The situation of Mrs. Bam was most melancholy, for most of the natives were prostrated by the fever, and the few who were not so exhibited a total want of commiseration for the poor widow, not only refusing to assist her in any way, but, taking advantage of her bereavement and want of protection, assailed her with continual demands for provisions and other things. Under these distressing circumstances, the few white people at the bay naturally did all in their power to alleviate the calamity. Having hunted out all the planks we could find, and broken up a few benches in the chapel, a kind stranger whom we casually fell in with at the station (and whose name I have forgotten) made a coffin, in which we laid the good missionary. Assisted by a number of Hottentots belonging to his congregation, we carried him to the grave, round which assembled the widow and her weeping children, Mr. Latham and his kind-hearted wife, another white man and myself, and, at the request of Mrs. Bam, Latham and I read the burial-service of the Church of England. It was a truly melancholy scene, and I could not help thinking as I stood over the grave: Such is the end then of those devoted men who, having sacrificed all for the cause of religion and humanity, are treated with such neglect, and even abused throughout the colony. As we could not leave the poor widow and children to the mercy of the savages, we took them to the bay, where they met with every kindness from Mrs. Latham, until they found an opportunity of returning to the Cape: and thus the mission establishment at Sheppmansdorp was broken up.

For a time we enjoyed the repose of a return to civilized habits after our long wanderings in savage lands; but the protracted delay became irksome, although the monotony of existence on this dreary coast was relieved by the extreme kindness of Mr. and Mrs. Latham, the only English residents at the bay—a hospitality extended, more or less, to all who land there. Mr. Latham is a man of considerable intelligence and attainments, sadly wasted on this desert soil; and his kind-hearted wife, having spent much time in Damara Land, had romantic adventures to recount, some of which might have enriched my narrative if I had her permission, and time and space allowed their insertion in these pages. In the society of these worthy people we enjoyed many happy hours; and in taking leave of Walvisch Bay I perform a pleasing duty in offering this public acknowledgment of their kind attentions, the recollection of which neither time nor distance will ever efface.

Walvisch Bay forms a very safe harbour, vessels of almost any tonnage anchoring within hailing distance of the beach. Although for the most part land-locked, the harbour is exposed to the north-west winds, but that point being protected by the coast, the anchorage is perfectly safe. At this part of the bay I have never seen a swell; there is no surf, the waves breaking on the beach being mere ripples. From the southward and westward the harbour is protected by a point of land jutting out northwards, called Pelican Point.

On our arrival at Walvisch Bay I was delighted to hear that a vessel was expected to touch there on her voyage to England, and hoped to embrace that opportunity of visiting my fatherland, and spending a few months in the old country. Weeks and months rolled on without any prospect of this project being at present realised, and both Edwards and myself were glad to take advantage of the arrival of a vessel, the Eblana, which, after discharging her cargo at the bay, would return to Cape Town. Embarking in this vessel on the 21st of January, 1856, the course held carried us considerably to the westward. The prevailing winds at this season being from the south, we could not make a direct course along the coast. We therefore turned west until, finding by our latitude that we should be able to reach Sandwich Harbour, where the captain had to take in some fish, we tacked about and reached that port on the evening of the 23rd. This also is a very safe harbour, with an outer and an inner anchorage, the latter, however, only open to vessels not exceeding fifty tons burthen.

Pursuing our voyage, we touched at Hottentot Bay and the island of Ichaboe, where we took in a quantity of guano, and, calling in succession at Angra Pequena and other small harbours on that line of coast, we cast anchor in Table Bay on the 1st of March, 1856.

Mr. Green's Account of Professor Wahlberg's Death.

"Upon my return from Lebèbè's, my own journey having been accomplished chiefly by water, and during which I had met Mr. Wahlberg on his way back to the lake, it was my intention to have retraced my steps to Walvisch Bay; but circumstances prevented the intended movement, and my friend, moreover, deciding upon spending the summer months in the deserts to the north-east of the Ngami, for the purpose of hunting elephants, and completing his collection of curiosities, and having expressed an anxious wish that I should accompany him, I was prevailed upon to do so. Accordingly we took our departure from the lake on the 22nd of November, but did not reach the land of elephants until the beginning of February. After having hunted these animals for some time, and with various success, Mr. Wahlberg one day (the 28th of February) left the wagons accompanied by two Damaras, one of whom, Kooleman, was his constant attendant on his shooting excursions, and was much attached to his master, who, on his part, I know, looked upon him as a good and faithful servant, as well as a brave hunter. He also took with him a Makalaka and a Bushman.

"The same morning I also set off, for the purpose of seeking elephants, though in a direction different to that which my friend had selected; for owing to my being subject to frequent attacks of fever, with which I was seized previous to our departure from the lake, and the long duration of Mr. Wahlberg's expeditions generally, I could not accompany him, but was obliged to limit my distance so as to enable me to return to my wagon the second or third day.

"On the present occasion Mr. Wahlberg had intimated his intention of not being absent long; and although this intimation proved incorrect, owing to his hunting excursions being very uncertain as to time, we did not entertain that anxiety for his return which we should have felt had his habits been more punctual. We knew that so long as he could find the game he was in search of, a month might elapse before he made his appearance. However, after waiting for him ten days in vain, I became uneasy, more especially as I could gain no information respecting him from the natives. In consequence, and my people as well as myself still suffering severely from the fever, I had determined on removing the wagons to a more healthy spot, when upon the very eve of my intended movement the startling intelligence of my unfortunate companion having been killed by an elephant was conveyed to me. The feelings of pain and sorrow which so shocking an event occasioned to all who were acquainted with him cannot be easily imagined, and will, I am sure, never be effaced from my memory. The following are the particulars of Mr. Wahlberg's death, as reported to me by the Damaras who accompanied him:—

"'We proceeded,' said the men, 'from the wagons in a

westerly direction, and on the day of our departure we struck upon the spoor of a young bull-elephant, which we followed until the third day, when we came up to him in company with three others, one of which master shot; another was killed by Kooleman. Thence we continued on the spoor of the two remaining beasts, one of which we fell in with and shot on the following day. The fourth morning we recovered the track of the young bull which we had taken up on the day of leaving the wagons. Not being able to come up with him before nightfall, we slept, as we had done on previous occasions, on the spoor. The next day, feeling hungry, and having managed to shoot a zebra, we camped for the night. The ensuing day, still continuing on the track, we reached a vley, where we bivouacked. Next morning we passed through a village situated on the banks of a large river called Tamalukan. The inhabitants were Bakoba, from whom we obtained some pumpkins, our master's provisions being exhausted. In the evening of this day we at last overtook the young elephant, which we found standing, together with another elephant, an old bull, in an open flat near a small vley. We approached them with difficulty. Our master and Kooleman fired three shots at the larger elephant, which then fled towards the river, where we soon found and overtook him. Mr. Wahlberg now sent us forward to turn the elephant towards a point where he took up a position in order to intercept him. We succeeded, and having fired a shot at him, he ran furiously in the direction of our master, but out of range. Mr. Wahlberg, accompanied by a Bushman from the werft we had passed through, then followed his spoor. Shortly afterwards, hearing the elephant trumpeting, we hastened to join our master, but had not proceeded far when we met the Bushman running in breathless haste towards us. We inquired for Mr. Wahlberg, and were told that the elephant had caught him! Hurrying to the spot indicated, we found only the mangled remains of our poor master, which the enraged beast had just quitted. There was no sign of life; indeed the body was so fearfully mutilated as to be scarcely recognisable. We carefully collected and buried the remains.'

"I deeply regret," Mr. Green goes on to say, "that, owing to the weak state of my health, I was unable to proceed to the fatal spot, but even could I have reached the place, at least twelve days must have elapsed from the time of the catastrophe, the distance from our wagons being very considerable.

"Mr. Wahlberg was a most determined and a most successful elephant-hunter, but he was far too adventurous, and his bravery throughout this dangerous hunting placed him in extreme peril with elephants on numerous occasions, and, alas! terminated at last so fatally.

"I had frequently endeavoured to impress upon Mr. Wahlberg the danger of 'foot-hunting,' but he always insisted upon its being the most safe; and, though I myself had never thought of pursuing such a course upon any former trip, nor had I ever met a European during my journeys in this portion of Africa who hunted elephants in any other way than by means of thorough good horses, from the success that attended my friend I latterly also became a foot-hunter. It was Mr. Wahlberg's opinion that he could always turn an elephant in his charge by giving him a shot in the head; but, alas! it seems that on this occasion my unfortunate companion had not even time to raise his gun to his shoulder ere he was hurled to the ground and pinioned between the tusks of the enraged brute. The rifle was discovered, broken short off at the stock by the elephant, as if the animal was possessed of the intuitive knowledge that it was the weapon employed for its destruction.

"Poor Mr. Wahlberg seems to have had a presentiment of his approaching fate. A letter from his head man, Mr.

Charles Cathcart Castry, addressed to Mr. Letterstedt, says: 'Some time previously to this awful event, Mr. Wahlberg came one day to me (Castry) and said, "If anything serious should befall me, I wish you to take my effects, collections, &c., to Mr. Letterstedt, my agent at the Cape." On asking him why he thought he should not return home alive, he referred to his having had several narrow escapes from elephants. Upon this, I remonstrated with him on his apparent recklessness, and begged of him to be more cautious and careful in his future dealings with elephants. To this he seemed to turn a deaf ear, merely remarking: "I cannot help myself; when I get sight of the brutes, I seem to lose all apprehension." ' "

Mr. Green's Narrative of the Defeat of a Treacherous Attack by the Ovambo.*

"Nangoro (king of the Ovambo) having refused to furnish my friends, the missionaries, with guides to conduct them northwards, they determined upon retracing their steps. Having made up our minds as to our future proceedings, it was resolved that we should take our departure for home on the 30th of July. On the previous evening we sent to Sjopopa, the Damara chief, who resides now at Ondonga, requesting him to inform the king that, as we had nothing more to do here, it was our intention to leave forthwith. But instead of conveying our message that same night, the chief postponed it until the following morning, just as we were about to start; and it is not impossible that this delay saved our lives. Before proceeding further, it will be necessary to state that, from the day of our entrance into the

---

* Mr. Green departed from Walvisch Bay, on the expedition here referred to, in April, 1857.—ED.

Ondonga to that of our departure thence, our people were seized with a fear of the Ovambo, amounting almost to a panic, which we could not understand. They would scarcely move from the spot without their arms, whether javelin or matchlock. We expostulated with them upon the absurdity of such measures; but they insisted on treating the Ovambo as a treacherous people, and even took upon themselves to caution both Mr. Hahn and myself against them, seeing that we mixed freely and unhesitatingly with the crowds. Our wagons were continually surrounded by hundreds of inquisitive spectators, all of whom, at least the men, never failed to carry with them their instruments of war. To approach strangers armed is a custom that I have never seen practised by any of the tribes in Southern Africa with whom I have come in contact; indeed, it is considered by them as quite contrary to all rule and etiquette. But not liking to construe it into any design of treachery, I attributed the practice simply to old and well-established habits. But not so our Damaras: they appeared at all times in momentary expectation of being fallen upon and massacred. Indeed, so conspicuous was their fear, that I became very angry, and threatened to deprive all my people of their arms if they did not desist; but it was all to no purpose. I may also mention that, two days prior to having formed our resolution of leaving the country, a large party of Ovambo, men and women, were engaged in dancing near the wagons, and as they were keeping up the conviviality of the night's amusements to a rather late hour, and, moreover, in a very noisy manner, Mr. Hahn sent one of his Damaras to beg of them to desist, as we were fatigued, and wished to retire to rest. This brought forth the following remark: 'These slaves, these dogs, that have come here to be killed, how dare they interfere with us?' This spoke volumes; and from that moment I myself, previously so sceptical, became apprehensive that their intentions were anything but friendly, and that our Damaras had more cause to be uneasy than we were aware of. However, I did not communicate my suspicions to anyone, but watched the Ovambo more narrowly.

"As already said, we had fixed upon the 30th as the time for our departure. Accordingly, on the morning of that day, when the first faint streaks of light appeared, we began yoking the oxen, and effected a start just as the sun peeped over the horizon. To guard against our exit from Ondonga being cut off, our caravan was thus arranged: The loose cattle and the donkeys were kept in advance, under an escort of all the available Damaras. Next came the four wagons, that of Mr. Hahn bringing up the rear. Behind were a number of the Ovambo, with Mr. Hahn himself walking in their company, engaged in conversation. Thinking it possible that treachery was meditated, I instantly hastened to the spot, begging the reverend gentleman not to trust himself amongst them, but to be on his guard. He took my advice, and, having seen him safely seated on his wagon, my mind was greatly relieved.

"I had ridden forward where the Ovambo were becoming very numerous, rending the air with their peculiar and wild war-cry, when I heard the report of fire-arms, and, on turning round, perceived the shots to proceed from some of the wagons. Retracing my steps without loss of time, I found, to my horror, a Damara servant of Mr. Rath's weltering on the ground in his life-blood. The cause was soon explained to me. The very party against whom I had warned Mr. Hahn, and which he had barely left, had committed this treacherous deed; and there can be little doubt that the villains would have served him in a similar manner had he not providentially parted from them when he did. The blow, which was probably intended as the signal for a general massacre, was struck by the chief's son, above alluded to, who stabbed his victim from behind, and then immediately retreated. The unfortunate man had just strength enough left to turn round and fire upon his treacherous foe, when he fell, expiring within half an hour after having received the fatal wound.

"This barbarous act was in itself enough to call for vengeance. Instantly turning my horse's head from the bloody scene, I rode forward to meet one man more daring than the rest, who was boldly advancing, poising his quivering spear above his head, ready to launch it at my breast. When within a short distance, I suddenly dismounted, levelled the rifle, and, whilst the man was hesitating—probably pondering on the best means of escaping my bullet—I slowly pulled the trigger with a steady hand, and the savage fell. This was the first fellow-creature that I had been compelled to kill; but though the deed, even when viewed in its most favourable light, must be considered a very responsible act, I felt, in this instance, as little compunction as I should have done in knocking over a savage beast. Remounting quickly, I took one glance at the fearful scene around; but rapid as it was, it would have been sufficient to make the stoutest heart tremble. Slaughter—indiscriminate slaughter—seemed inevitable; but I was determined that our destruction, if such was to be our fate, should be dearly accomplished. Fortunately, I was well armed: I carried my revolving rifle in my hand; a six-barreled pistol and your stout hunting-knife were stuck in my girdle; whilst close at hand I could command a sword —a weapon I could use not indifferently at close quarters.

"Seeing the enemy gathering in great masses on every point, I gave orders to form the wagons into a square; but Mr. Hahn, more wisely, suggested that we should continue to journey on, but leave our old route, and strike away to the open plains. This manœuvre was immediately executed, and the caravan moved forward.

"The battle, or, as I may not inaptly call it, the struggle for life, now became general. Arrows were flying thick and fast around us, and projected with considerable skill and precision. The missiles appeared to be especially directed towards Mr. Hahn and myself, who were regarded as the chiefs of our party; and it was probably thought that if they succeeded in despatching both or either of us, they would make short work of the remainder.

"Our fire was not a brisk one, as, of the number among our party who carried guns, only ten individuals, including ourselves, could use them with any degree of accuracy. Besides, I saw at the first glance that a slow and steady fire would prove more effectual, and tend to frighten the enemy far more than a volley of shots fired at random. The Ovambo never understood till this memorable day the deadly effects of fire-arms—nay, indeed, they had often ridiculed the guns whilst we were at Ondonga. I was much pleased with the cool and steady behaviour of our party, excepting two or three Damaras. I am sure not one of us thought this day that he should see the light of the morrow.

"During our retreat we repeatedly halted, more especially when our adversaries approached too close, in order to convince them that we were not running away; and, even in the face of several hundreds of the savages, we drew the wagons up near a vley to fill our empty water-vessels. While some of our people were thus employed, I uncased my elephant-rifle, and threw some shots over the heads of the nearer parties among some distant groups, the range being about a thousand yards and upwards. Whether any of the Ovambo were killed by these discharges we were unable to ascertain, but, at all events, it had the effect of putting many of them, both far and near, to a precipitate flight; and somehow I fancy that after these shots they began to have a notion that they could never get far enough away. The last man killed

showed himself too daring, and was shot through the forehead
by young Bonfield, this being the eleventh we saw actually
fall; but there is no doubt of many more having been killed.
Indeed, it was all but impossible to ascertain the number of
their losses, since, at every discharge, they all dropped simultaneously flat on the ground, so that, had some been killed,
we might not have perceived them fall from the shot. On
our side, I am happy to say, not one fell, nor was even
wounded, although more than one of us had narrow escapes
of being struck by the flights of arrows aimed at us.

"The fight lasted fully two hours, during which time the
enemy had constantly received strong reinforcements. Upon
several occasions they appeared ready for a general rush, but
at such times I directed my fire at the leading and most
daring men, who invariably paid for their temerity with their
lives. Could the Ovambo have succeeded in closing with us,
we must inevitably have been destroyed to a man, for the
odds against us were fearful—certainly not less than five to
six hundred spears, and these almost completely encircling
our handful of men, a great portion of whom were occupied
in attending to the wagons and the cattle. It seems really
incredible how we could escape from such a perilous situation, and it can only be attributed to the assistance of an
all-powerful and gracious Providence.

"From the frequent losses they met with on every occasion
when they attempted to advance, the enemy at last began to
show signs of fear, and eventually dispersed; so that about
three hours after the commencement of the action not a soul
was to be seen. Nevertheless, we continued on until we
reached a small water, where we outspanned, to allow the
cattle to drink, and to bury the Damara who was killed
before the fight commenced in real earnest.

"Having changed the span of each wagon, we again proceeded on our journey, keeping the open plains, fearing to

risk the probability of an ambush where any trees were visible. At midnight, however, we came to a halt, and prepared ourselves for a night attack, as I fully expected we should be followed up. In this position of affairs, I felt a nervousness to which I had been a stranger in the heat of the struggle; and I am free to confess I never remember hailing the return of day with greater joy and satisfaction than I did after keeping watch during this night. From this spot we made forced marches, travelling during the greater part of the day as well as the night, but we did not reach water until noon the third day from our leaving Ondonga. This constant toil, without water, naturally very much distressed our oxen; and as to ourselves, being compelled to keep a strict look-out for fear of a surprise, we were scarcely less fatigued. However, the Ovambo, having paid dearly for their foul treachery, did not seem inclined to renew our acquaintance; they had discovered, but too late, that when necessity compelled us to fight we were not 'women'—an epithet they loved to apply to us whilst at Ondonga.

"On Sunday, the 2nd of August, we reached a large vley on the Otjihakondoa-Tjomutenya, where we halted. Here a party of thirteen Ovambo passed us, carrying copper ore: they were of Otjipanga's (the brother of Nangoro) tribe. We called them, and gave them a full account of the treachery of their countrymen, informing them at the same time that, now they were in our power, we could destroy them all if we chose; but it was not our business to kill people at any time. We also pointed out the difference between ourselves and the red people, saying that, were our party Namaquas not one of them would survive that day. We further gave them a message to Otjipanga, with a full explanation of the conflict having arisen from the treachery of Nangoro's tribe. The men were in a great fright, and did not wait to be told a second time that they might depart.

"On the following morning, Mr. Hahn's guide, who had left us at Ondonga to pay a visit to his relatives, who were living amongst Otjipanga's tribe, made his reappearance. We had much cause to rejoice at his safe return, as we greatly feared that the Ovambo would kill him, knowing him to belong to our party. The news he brought was not of much consequence, except that the son of Nangoro, who stabbed the Damara, was among the slain, the guide having seen his body together with that of another of the victims. These two were probably killed in the first discharge by the man who was stabbed, and who carried a double-barreled gun. If my conjecture be correct, the poor fellow amply revenged his death. This made the loss of the Ovambo amount to thirteen killed, of whom we were certain. I shot one of the head men at the commencement of the affray: I fancy he was Nangoro's brother-in-law, but am not sure. The guide further stated that a party of Otjipanga's people arrived in company with him at Ondonga on the day of the battle, to greet us on the part of their chief. They were accompanied by four men from Debate, who immediately, it is said, recognised me. The tribe of Debates are called Ovaguembe, and are included with the Ovambo nation.

"The messengers of Otjipanga appear to have been very much grieved at the treachery which occasioned the fight. It is stated that they held a meeting with Nangoro, the result of which appears to have been a resolution to send us cattle as a peace-offering on the part of Otjipanga. The Damara said that he was to have accompanied the messengers, but, on their stopping at a village for the night, he made his escape.

"And now, my dear Andersson, you have the principal details of one of the most marvellous escapes from a set of treacherous 'friends' that has perhaps ever been recorded. At any rate, I trust sincerely it may never again be my lot to act one of the characters in a similar tragedy. There is no doubt that, had we remained longer in the country, the natives would have surprised us in the night, and butchered us to a man.

"The only other incident of note which occurred during this unlucky journey* was occasioned by a brute of a lion, who followed and harassed us excessively. He succeeded at last in carrying off, from near our wagons, an unfortunate Bushman chief who had come to visit us. I followed the horrid devil, but failed to inflict the summary justice he so richly merited. This was a cowardly lion; he would not make a stand, nor even show fight to our dogs, who came up with him on several occasions.

"As regards sport, I have had very little. I have managed, however, to 'bag' a few elephants, and might have killed a good many, had I devoted myself to hunting, instead of visiting the Ovambo country."

* Mr. Andersson does more justice to one result of this abortive expedition, which failed in its main object, that of penetrating the Cunené river, than Mr. Green, in his modest reference to its incidents, has done to himself in omitting to mention his discovery of the great lake Onondova.

"After such a tremendous lesson," observes Mr. Andersson, "my friends, as may well be imagined, gave up all further hopes of reaching the Cunené, and forthwith retraced their steps to their respective homes—Mr. Green lingering somewhat behind his fellow-travellers.

"I have omitted to mention one interesting fact connected with this expedition, and which, in some measure, redeems the credit of the undertaking. This was the discovery of a fresh-water lake called *Onondova*, which the explorers actually stumbled upon; for though they had people with them perfectly well acquainted with the country, they were not aware of its existence till they actually and incidentally caught sight of the water. This lake, as far as it is possible to judge either from the west or east—and I have been within a couple of days' journey of it—is situated about latitude 21°, and longitude 19°. The travellers did not go round it; they merely saw it at its eastern extremity; but water appeared as far as the eye could reach all round, and they estimated its circumference at from 25 to 30 English miles. Mr. Galton and myself, in the year 1850, actually passed within one day's march of this superb inland sea, without—such is the difficulty of obtaining information from the natives—having the slightest suspicion of its existence."—*Andersson's Okavango River*, p. 5.—[Ed. B.]

CHAPMAN'S TRAVELS.

JOURNEY FROM WALVISCH BAY
TO
THE VICTORIA FALLS
AND
SINAMANI'S TOWN, ON THE ZAMBESI.

(1860–3.)

CHAPTER XVI.

Preliminaries—A New Expedition—From Cape Town to Walvisch Bay—Fisheries—Discouraging Prospects—Oesip—Geological Features—The Schwagoup River — The Berg-Damaras — The Chobé River — Intense Heat—Otjimbengue—Cattle-stealing—Missionary Labours.

THE journey now about to be described was undertaken as a preliminary step towards the ultimate execution of a design which the author had long cherished, viz. the establishment of a line of commercial stations across Southern Africa from sea to sea. From an early period of my experience of travel in the African interior, I imbibed a deep conviction of the important results, commercial and social alike, which would ensue from the formation of an organised chain of communication, such as could only be maintained by the agency of a trading company established for the purpose between the east and west coasts of the continent, and the efforts of many years have been perseveringly, though hitherto unavailingly, devoted to this object. When, in the course of the travels described in the preceding chapters, I first reached Linyanti (in 1853), I was altogether unaware of the extensive views entertained by the great missionary traveller who was at the time in its vicinity, and with whom, though failing to meet him in person, I was so fortunate as to hold some friendly intercourse by way of correspondence. It was not until my arrival at Walvisch Bay, at the close of a lengthened journey, commencing at Natal, on the east coast, and extending in various lines of route over a large area of the intervening space, that I heard with astonishment of Dr. Livingstone's having reached Loanda, several degrees to the northward. It was with still greater surprise, and with unbounded admiration for the untiring perseverance and wonderful self-resource of that distinguished explorer, that I subsequently learnt the completion of his great enterprise by the return journey to Linyanti, and thence, by way of the Zambesi valley, to the shore of the Indian Ocean.

Although anticipated in the fulfillment of one part of my favourite project—the establishment of a communication between the opposite shores of the continent across the hitherto unexplored deserts of the interior—I yet clung to the idea of making a full examination of the middle and lower Zambesi, with a view to testing its capabilities for navigation, and for becoming a highway of commercial intercourse. The insurmountable obstacle which the tsetse opposes to wagon travelling beyond a certain distance to the northward of Lake Ngami gives additional importance to any facilities which may be offered by inland water communication; and, with the imperfect knowledge then entertained as to the character of the Zambesi below the Victoria Falls, I cherished the hope that, by reaching the river at a point at or near the junction of the Gwai or Quagga (the existence of which stream I had the good fortune first to have made known*), I might embark on its channel and descend it to the ocean.

Failing in the attempt to form a commercial company for the accomplishment of my larger design, but sanguine as to the feasibility of the lesser object just referred to, I determined to undertake the cost of such an enterprise out of my own resources and at my own risk. While full of this project, I was so fortunate as to renew at Cape Town my acquaintance with Mr. Thomas Baines, the well-known artist and traveller, who had been attached in the former capacity to Dr. Living-

* During my journey of 1854, when accompanied by Mr. Edwards.

stone's later expedition (1858), from which he had separated himself. Mr. Baines was, however, earnestly desirous of visiting the Zambesi, and especially of employing his pencil in delineation of the wonderful "fall" to which Dr. Livingstone had first drawn the notice of the civilized world, and to which I had determined, at whatever cost to myself, to carry my photographic apparatus. He was thus prepared to enter cordially into my views, and to become my companion in the extended journey which I then contemplated, and which is described in the succeeding pages. The many admirable qualities of my proposed fellow-traveller were well known to me, and his companionship and assistance promised to be of the greatest service.

The preliminary arrangements between Mr. Baines and myself were soon completed. The only conditions which I made with my friend were, to abstain from publishing, to assist me in taking observations and laying down a map of our route, and in making an occasional sketch or two from time to time. We agreed to work together for the benefit of science, and to forward our joint observations to the Geographical, Botanical, and other learned societies, as occasion might occur.

In all my speculations of sailing down the Zambesi I had always feared one great difficulty—the conveyance of boats for the purpose, as it was probable that nothing less than a couple of spare wagons would suffice for that exclusive service; but in the course of conversation with Mr. Charles Bell, the surveyor-general of the colony, he showed me, amongst many other models and ingenious contrivances of his own, one of a double canoe, or two half-canoes, of tarred canvas, fitted and made buoyant with inflated bladders. I conceived that this, on a large scale, would answer our purpose well, provided the canoes were placed far enough apart, and able to support a deck or raft sufficiently large to carry several tons of luggage, ivory, &c. &c., down stream. Baines had also a somewhat similar plan of his own, involving two whale boats to be built of sheet-iron, as being more durable. My own ideas were always, as they still are, in favour of a single boat, to be built on reaching the banks of the stream. I thus had it in contemplation to purchase in Cape Town a sufficient quantity of galvanized iron sheeting, and take it up to the Zambesi Falls in the bottom of my wagons, where it would occupy no great space; thus Baines might set to work, with what assistance I could give him, and put the whole together over a skeleton frame of wood. After some experiments, however, Baines altered his opinion in favour of two boats, and of the much more expensive material, copper. He thought he could complete the whole in such small sections as would admit of their being stowed within the wagons, and filled with merchandise, &c.; and as he wished to have the fullest liberty of action here, I consented to his building them after whatever fashion he chose, and at his own expense, so long as they were suited to be stowed in wagons without loss of much extra room. It was thus that our journey was inaugurated, Walvisch Bay being our destined point of departure for the interior. I proceeded in advance, in order to make various preliminary arrangements, leaving my intended companion to follow at his further leisure.

From Table Bay we had a smart run of five days, between the 9th and 14th of December, 1860, to Sandwich Harbour, the weather being unusually fine for the season, with favourable winds. On Sunday, the 17th, we cast anchor in Walvisch Bay, where I had the gratification of finding that my anticipations—formed on the occasion of my previous visit to this spot—of its becoming the seat of a fishery on an extensive scale had already been realised. A party of men in the employ of my friend, Mr. Wm. Latham, had just taken 650 large fish, each upwards of 100 lbs. weight, and another party had captured upwards of 2000 rays and sharks. The only use made of them consists in taking out the livers, for the sake of the oil. Having had that yielded by the sting-ray and the black shark analysed, it has been found to contain as much iodine as cod liver oil, and its efficacy in pulmonary and other complaints has been fully proved.

We had a present sent us of some fine soles, milk-fish, and leather-fish. Sprats are sometimes sent in this way to catch a mackerel, and a flask of gin or bottle of French brandy is always welcome. The large kabeljau fish, which, when cleaned and dried, weigh eighty to the ton, realise in the Mauritius from £28 upwards. A few years ago, about Christmas, or a little after, it was not an uncommon thing to see half-a-dozen whales sporting about this bay, and I have known two or three to have been washed up dead. Some whalers from Table Bay have occasionally done very well on this coast. I have counted at one time as many as eighty whales on a calm day sporting round our ship, when becalmed. Lampreys abound in the lagoon, where they burrow in the mud after the tide recedes, occasionally peeping out of their holes and basking in the sun. A species of mullet, or herring, is also taken in large quantities, and salted down in casks of a couple of hundred-weight each.

Sharks are plentiful in the bay. I have speared as many as twenty-two in a quarter of a mile's walk along the beach, scarcely up to my knees in water. The black shark is a very ferocious animal, and this makes sea-bathing on this coast a most dangerous amusement. Some prodigiously large sharks have been killed in the bay at various times; the jaws of one which I saw could easily take in a man of even more than the ordinary proportions. From the liver of this shark, the black or man-eating species, upwards of 200 gallons of oil were obtained. They glide up and down near the beach, occasionally showing their flukes above the water, reminding one very much of the crocodiles in the rivers of the interior, and they are not unaptly distinguished by the fishermen by the sobriquet of "alligators." The bone-shark is a very large but exceedingly harmless fish, sometimes attaining nearly the proportions of a whale, but living by suction. I have seen some of these upwards of thirty feet long.

As my friend Latham was not at home, we found some difficulty in getting our goods landed without the assistance of his people, having to carry or roll the heavy packages up to the store ourselves, and this while working against time also, as the tide was rising fast. The landing being accomplished, we sat down to listen to reports from the interior, and these were extremely discouraging. The lung-sickness was making dreadful havoc, for which the natives had already, it appeared, passed sentence of death upon me, laying to my charge the introduction of this disorder into their hitherto uncontaminated country. We heard also that a party of Hottentots of the very worst description, the Vilschoen-Dragers, who cut off the heads of some Lake people who came to them on a friendly embassy from Lechulatèbe, and in consequence of which the Bechuana tribes call them "Makaula Hogo," or "head-off-cutters," had plundered two cattle stations of Mr. J. J. Wilson and his partner, Mr. Castray, of several hundred head of cattle, and shot down in cold blood the unresisting Damara herdsmen, with their women and children. The drought had also been so severe as to render it a matter of doubt, if not of utter impossibility, that we should be able to reach Otjimbengue. The great scarcity of grass and water was unprecedented, and it appeared that, even if we could succeed in reaching Otjimbengue, there would be no possibility of proceeding beyond, since the Hottentots were said to be averse to travellers passing from this side in the direction of the lake, for fear of having the lung-sickness introduced amongst their cattle, and all the tribes had determined to support the cry of " No thoroughfare " by every means in their power. Above all, to complete the catalogue of disasters, my cattle were said to have all died of the murrain. Pretty prospects these with which to commence a journey across the African continent!

I had, however, found by experience that in this region one must not rely implicitly on the rumours abroad, and resolved at once to go up as far as Otjimbengue, where I had already established a cattle station, to see after my own affairs and judge for myself as to the state of the country. At this moment I received an offer from Mr. Andersson of carrying my bedding, &c., in his wagon. This proposal I thankfully accepted, arranging to walk beside the wagon myself, and taking with me a single attendant, to render any aid which my then weak state of health might render necessary.

We crossed the plain which intervenes between the bay and the station of Oesip in twenty-one hours, without unyoking the oxen. The first three miles of level, muddy beach, at the confluence of the Kuisip with the ocean, we passed without much difficulty; but then came three and a half miles of deep, heavy sand in the dry bed of the Kuisip river, which winds between high shifting sand-hills overtopped with narra-bush and dabby. In a thick crust of a very fine sediment—the consolidated deposit of alluvium from the interior—we find very distinct impression of the tracks of birds and beasts, men, women, and children, of seemingly old date. The spoors of rhinoceroses and elephants which have been observed here have the appearance of being quite recent, although it is well known that no elephants have frequented this country for more than twenty years. I cannot but think that this fine sediment might be applied to some useful purpose.

It was now midsummer in this hemisphere, but the nigh

fogs, aided by northerly winds, made it exceedingly cold at night, though just under the tropic. Crossing the plain in twenty-one hours seems a very summary account of this truly irksome journey over a dreary waste of sand in the darkness, and through the thick cold fog, without grass, spring, or fountain. The tale of personal anxiety and suffering would, however, be tiresome, and must be left to the reader's imagination. Perhaps the most painful incident of the journey is the cruelty one is absolutely compelled to inflict on the poor oxen in order to accomplish it. Sometimes, when they have been driven the last 70 or 80 miles without food or water, they have been unable afterwards to take us on more than six miles in as many days. It often happens that from one to a dozen oxen faint and die on the road under the exertion, and we have to lie awake all night and watch, for fear of losing some of our loose cattle and sheep. The journey must be accomplished during the hours which intervene between 3 or 4 o'clock in the afternoon and early sunrise on the following morning—it being absolutely necessary to give up all thought of farther progress by the time the sun is an hour above the horizon, otherwise your oxen are done for, and your people likewise. By dint, however, of sharp and constant driving, when the load does not exceed 2000 lbs. it can be done well enough, but we have to keep strict watch on our men, who, notwithstanding that they generally put pieces of stick across their eyes—from eyelid to eyelid—to keep them open, will sometimes drop down overcome with fatigue, or even fall asleep while walking. Their bleared eyes, when thus pegged out with sticks, and their dust-covered hair and faces, seen by the light of our wood fire, look truly diabolical.

At Oesip I made a collection of geological specimens for a juvenile museum. I also collected some for myself. It is a favourable spot for the purpose. Most of the rocks here bear traces of volcanic action; indeed, the whole country is charred and seared, and the hills—covering extensive tracts—assume the appearance of a stormy cross-sea, whose gigantic and tumultuous black and crested waves had been checked suddenly in the midst of their fury. Broad veins or dykes of black, metallic-looking basalt stretch in chain-like walls from peak to peak, the softer rock having crumbled from its sides or been washed away. The dykes cross each other at right angles, or branch out of one another in serpentine veins, contrasting strongly with the pale gneiss; and the hills of grey and spangled schist are streaked in the same manner with wavy, serpentine, and parallel strata of white mural quartz from one to twelve or fifteen feet in thickness, imparting to the whole scenery a picturesque appearance, and compensating in some measure for the great scantiness of the vegetation. The hills are sometimes torn and rent asunder in the most fantastic manner, and it requires but little aid from the imagination to realise the semblance of towers and spires in the huge square boulders, or the tall sharp splinters of rock cutting their bold outlines in eccentric angles against the clear blue sky. In parts the honeycombed appearance of the cavernous rocks was in the highest degree interesting. They had been eaten into by wind and weather until the pretty rock-pigeon, together with bats and owls and hawks, could find a safe shelter in which to rear their young broods; while elsewhere the hollows enclosed between or beneath the huge projecting boulders formed the abode of the cony, the bat, and the vulture. Meanwhile the disintegrated gneiss and detritus washed from the hills fill up the valleys, and seem to be forming a fruitful soil for the future. Under the shade of some of these weather-worn rocks at Oesip a strong nitrous exudation may be found. I have gathered here about a bushel of fine white salt in an hour's time, and I picked up a piece of sal-ammoniac in the same neighbourhood.

Notwithstanding the usual dreariness of these plains, we saw several gnus, springboks, ostriches, and jackals; wolves (*i.e.* hyenas) were also heard. It is a mystery to me what these animals live on, or what they seek so far away from the river, where, this year at least, there is neither grass nor water to be found.

From the outspan at Oesip I walked down to the Schwagoup (or Swakop) river, distant about four miles, whither our oxen also were obliged to be sent for both grass and water. From this outspan, which seemed just at the northern border of the plain, I followed a long narrow valley, with a steady decline, for about an hour and a quarter, the hills and rocks on either side being bare as a burnt brick. The valley became more broken and rugged at every step, until suddenly rounding an angle I was most agreeably surprised at finding myself in a pleasant green spot, an oasis in the wilderness, enlivened by the notes of a multitude of pretty-plumaged birds of the finch tribe, the sweet cooing of the turtle and ringdove, interrupted occasionally by the somewhat discordant cackling of pheasants and guinea-fowl.

A deep valley between high hills, and splintered pinnacles of schistose rocks, several hundred feet high, forms the channel of the Schwagoup river. The bed, consisting of white sand about 300 yards wide, and level as a floor, frequently exhibits a saline efflorescence, or other sign of moisture, where the water is nearest the surface; and in this bed of sand the thirsty traveller has often to dig to the depth of three or four feet in order to obtain a supply of water for his thirsty team. A little to the eastward, however, a spring percolates in some seasons through the sand and flows for a short distance over the surface. The banks are studded with the large and majestic anna-boom, an enormous acacia, casting round a deep and grateful shade. There are also melancholy-looking ebony trees, which resemble the weeping willows in growth, and the camel-thorn abounds. The tamarisk, or dabby-bush, forms sometimes an inaccessible thicket, matted and overrun as it is with the too exuberant burr-grasses—the great pest of this country, though the seeds are eaten by the natives as a grain. The "gona, or soap-bush," from the ash of which, as well as the brakbosch, the ley for soap boiling is made, also abounds. A great variety of other saline plants and bushes, peculiar to a soil impregnated with nitre, luxuriate near the river-banks; more especially several kinds of pulpy-leaved plants called "slaai-bosch," or salad-bush, of the same genus as the portulaca. One kind of slaai-bosch, with a very large fleshy leaf, seems, even during the greatest heat of the day, to be literally covered with sparkling dewdrops, but on feeling for the moisture you are soon undeceived, and what appeared dew on the surface is firmly secured under a thin transparent film, which is raised throughout the surface of the leaf, in small and brilliant globular blisters that are pleasing to behold. These sappy plants are found in the most sterile parts, high up amongst the rocks, and I believe may safely be eaten. John Cluett, my present servant, who is an old stager, is in the habit of washing his hands and face and clothes with the juice of the plant in preference to water, as it contains some property which answers as a substitute for soap. He further assures me that skins first damped with the juice of this plant, and then dressed, are rendered "as white as milk, and as soft as silk." All these plants grow abundantly, even near the bay, in the river-bed. The grasses are coarse and prickly, such as we usually find wherever a nitrous exudation is visible on the surface of the soil. A few gigantic aloe trees, and the poisonous euphorbia and cacti, exist, though but sparsely interspersed through the distant heights, rearing their star-like crests, and standing out in bold relief against the clear blue sky.

Having spent Christmas Day at Oesip, we proceeded over

the next trek of about eight hours to Davikip, another outspan far removed from any water, and situated between immense spherical hills of smooth granitic rock with scarce a fissure, and huge boulders of gneiss and granite, streaked with veins of pure quartz, masses of black and white mica, and large lumps of a black coal-like mineral, often imbedded in the whitest quartz. Within half a mile of our camp an immense mass of granite, hanging from a smooth hill of the same formation, seems as if a thunder-clap could easily dislodge it.

For the first hundred miles from the coast the country generally presents a wild and sterile appearance, more especially so for the first 70 miles. The red baked earth and barren rocks are seldom, perhaps only once in ten years, cooled by a grateful shower; an unhealthy fog, perhaps detrimental to the growth of cereals, prevails during half the year. There seems to be a prevalent opinion that this country, in common with South Africa generally, is visited by a succession of wet and dry seasons, and in this manner some of the older inhabitants account for the perfect desiccation of fountains which were formerly large running streams, inhabited by the hippopotamus; I have myself found evidence at Ghanze which goes far to corroborate this theory, in the presence, twenty feet below the surface, of Bushman utensils—a tusk of ivory and a ladder—where it is well known that, until lately, there has been no need within the lifetime of any living native to dig for water.

The deficiency of regular rains near the coast is no doubt owing in part to the absence of vegetation, but more particularly, perhaps, to the winds. These always blow with greater violence from the sea during the rainy season, just after mid-day, and with sufficient force to check and drive back the thunder-clouds, which, as a rule, generally come from the east in the afternoon, dispelling them in circles over and around Otjimbengue, and back again into the interior.

During the dry season the grass seeds which are blown about the plains remain in a dormant state, the dew not being sufficient to effect germination; but when a shower or two, in some favourable season, does fall, the plains speedily exhibit the appearance of an immense corn-field, and become clothed with a profusion of wild flowers which can scarcely be excelled. I have noticed in the neighbourhood of Hykamgoub some very large bulbs of the amaryllis and others, as well as another large woody root or stem, something like a bulb,* having a rough bark like the oak, with long, broad,

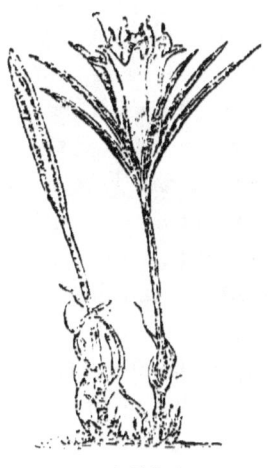

AMARYLLIS SP.

wavy coils of fibrous leaves, fifteen or twenty feet long, with scarlet flowers, and seed pods like small fir-cones. These plants, which have a very long tap root, weigh sometimes upwards of two hundred-weight. I observe it has but two leaves, which, however, are often split up, imparting the appearance of there being several. The top of the plant is flat, of an oblong form, and one leaf springs near each edge out of a longitudinal rent, and the cones or flowers

* This plant is called by the Damaras, Otjetumbo Otjehooro.

grow on branching stems along the margin and sometimes under the leaves. The circumference of some of these root-stems is from six to eight feet at least.

From Davikip, three and a half hours the next day brought us to the "Reed," where we hailed with delight the merry chirping of the birds, of which we shot a great many, and some of exquisite plumage, such as the lustrous *Lamprotornis leucogaster*, and Honey-suckers (*Nectarinia fusca*), three varieties of woodpeckers, small green and blue paroquets, with red bills, and a larger grey one with yellow shoulders, fly-catchers in great variety, chatterers, brilliant bee-eaters, and butcher-birds. We also shot some guinea-fowls and pheasants. Andersson shot a little steinbok with very small shot. I amused myself by shooting a few birds for Andersson, who skinned and stuffed them almost as quickly as I could procure them. I saw here a rabbit of a pinkish or pale violet colour, but had no gun; I have since tried in vain to get another glimpse of this animal, perhaps new to science.

We next proceeded to the junction of a river coming from Onanies, or, as it has since been facetiously called, "Old Nanny," outspanning at Onanies Mouth. A trader's cattle were stationed here in charge of some Berg-Damaras, who made us an acceptable present of some coffee and milk. These people are a much cleaner race than the Beeste-Damaras, who never wash their milk pails, for fear of their cows drying up, nor do they go to the trouble of taking out drowned flies; hence their sour milk is invariably tainted with the offensive odour of utensils never cleaned for years together, or discoloured by the swarms of flies.

During the night we were occasionally disturbed by the falling of pods, displaced by the gentle wind which played through the trees. Insects flitted and boomed about the wood fire, and bats whirred all night in pursuit of them, while the little owlets (*Strix arcadia*) threw their loud, whistling notes, in tones which rose higher and higher on the still air of night, with the regularity of a scale in music, until they could go no higher. I collected some beetles and butterflies during the day, and at night one very large wood-beetle, which I pinned just over my head against a tree. It had not been long there before it was attacked by several small bats. At any other time I should have been glad to see a family of bats at supper, but as my specimen was a rare one, I could not indulge their taste.

From Onanies Mouth we travelled over a series of rough and flinty undulations a few miles farther up the river, and then struck off to the right; and having ascended a considerable distance, with infinite toil, over the sandy bed of a narrow rocky glen, the sides of which are a study for the geologist, where we occasionally disturbed a number of dassies, we emerged at length on to a plain. By a very winding, round-about way we reached, after another six hours, Wilson's Fountain. Here I photographed some botanical specimens, and cut down one of the aloe trees, wishing to use some of the wood, as a substitute for cork, to line an insect cabinet. Between Wilson's and the Tsoubis we saw zebras, koodoos, springboks, klipboks, steinboks, pouws, and some rock-rabbits, or dassies. The excrement of this little animal, which is generally deposited, together with the urine, in one particular spot, is with the Hottentots a great antidote against dyspeptic complaints, constipation, indigestion, &c. The spoors of giraffes were here plentiful.

From Wilson's Fountain, having cleared a mile or two of flin ground, covered for the most part with a variety of euphorbias, or milk-bushes, we reached more level ground. One kind of these milky bushes has a very peculiar tuberous root, resembling in colour and form the granite rocks on which it grows. Some of these roots must be half a ton in weight. The hills and plains on either side are covered with tall, wavy grasses, and studded with the aloe tree and a peculiar gouty-looking sterculia, or scattered over with fields of shattered granite. Along the south of our track, and parallel with it, is a higher range of granite mountains, thickly strewn with blocks of the same formation, which may have crumbled from its now smooth and bare summits. One point in particular, which has at the top a deep niche between two pinnacles, I have ever called in my book of bearings the "twin turrets." It is one of the best landmarks in the country, being visible from near the coast. This hill is called by the Hottentots "Annison's Ears." My friend told me he believed it had been thus designated by the natives in honour of his explorations, but on subsequent inquiry I found it means simply "Owl's Ears!"—Annis meaning a bird, or owl, which this hill much resembles in shape.

We passed through two pretty wooded ravines,* studded with anna-wood and camel-thorn, and descended through a rocky country to the Tsoubis, a river coming from the south, and flowing into the Schwagoup west of this place. Just below the outspan are conical hills, formed apparently of detached granite rocks, piled one above another. Near the base is a brackish fountain, where we procured water for ourselves and cattle, the running fountain of former days, near the outspan, having, to my own knowledge, dried up within the last few years.

Having had to walk the greater part of the distance from Walvisch Bay, night and day, I felt thoroughly exhausted, and was not sorry to enjoy here a day's rest. The heat gradually increased towards noon, when it could not have been less than 112° in the shade. It was a fearful day. Panting, and hiding wherever I could get from the sun's

* All the rivers in the country are periodical. In the rivers one can always find water by digging, but in the ravines the water lies too deep for digging.

piercing rays, I could not evade the devouring blast, which seemed to scorch up our entrails. The wind, which had travelled over hundreds of miles of baked earth, poured upon us its accumulated heat in blasts that might have issued from a furnace, and the dense clouds of fine dust, which swept down like a torrent, were truly stifling. We could not see one another, nor dare we open our mouths. The little rain which followed was just sufficient to lay the dust, and to bring out innumerable small beetles from their hiding-places, while pouws (bustards) and korhaans (*Floricans*) came from every direction, guided by unerring instinct, and pounced upon them.

Finding that neither water nor grass were to be had for the cattle, we had to send them back about five miles a little to the north of the road. Nine and a half hours more of trekking brought us next day to Otjimbengue. The intervening country, which gradually became more thickly wooded with mimosas and other thorny bushes, is still frequented by the giraffe, whose spoors were visible. On nearing Otjimbengue we passed near several conical hills of granite rocks, which are interesting from their peculiarity. On the extreme summit of one of these hills stands an immense block of stone (as if it had been placed there by giant strength), like an inverted cone on a very small square pedestal. It looks for all the world as if a gust of wind could blow it over.

At Otjimbengue we learnt that the inhabitants had been startled on the 22nd of November by an unusually severe shock of an earthquake. The rumbling by which it was accompanied seemed to travel in a N.NE. direction. The peaceable inhabitants of Otjimbengue had also been startled by the depredations of some Hottentots, who had lately plundered their neighbour, Mr. Wilson, of his cattle. These Hottentots were of the tribe of Vilschoen-Dragers, living on the Visch river. They had previously been to Bokberg, where they shot down a number of Berg-Damaras and carried off large flocks of goats and sheep; but some of the Berg-Damaras who had escaped waylaid them, and, poisoning the waters in their path with the milk of the euphorbia, succeeded in destroying all their booty, and horses as well. Disappointed in this affair, they came upon a cattle-post of Mr. Wilson's, and, shooting down six or seven Damara herdsmen, carried off all the cattle in their charge—about 500 head. They then very coolly went to Mr. Wilson's house, where there was only one solitary white female at home, and sold her some of her employer's cattle; and, behaving in other respects with all due propriety, gave not the slightest cause for suspicion: it was some days before Mr. Wilson found out what had happened, and it seemed then too late to follow the marauders up.

Such being the insecurity of property in this country, I took the liberty of urging on Mr. Wilson the necessity of exerting himself in this matter, which he seemed to take rather quietly, having since the occurrence sent out his partner, Mr. Castray, and servants on trading excursions 300 or 400 miles up the country, instead of trying to recover his cattle. All the white inhabitants in this neighbourhood are willing to aid him in any steps he might take, and are greatly disappointed at his indifference, justly considering that a bold step now might prevent future aggressions of the same kind, and perhaps save in the end a great deal of bloodshed.

The cattle of the Africaners have been dying very fast of lung-sickness. Jonker's own wife and daughter, report says, have been shooting the cattle belonging to the chief himself, and killing some of the Damaras in charge of the station. These Hottentots are a despicable race; no one of them can put any trust in others—a chief in his subjects, a master in his servants, a husband in his wife or his daughter. The same want of good faith also makes the tribes distrustful of each other.

At Otjimbengue hardly any rain had fallen during the last two years, in consequence of which there is not a blade of grass to be seen, and the bare trees and bushes appear miserably scorched. The thermometer, after sunrise, stands already at 80°, at 2 o'clock, 106°, and at 10.30 at night, 98°. This in the thatched houses; they feel so close and uncomfortable that we are obliged to wet the floors several times a day with a watering-can, and pour several bucketsful all round under the verandahs. The water, too, is so tepid and nauseous that, unless bottled and hung in the air in a flannel bag, it is not fit to drink. The natives, as may be supposed, suffer less than white men from this high temperature; indeed, they seem to enjoy it. We seek the shade during the heat of the day; the Hottentots and Damaras are often seen sleeping in the sunshine, with their faces turned upwards, without any bad results; while I feel convinced that ten minutes of such exposure by a white man would bring on a sort of sun-stroke. With a similar contrast as regards enduring cold, if you happen to take pity on one of the natives in crossing the foggy plain near the bay, on a cold night, and supply him with a rug or duffel jacket, he will carefully stow it away, and bring it out and wear it only when the sun becomes powerful.

I visited my friend Mr. Hornemann at the mission station of Richter-Veldt. He was, as usual, working himself to death; and I fear will find out that people cannot work within the tropics as they can in Germany without the risk of material injury to the constitution.

CHAPTER XVII.

At Otjimbengue—Loss among Cattle—Return to Walvisch Bay—Bed of the Schwagoup—A Buffalo-hunt—Stories of African Adventure—Lion-killing—M. Gérard and his Exploits—Wreck of the Canute—Walvisch Bay again, and return to Otjimbengue—Meteors—Shocks of Earthquake—Journey farther Inland—Change in the Aspect of the Country—Hot Springs of Barmen—the Schwagoup—Jaager Africaner.

I DISCOVERED, to my great disappointment, that out of 148 oxen, which ten months ago I left at Otjimbengue, only seventeen were now alive. One of my wagons, by my own directions, had been bartered for oxen, all of which had also died. I was obliged to borrow and purchase some wherewith to return to Walvisch Bay with the one remaining wagon, in order there to purchase a second. My first care was necessarily to bring up from the bay some trading goods to barter for additional oxen. On the return journey I lost eight within the first four miles, which it took twice as many days to accomplish. In company with Mr. Dickson, who lent us timely aid in lightening our overladen vehicle, we rested on the way at Hykamgoub, a beautiful fountain in the bed of the Schwagoup, to refresh our exhausted cattle. Hearing opportunely that buffaloes had been seen a few miles lower down the river, six having been shot some days before, it was resolved to make up a party and pursue them. In the meantime some gnus, ducks, rails, water-hens (or coots), and snipe, were shot near the fountain, all of which we demolished with true hunters' appetites.

Having yoked some oxen to a light cart, into which I had

stowed my photographic apparatus, we proceeded down the bed of the Schwagoup, between towering perpendicular mountains of granite rock. These rocks, in and near the river, are all rounded and smoothed by attrition, but away from the river the outside crust of the granite rocks peals off in rotten and crumbling scales. Others wear away very unevenly, leaving in some places cavities, in others huge projections. Some of these have been worn into the resemblance of human or other figures, and one in particular is such an exact representation of a couchant lion on a square pedestal, that one would not, but for its greater size, be able to distinguish the rock from the living reality.

After three hours we reached a fountain called Kannekompdis, the site of the former residence of a Mr. Dixon, whose family, having all grown up in this country, could relate some most interesting stories of lion-hunting as it was in the days of old, when the lions were bolder than their degenerate successors of the present day. Next morning we proceeded to Nonidas, another fountain, or rather a marsh, in the river-bed, within two or three miles of the sea. Here the general hilliness of feature gradually gives place to mere sand-hills and sandy plains. During two long days we strove to drive a troop of buffaloes out of an extensive reed-marsh, sometimes up to our waist in water, at others firmly caught between the tall and matted reeds, in which position we were often in danger of being overrun by the buffaloes as they were driven about by the dogs. In vain we strove, with the assistance of the dogs, the holloing of men, the cracking of whips, and discharging of fire-arms, to dislodge the game. The wary beasts would not break cover; and though they often passed within five or six yards of us, we could not see them, owing to the denseness of the reeds, which they crushed down with a great noise. A party of Topnaars from the south of Walvisch Bay now joined us, bent upon varying their narra diet with buffalo steaks and marrow-bones. Taking advantage of this accession to our beaters, we drove them into the reeds with whips and guns, after our oxen and dogs, bidding them make as much noise as possible, while we watched at different points on the margin, catching occasionally distant glimpses of the game as they fled from cover to cover to the end of the marsh. Then they rushed back again through the ranks of the Hottentots, one of whom, bolder than the rest, took courage, and, creeping in towards the crackling reeds, observed a dark object before him, at which he instantly levelled and fired. Down came the animal with a groan, and out rushed our hero, declaring he had finished it. Not altogether trusting the report, the spot where the beast had fallen was approached with a caution that appeared almost ridiculous, no one seeming particularly anxious to occupy the post of honour at the head of a party in search for a dead buffalo; and it occupied half an hour before the fact was ascertained that our friend had certainly slain one of our own oxen! Not much disheartened by this circumstance, we made renewed efforts to dislodge the buffaloes; but the Hottentots persisting in firing ball wherever they heard the reeds breaking, we found that we were in great danger of sharing the fate of the poor ox, and gave up the chase. The poor fellows, afraid of being punished for their mistake, and unable to bear up any longer against hunger, as they never carry food with them, went off in a great hurry for their narra-fields. We felt confident that after the worrying received by the buffaloes they would certainly abandon the marsh. This was what we particularly desired, as they had no other retreat but up the Kaan river, which joins the Schwagoup, where we should fall in with them. We therefore left a troop of more than a hundred gnus quietly grazing or resting on the plain, and, passing some gemsboks and ostriches on our way, proceeded to head them at Kannekompdis, having first ascertained that they had left the marsh during the night. Making a short cut, under the guidance of one of our party who was intimately acquainted with the ground, we passed the spot where our game lay concealed by a bed of reeds. Some of our followers were instructed to fire these reeds; after doing which they climbed a rounded mass of granite, 100 feet high, from which they rolled down huge pieces of rock amongst the hiding buffaloes; but neither rock nor fire would make them move, until one of the Damaras, catching sight of a cow, shot her. Then the rest of the herd, already scorched by the flames, sulkily trotted out of the thicket on to the plain intervening between our perch on the rocks and our cart, which was now coming up. We had become so impatient that, on sighting the buffaloes within a few hundred yards of us, I descended the rock in order to make a stalk in upon the troop, although B—— insisted they would certainly pass within a few yards of our ambush. I had scarcely descended when, to my amazement, I found myself face to face with three large bull-buffaloes. To fire at the biggest and retreat behind a rock was the work of an instant; but my smooth-bore gun had been loaded with hollow conical bullets, and the second shot I fired at the retreating herd missed them altogether. Away they scampered across the river-bed, a quarter of a mile wide, the wounded bull lagging behind, until they entered a large patch of reeds, from which our wounded buffalo never departed more. It appeared to me that the opposite side of the river would now be the best position, as the rest of the herd would be afraid to pass where we had just been firing; but my friend B—— knew otherwise, for, owing to the cunning habits of these animals, and the circumscribed limits of the cover, they always make the shortest cut from cover to cover.

I had not rested long in my new perch—about twenty feet

up in the rocks—when, while listening to some of B——'s stories, on looking up I saw, to my great surprise, other listeners close at hand. In fact, six buffaloes were standing within fifteen yards staring at us, but before we could bring our guns to bear, they were on the move, running past us. Giving my companions notice of which animal I intended to shoot, so that they might not aim at the same one, I fired at the largest cow, the bullet passing right through her and out on the opposite side. B—— fired next, but, missing his mark at another full-grown cow, he bowled over the youngster which was galloping by her side. The rest of our little party, never having seen such large game before, were so confounded that at first they forgot to fire, and, when they somewhat recovered their presence of mind, shot wildly, and missed their aim.

Having eaten nothing since the previous day, and it being now near sunset, we lost no time in cutting off and cooking some portions of the buffaloes we had killed, intending on the morrow to follow up the remainder of this herd, and shoot a few more at Hykamgoub reeds, where they were sure to take shelter during the night. But our horses had run away while we were killing the buffaloes, and it was late the next day before they were brought back. It was probable that the buffaloes had meanwhile escaped up the Kaan river. The tame process of buffalo-shooting in this country had, moreover, lost all its interest and excitement, and, there being no chance of having a spirited dash after them on horseback, or even of obtaining photographs, I began to wish myself back. Not forgetting, however, that we had not broken our fast the whole day, we got up a stew of the liver, heart, kidneys, &c., well seasoned with pepper and salt, and this, with the addition of biscuit and coffee at 10 o'clock at night, put us all in good humour, and we sat up till a late hour telling our own adventures, or listening to those of others, and then lay down on the hard flat rock to sleep, with only a blanket under us.

The strange stories one sometimes hears in South Africa lead one to the conclusion that it is dangerous to give implicit credence to all that is told. Some of the old inhabitants of the colony, as well as of this part of the country, tell of wonderful things that we of the present generation never have the good fortune to meet with, many of them evidently being but chimeras emanating from their own imagination, and retailed from father to son with compound interest. I can believe in the *possible* existence of a sea-serpent, though the idea has been scouted as ridiculous; but I must hesitate to believe in snakes exercising a fascinating power upon man (though I have witnessed a somewhat similar influence exercised by them over birds). But what can be said of the wonderful feat attributed to the hoop-snake, which, when it chases you down-hill, bends its body into the form of a rigid hoop by putting the end of its tail into its mouth, and then rolls after you? what of the charmed cap of the snake-charmer, an old Hottentot in Namaqua Land, who is said to boil his dirty old night-cap, which he has been wearing for half a century, and a decoction of which, taken internally, is said to be an infallible cure for the most venomous snake-bite? A blow from this man is said to produce immediately a swelling on man or beast, and a scratch would terminate fatally in a few minutes without the administration of the above antidote. It is said that this man swallows the poison-glands of all the snakes he gets, and that all kinds of venomous reptiles nestle harmlessly in his bosom. I have been thinking that the story of the dassie-slang, so prevalent amongst the Boers of the Trans-Vaal, must have originated in the imagination of some half-blind individual, who perhaps had seen a rock-snake or python seize a dassie — their usual prey — by the hind quarters, and glide with it into the nearest hole, there to swallow and digest it.

I have just been reading Jules Gérard, and have puzzled my brains very much to discover the advantage of such close contacts as that gentleman allowed the lions to gain before he fired at them. I consider this to have been unnecessary; in fact, an obstacle to success, and dangerous in the highest degree. I hold that no man's ocular faculties can be equal, much less superior, to a lion's at night, and my only wonder is, that M. Gérard's statements on this subject have escaped the criticism of scientific men, and that not one of them has ever thought of testing this wonderful freak of human nature. It would have been no discourtesy to M. Gérard to have asked him for a single exhibition of his marvellous powers. M. Gérard certainly makes the most of everything; he strains every nerve to show that the northern lion is more dangerous than the southern. Of the danger of shooting these animals, his English editor, if I remember rightly, tells us that this is proved by the circumstance of Gordon Cumming not having nearly so much trouble in killing his lions as Gérard had. The fact is, that Cumming, who is well known out here to be but an indifferent sportsman, has killed three times as many lions as Gérard did, and has talked less about it. No Englishman would indulge in such gasconade. I could produce a man who has killed no less than fifteen lions in a single night, and perhaps hardly ever afterwards recurred to the subject. And while I am writing, I have been in conversation with a little woman, only four feet high, who, when a girl of about twelve, wounded a lion in the doorway of the house with a discharge of pebbles from a blunderbuss, in the absence of lead, and who had to keep watch every night over her mother and little sleeping brothers and sisters, while the father was absent on a journey. Of the fact that lions are as daring in this country as in any other, we have had sad experience in the number of people that have been dragged away from out of their houses and from their fire-places, in spite of every resistance. Still, it is sometimes the amusement and sport of young lads in the Trans-Vaal to make an ambush or skaarm, and waylay the lion and the rhinoceros at night, and I have before me an instance of daring such as would, no doubt, astonish many persons, in the person of my friendly host's eldest son, aged nine years. A short time ago he followed up a lion that during the night had killed a horse on the homestead, and fired a charge of buck-shot into him. I have known more than one man who had killed upwards of a *thousand* elephants, and you scarcely ever heard a reference to it.

There is throughout M. Gérard's wonderful book a carelessness of expression, which, were one inclined to cavil, would lay him open to severe criticism. For instance, he goes out to shoot lions one night, when it was so dark that he could not see two yards before him, nor even the barrels of his gun, and yet he kills or wounds his lion at a few yards! He can see the eyes glaring like two balls of fire, &c. Now, I know very well that an animal's eyes can be very vividly illuminated by the reflection from another light, and *that* only when you happen yourself to be stationed between the object and the light. Under no other circumstances can I believe it possible. Again, a lion gets three shots behind the shoulders, and lives three days afterwards : this is quite possible, but it seems a very vague account. With M. Gérard the shoulder probably reaches as far as the tail; with us in South Africa, " behind the shoulder " always conveys the idea of a mortal wound, somewhere about the region of the heart, liver, or lungs—a wound which the animal could not survive longer than a few hours. There are other people equally careless in their descriptions : thus I can hardly imagine a man's losing his self-consciousness so far as to make either himself or others believe the fact of a rhinoceros standing over him while the saliva was dropping from the beast's mouth on to his own face, the animal being at the time unable to smell, even granting that he could not see him!

Having still a great quantity of goods to bring up from the bay, I returned for another load. Lions were reported to be very plentiful and troublesome on the road; and one, a large male, was even bold enough to follow on my spoor to Tsoubis outspan, at mid-day. Leaving my man to pursue his journey to Otjimbengue, I started for Hykamgoub by the new road, and, on reaching the outspan at midnight, learnt from some people there that they had just been visited by six lions, who had killed one and wounded another of their dogs, within three or four paces of their wagon. On reaching Wilson's place, I met the post-boy—a Damara—who travelled on foot from Walvisch Bay, and who brought the painful intelligence of the loss of the Canute, somewhere near the island of Ichaboe. I lost in the wreck upwards of £300 worth of goods, among which were all my provisions and all my beads—the only currency in the interior, and the only article I had with which to buy food from the native tribes. I lost besides many little requisites, and some books and instruments which would have been very useful to me.

From Mr. and Mrs. Latham, who barely escaped with their lives, we heard all the particulars of the disaster. The ship was wrecked on a desolate coast, and was soon plundered by a set of lawless ruffians then stationed at Ichaboe, the possession of which was at the time disputed between certain mercantile houses of the colony. Mr. and Mrs. Latham had, with much personal suffering and anxiety, made their way along the coast, eventually reaching Walvisch Bay. I was truly glad to learn that my intending fellow-traveller, Baines, whom I had for some time been anxiously expecting to join me, had not taken his passage in this unfortunate vessel. Arrived again at Walvisch Bay, I heard that, notwithstanding the great quantity that had been caught on our landing at this place, no more fish had been taken since Christmas. It appeared that there was a peculiar epidemic among the finny tribe; thousands were washed up dead on the shore, and the water had a reddish colour, without any apparent cause. There had been no extraordinary winds in shore to disturb the sediment at the bottom of the sea, nor was the colour at all traceable to sand or mud. It is observable, however, that the spring tides had never been known to be so high as just before. At night-time the sea flashed with phosphoric scintillations. A similar phenomenon had occurred eleven years ago, when thousands upon thousands of fly-blown fishes lay scattered all over the plain, breeding a pestilential atmosphere that was quite unendurable.

On the return journey to the interior we had hard work between the Hykamgoub and Otjimbengue, the oxen dropping in the yoke from starvation, poverty, and exhaustion. I was obliged to bury 700 lbs. weight of lead, which could only be with difficulty recovered several months afterwards; and we often had to unload some goods and make two trips to bring them up over heavy places.

The usual outspan at Onanies Mouth is a very pretty spot. The white river-bed is here studded with majestic anna trees. These have not the tall, graceful appearance of those we found in the bed of the Schwagoup, but exhibit more outspread branches, the lowest of which grow downwards until they touch the ground. When the periodical floods rush down from the Onanie hills these branches are immersed in the water, and I have observed on them the marks of the drift as high as ten feet above the ground. The surrounding reddish hills, though wild and naked, are thinly sprinkled over with grass, and add to the beauty of the scene by their striking contrast with the bright green foreground of ove'-ngambu bushes and gigantic anna trees, the latter overrun with parasites, which cluster and droop round their trunks and branches.

Owing to the great heat in the river-bed, the open ground above forms the more agreeable resting-place; but even this has its inconvenience in the shape of creeping reptiles and insects. I was bitten by one of the latter, which gave me acute pain at the time: thinking it was only the bite of a scorpion, and would soon wear off, I tried to forget it, but in vain. The swelling spread from the knee into the thigh, and I was soon startled by sharp pains, like the piercing of red-hot needles, flying all over the body, but felt especially in the breast. The swellings and other symptoms became quite alarming. Tobacco-oil and poultices were successively employed, but the pains did not cease, nor did the swelling go down for several days. I have never been able to find out what my tormentor was, but, my friend Mr. Eckard says there is a very venomous small black spider which bit a Namaqua woman six months ago, and she has been quite paralysed ever since, and is not expected to live. Another nuisance from which we suffer in sleeping under the trees is the hairy caterpillar already mentioned.*

This last-named insect is very abundant in the early part of the rainy season: on coming in contact with the flesh, the hairs, which are barbed, detach themselves, and, penetrating into the flesh, create a most intolerable itching. Persons suffering under this torment being frequently unable to resist the inclination to scratch, ulcers are induced and sometimes end fatally.

I observe that the natives here have the same custom that prevails amongst the various Kaffir and Bechuana tribes, of raising monumental piles of stones in memory of their

* See *ante*, p. 285.

departed chiefs. These piles, by added contributions, sometimes attain to very large proportions. The Damara groves can be easily enough distinguished when recent, there being generally a pile of bullocks' horns heaped over them, or suspended between the forks of the nearest tree. But the Namaquas and Berg-Damaras have a more peculiar custom of adding their mite of rock to a pile which they build at every crossing of a road they come to when they go out hunting. This is done to insure them luck. We find many of these conical piles all over the country, and, as nobody lies buried there, they are very probably intended to propitiate the powers that are supposed to aid the hunter. They give distinguishing names to these piles.

From Onanies Mouth we hurried on for Otjimbengue, taking up Mr. and Mrs. Latham as passengers from Wilson's Fountain. I saw in the evening a great many meteors, some of them among the most beautiful I ever observed; most of them fell almost perpendicularly. The moon was shining brightly, but the light it shed was more than once eclipsed by the descent of large meteors, leaving streaming showers of *flame* — not merely sparks — behind them. At about 10.30 P.M., an unusually loud report was heard, like the discharge of a small cannon, or the explosion of a powder magazine; yet no one had fired a gun. Wilson questioned us about it, and we inquired of him. There was no one living or likely to be about here, and we ascribed the report to the meteors. I have before heard them burst with a similarly loud noise.

As we outspanned at midnight at the Koppies, or conical piles of rock, near Otjimbengue, we both heard and felt three very distinct shocks of an earthquake, and, on reaching Otjimbengue next morning, heard that three other shocks had occurred during my absence at the bay. These, taking place so frequently as they do, are thought little or nothing of—the arrival of a wagon is of greater importance. I was told that a few nights since the rumbling of an earthquake roused some of my friends here from their slumbers; they mistook it for the rolling of a wagon, but, when they found their mistake, returned in doors, exclaiming, in answer to the eager inquiry, "Whose wagon?" in a tone indicative of intense disgust, "No wagon at all; only another earthquake!"

On arriving at Otjimbengue, I found some excitement prevailing amongst the white residents there on account of the recent stealing of Wilson's cattle, and the difficulties interposed by the natives, on the plea of guarding against infection, to cattle being driven through their district. Mr. Andersson proposed our going in a body to Zwartboi's, and making a hostile demonstration against that chief, who had promised to do his best towards the recovery of the lost oxen; but differences of opinion were entertained, and no common plan of action was agreed on.

In consequence of the great and almost unprecedented drought, I had been detained a long time in getting my goods and outfit brought up to Otjimbengue; there was also no intercourse or thoroughfare allowed us by the neighbouring tribes, for fear of our introducing the lung-sickness among their cattle. They had sent us warning after warning not to make the attempt, as in so doing we should endanger the loss of our lives and property. I had heard of their shooting down a whole team of oxen, drawing the wagon of one of their own friends, for fear of allowing this scourge to find its way into their herds of healthy cattle, these constituting their sole wealth and source of subsistence. This quarantine, however, was becoming so irksome—my expenses heavy, my stock diminished, and my people and companions disheartened—that I resolved, in spite of all the difficulties that threatened, and although Baines (who was to accompany me) had not yet arrived, that, rather than delay here any longer, I would make an effort to get through ; and, if I succeeded, would send back for Baines, when I should hear of his arrival at Walvisch Bay. I had a difficult undertaking before me—one in which nobody believed I could succeed—that of getting my wagons and party through the Namaqua tribes to the lake and the Zambesi. Even Dr. Holden, who had left his wagon and property at the lake, turned from his course thither in quite a different direction, despairing of ever accomplishing such a journey in the teeth of the opposition threatened by the intervening tribes. But I had the one object, for which I had come thus far, too much at heart to abandon it without, at least, making a trial. I felt thus determined, if possible, to accomplish my purpose, although new, rich, and inviting elephant-fields had been discovered to the north and to the eastward of Ovambo Land. My love of sport, always great, was now blended with a strong desire to carry out a plan that embraced objects of more importance than the mere indulgence of my ruling passion in the chase of elephants, however numerous. I was now more than ever bent upon examining and making known the resources of extensive and, as yet, perfectly unexplored tracts of country between the two oceans, and where, in course of time, numbers of people now leading a life of indigence, as well as much capital open to the demands of any promising enterprise, might find lucrative employment.

In prosecution of this design, I started from Otjimbengue on the 2nd of April, 1861, with inoculated cattle, my object being to get through to Amraal's by means of oxen thus treated, as such cattle were generally supposed to be safe ; but, as it had recently been proved that it was not impossible for an inoculated ox to have the disease, the natives were so prejudiced against even these, that no bribe would induce them to allow me to pass with them. Still, if I could only manage to take these cattle as far as Amraal's, I might purchase others there, and so pull through; but it turned out that, after getting only 75 miles farther inland, I was obliged reluctantly to abandon all hopes of proceeding beyond by means of the forty-five oxen which I had with great difficulty, and at high rates, already procured. The old grievance of my having first introduced the disease into the country was revived, and many unmistakable and far from pleasant insinuations and threats were constantly made within my hearing, that to have my throat cut or to be beaten to death would be but a fitting penalty for one who had brought such a calamity on their country.

The 75 miles already spoken of were not accomplished without severe toil. In our first day's journey from Otjimbengue we found the country was still as parched and bare as it was to the west. We had extreme difficulty in getting over it with our famished cattle, for which we could find neither grass nor water, in a country where we have seen grasses and herbs growing in the greatest exuberance when rains favoured the earth at the proper seasons, and where now the cattle grubbed with their snouts in the dust for the roots of the grasses that had long since been devoured or had crumbled. It took four days to accomplish what, under ordinary circumstances, would have been the work of one; and even this not without inflicting the most painful and apparently barbarous cruelties upon our jaded and suffering oxen.

When, however, we had accomplished the first 30 miles, a glorious contrast in the face of the country burst on our view. The earth, which hitherto had been baked and barren, was here covered with a profusion of soft green grass, bright flowers, and odoriferous herbs—a prospect as exhilarating as it was unexpected. The acacias and mimosas, which before showed a scanty, drooping, withered foliage, shrinking and closing their leaves after 9 A.M., from the effect of the sun's heat from above, and the reflected heat from the red and baked earth beneath, were here expanded by grateful showers to their full freshness and greatest vigour; birds of the gayest plumage sung and chattered in their merriest notes; while moths, bees, butterflies, and beetles, in endless variety and of the most brilliant colours, fluttered and buzzed and boomed about the odoriferous mimosa blossoms, and the red, clustering, parasitical flowers suspended from the boughs—thus making a paradise on the very borders of the sternest sterility. Our cattle snuffed the delightful gale, and seemed to have imbibed another and a new spirit, trotting and frisking before the ponderous and heavily-laden ox-wagons, which they now dragged like a toy, crushing remorselessly whole fields of beautiful wild flowers (*Amaryllidæ*) in their merry career, and snatching occasional mouthsful of long sweet grass which invitingly lined the road-side; indeed, it was quite a holiday to man and beast. The effect of such a change on man, as well as on the lower animals, is of the most invigorating description: one's hopes and energies suddenly revive, and the most sanguine expectations of ultimate success are re-awakened.

Our Damaras here collected an odoriferous herb very like the shamrock trefoil, but possessing a powerful aroma, with which they perfume themselves. Nearly all the native tribes that I am acquainted with have, in common with more civilized beings, a predilection for perfuming themselves, but generally with odours disagreeable to our senses, or unappreciated by the white man. One secret which I have learnt in connection with this is, that the Namaqua and Bush women use these perfumes chiefly to destroy or neutralise the powerful and disagreeable odours secreted by their own persons. These herbs, of which there are a variety, they call bucho.

They carry them, reduced to a powder, in tortoise-shells, and use them when required. Some of the paints and powders used by the Namaqua women are derived from certain lichens, growing on stones, and emitting a strong aromatic odour. They also use a kind of saffron, of which there are two or three varieties found in the country; also a small bulb, and other aromatic plants. They paint their faces with a paste composed of charcoal and fat, on the cheeks and over the eyebrows, and with dry redlead round the eyes. The missionaries have remonstrated with them on the unseemliness and vanity of this practice, which they defend on the ground of its being an antidote for ophthalmia. Dogs and Damaras are equally fond of the odour of the polecat-musk, with which the latter anoint themselves; and the former delight in rolling themselves over the carcase of a dead polecat, or on the spot where a live one has lain, in order to impart the odour to their bodies.

I made a good collection of grasshoppers, silver-spangled caterpillars, and a large edible beetle, which, preserved in spirits, I forwarded to the South African Museum.

At the next outspan, under the shade of some large camel-thorn trees (*Acacia giraffe*) and anna trees, we found ourselves close to a large termite nest. The mound was fully twenty or twenty-five feet high, and about thirty or forty round the base. I regretted much not being prepared to photograph it. I also found, amongst the rock, a very beautiful, though stunted, variety of the China rose, the root of which is esteemed a capital poison for anointing spears and arrows, and the odour of the flowers is also said to be poisonous.

One of my young companions, who had been taking a stroll, returned with a countenance full of importance at some grand discovery he had made. He was no sooner seated than he very confidently assured us, as he seized his journal, preparatory to notifying the fact, that this valley had been covered, at a period not very remote, with an alluvial deposit of at least thirty feet in depth, traces of which, he further assured us, were still visible in numberless pillars of this earthy covering, fifteen to twenty feet high, which were still left standing. When, after enjoying a hearty laugh at his expense, I explained to him how these pillars had been raised by the industrious labours of the termites, he was rather annoyed at the failure of his theory.

At Grey's park we found a most lovely spot of level green sward, dotted with immense patriarchal anna-wood trees, and the gigantic and far-spreading *Acacia giraffe*. This natural park being quite free of underwood, and covered with a carpet of short grass, forms a most delightful camping-place. One is led to wonder at finding such lovely spots almost without an inhabitant, a name, or an owner. In the background, high, and somewhat rugged, purple and grey hills loom above the trees, or gradually fade away in the distance, till they blend with the sky.

Our next trek was over stony ground to Little Barmen, the first of a series of hot mineral springs. Its principal chemical constituents seem to be iron and sulphur. The greatest heat by my thermometer was 143°, while the temperature in the shade stood at 82°. Water boils here at $205\frac{9}{10}°$.

In the evening we reached Great Barmen, or Otjikango. Here we found another hot spring—the greatest heat of the water being 144°, the temperature in the shade 70°, and water boiled at $205\frac{1}{10}°$. The geological features here are peculiarly interesting: the columnar peaks, apparently of gneiss, and the long walls of the same formation at the back of the missionaries' dwelling-house, look like natural fortifications. Some date trees, planted by Mr. Hahn, were in full bearing; tobacco, onions, potatoes, beets, water-melons, pumpkins, and other vegetables, flourished beautifully, and wheat and barley are also grown. I would gladly have preserved a memorial of so interesting a spot, but my cameras were broken, having been upset by the wind. I delayed here some hours to make a new camera out of the wreck of the old one, and, as soon as finished, brought it into use, but the wind blew so strong that I found it impossible to get a picture: at last, camera, chemicals, and all, were again upset, and I lost another nitrate of silver bath, some varnish, and a half-gallon bottle of glacial acetic acid, which spread over and destroyed nearly all the medicines and chemicals I possessed. These were things that, unfortunately, I could not replace.

Groups of native (Damara) huts were scattered round the station, and men, women, and children assembled round the wagons to look at some natural curiosity which, they understood, I had introduced into their country. Their eager looks and exclamations were, however, all directed to my servant, Laing, whose fiery-red hair and whiskers, such as were never seen here before, astonished the natives out of all propriety.

I availed myself here of the services of a number of half-starved native children, whom I sent out, each with a tin box, to collect insects, while I kept a large boiler of meat constantly on the fire, from the contents of which they were rewarded. The children soon became very expert at their work, and I can recommend this as a very expeditious way of collecting specimens.

Four hours beyond Barmen, in a southerly direction, we struck the Schwagoup river, at the junction of its two main streams. Here our progress was for a time arrested by the fury of the stream, which came roaring down in a muddy torrent, overwhelming trees, and sweeping down logs and islands of drift. In a few hours, however, the river, owing to the quick drainage of these periodic streams, showed a depth of only three feet when we made the attempt to cross. The first wagon had no sooner entered the river than both oxen and vehicle sank in the quicksands, and were nearly overwhelmed in the mire—both the wagons in great danger of being overturned by the force of the stream. Two teams were at once put before the first wagon, while the other was left to take its chance: after floundering about for more than two hours we emerged in safety on the opposite bank, and three hours farther on halted for the night.

Early in the morning of the 12th of April one hour's travel brought us to the Dabby-bush, or Quarantine Camp. Meeting with nobody there to oppose our progress, we marched on two hours farther, to a place called Otjehorongo, and then, not wishing to run any unnecessary risk of quarrelling with the native Africaners, sent back our cattle to the Quarantine Camp. It was well we did this, for the people had given notice everywhere that they would shoot all cattle from the tainted country; and a few moments after a party arrived, who informed us that we had already exceeded our limit by five miles, and were rather astonished at our rashness.

Leaving my wagons and companions at Otjehorongo, I proceeded next day to Windhoek, about 35 miles, in order to have a personal interview with the chiefs and people. I was well received by old Jaager Africaner, who acted as paramount chief during his brother Jonker's absence; but the people in general cast furtive and uneasy glances at me, as if fearful that I had brought the contagion in my pocket. I became rather alarmed at the non-arrival of two Damaras whom I had outstripped, and who carried my blankets and pistol. However, both Jaager, who spoke Dutch, and his wife were very civil. He offered me a clean wagon to sleep in, in preference to a hut, in which he feared I should find no rest, in consequence, he said, of the innumerable insects which infested their houses, and which he facetiously and delicately called locusts. Being tired, however, I stretched myself on the hard boards of the wagon, spending a cold and wretched night, during which I missed the comforts of my own wagon very much.

I had laid my case before the chief and his councillors during the evening, and after a long palaver, which foreboded no good luck to me, they retired. Next morning I exercised my greatest eloquence, and all the ingenuity I could command, and having at length touched them in the tenderest point, their avarice, I succeeded by adroitly putting the question, not in the form of a petition, " Will you let me go through ?" but " I am going through to Amraal's country, and have come to make you an offer for the loan of two teams of oxen before I send mine away altogether. If you do not care about the job, I shall leave my wagons, and proceed inland to Amraal, only 200 miles, on foot, and can then reckon upon obtaining the loan of healthy cattle to fetch my wagons."

At length I gained my point, not until after a good many palavers, or *raads*, in which I laboured under the great disadvantage of being unable to understand all that was said, and some friendly gossip with the women, in which they were well informed of many nice things in the wagons, in order to excite their curiosity; nor was I at all scrupulous about promising a present of a dress or a shawl here and there, in case I should be passing their way. They, therefore, were not backward in begging that I would come past their village ; and thus the battle was won, and the loan of thirty oxen promised. It was not my interest to stick at trifles, and the price promised to be paid was in itself sufficient to have purchased an equal number of oxen. But the point was gained, and the road which had been so long shut to travellers was successfully opened. It took two whole days of palaver to bring the affair to this happy termination, during which time many *raads* were held, and there was no lack of hard things said at me or about me.

The women, rejoicing in the misfortunes of poor travellers, which compelled them to hire oxen upon their own terms, now struck up a dance for my edification, which for unseemliness can hardly be equalled by any I have ever seen. I made a rough sketch of this scene, which I kept till Baines joined me, when I got him to make a drawing from it.*

The original Hottentot, or Namaqua dance, somewhat resembles that of the Bechuana and Bushmen, a monotonous singing, stamping, and clapping of hands, together with some uncouth gestures. The dance got up for my edification is chiefly memorable for the extreme inelegance of its figures, and the vulgar attitudes which the women delight to place themselves in. About a dozen of these followed one another in a circle, in a hop-skip-and-jump sort of manner, singing, clapping their hands, and lifting one leg high in the air, throwing a bitter melon through under it for the next following to catch, and so on. The vulgarity and unseemliness of this dance are much increased by the almost nude forms of the women, whose enormous pendant bosoms dangle about in a most disgusting manner. The men pride themselves now-a-days in playing a sort of reel, called "Valshe Rivier," which they have learned from the colonial Hottentots, to which they dance, and seem never to tire of it.

It having been arranged that all my wagon gear was to be burnt or destroyed, and that the native oxen were to fetch my wagons from Otjehorongo, I hired horses from Hendrik Jaager, son of old Jaager, a tall smart-looking fellow, with a defect in one eye, who gave me a guide to take me a short cut through the veldt. But, as he candidly told me, he had purposely given me one whom I could not understand, and who could not understand me, in order that we might not conspire and humbug him. I found what Hendrik said only

* This drawing has, I find, been published amongst the illustrations given in Mr. Baines's own work.—J. C.

too true, for this fellow's whole Dutch vocabulary consisted of two of the vilest oaths in that language, which he vented alternately as his hat blew off or when his horse stumbled. He was one of the most ruffianly, cut-throat-looking beings I ever beheld.

On returning to the wagons, I met the welcome arrival of an express from Walvisch Bay, bringing letters from the Cape.

CHAPTER XVIII.

Journey towards the Lake—Mineral Springs of Windhoek—Difficulties of Progress—The Namaqua Chief Jan Jonker and his Family—Elephant's River—Droogevley—Adullam—Further Impediments—Edible Roots—Gobabie, or Elephant's Fountain—Amraal—A Novel Mode of Duelling—Lions and their Victims—Amraal again—Want of constituted Authority—Commerce with the Cape—Namaqua Habits, Manners, and Superstitions.

HAVING made arrangements with my brother for the conveyance of our boats and provisions as far as the lake, or to the Zambesi, if necessary, we parted company. He had been very unsuccessful in his search after elephants, had lost some 500 head of cattle by the murrain, and had not decided what course he should next pursue, so that I easily persuaded him to follow us up as soon as he could arrange his affairs.

In strolling about the hills near Otjehorongo after koodoos, I fell in with strong surface indications of copper ore, on a conical hill about one mile to the north, and brought away some rich specimens.

Jaager's cattle at length arrived, and the sentence of burning all our gear and tackle was commuted to simply painting or tarring; so we got on our way at last, and the cattle, being strong and large, jogged merrily along while the road was fair; but soon we had to cross a sandy ravine, in the bed of which we stuck as if we had grown there. There we broke the pole of one wagon, but were under way again next day. From Otjehorongo two days through a hilly country, dotted with mimosas, &c., and river-beds, whose green banks were abundantly clothed with tall and shady trees, chiefly acacias and mimosas, brought us to Windhoek, where a series of hot mineral springs bubble out of the side of a high sloping hill. These fountains being situated at the foot of the town, and not very far apart, all flow down the slope to the westward, and forming a junction lower down the valley, by which time the waters are already cool, flow northward into the Schwagoup, just below Eikhams, or Warm-water, another mineral hot spring, and the site of a deserted mission station. The warmest of these springs I found to be 162°, though there are some farther south which the natives assure me are much hotter, and fit to boil flesh in. The water is, however, quite hot enough to be unbearable; birds and wild animals venturing in are scalded to death, and dogs and jackals in particular.

The weather here was extremely cold for this season of the year. The thermometer was scarcely ever higher than 40° during the day, and wind and clouds prevailed. The contrast between this sort of weather and what we experienced during the latter weeks of March may be easily imagined when I state that in the coolest shade, at mid-day, it was never less than 95°, but more frequently at 112°.

I delayed here a few days for the purpose of trade, and then prepared to depart; but, to my great annoyance (I cannot say astonishment, because I always expect these people to act dishonestly), I was informed we could not proceed, as we had not paid, in the first place, the hire of the oxen thus far, nor had I arranged about the price from hence onwards. The oxen, meanwhile, had been removed, so that, as they knew very well, we could not get away. They had calculated that I would, in a fit of despair, succumb to any terms they chose to propose; but in this they were mistaken. I went to my wagon and quietly played at chess, or, when the weather would permit, amused myself with photography, or in collecting and preserving botanical specimens, or taking observations, and not taking the slightest notice of those who were constantly sent down, as I supposed, to see how the wind blew, or, in other words, "what our plans were."

At length, finding from my people that I was determined to proceed to Amraal's in search of assistance, they came to hold another raad, or great talk, but not in the same violent strain as before. They acknowledged their wilful error in having purposely misunderstood the terms of our agreement, and were informed that they would now get nothing for what they had done if they did not fulfil their contract, and lend the oxen as far as agreed on. They softened down wonderfully upon this, maintaining, however, that they were very honest, innocent people; that certainly some of them were poor, and wished only to see what they were to get. As their tone was this time quite humble, and unaccompanied by threats or menaces, I agreed to pay them in advance, but with the condition that the oxen should be first handed over to me, and should remain with me as my property until the due fulfillment of the stated work, while, in case of any after-treachery, I should be at liberty to shoot every ox. Delighted to be rid of such company, I now lightened my wagon of some of its weight, and took a few photographs before leaving, which so pleased and interested them that they had neither time nor inclination for throwing any further impediments in our way.

We had not, however, proceeded far before we encountered a number of Namaquas (Africaners) with long sticks, running to meet and drive us back, saying we were conveying the contagion, and, in spite of our remonstrances and passport from their chief, they disputed his authority, and dared us to proceed. They now became furious, yelling and screeching in a very ridiculous manner, and loading us with the most fearful imprecations. We halted at the base of the Awass mountains, having been requested to wait there, as numbers of the people at Eikhams intended following us with cattle to trade; but, having now given them ample time for that purpose, and feeling still insecure until a little farther off, we started again at midnight, and travelled a considerable distance before unyoking. Here we felt as if in quite another atmosphere, pure, fresh, and bracing, and ready even to give battle if necessary. We were now in the highest part of the mountain, and directly in the road. Here water boiled at $199\frac{1}{2}°$. Some portions of the hilly range to the southward are probably 500 feet higher.

Messengers overtook us to say that several people were coming after us to trade; but they had been already such a source of annoyance that we felt no regret in giving them a long and fruitless chase after us. Being now rid of these people, I could with safety leave the wagons for a stroll among the hills running parallel with the road. The country being favourable for small game, my particular delight, I succeeded in bagging three or four little steinboks every day, and perhaps also a brace or two of fine plump bustards or pouws. One day I killed, on the Elephant river, no less than eleven fine antelopes, among which were three pallah-bucks. These, with two large pouws, formed an excellent bag for such a country as this. Guinea-fowls and red-winged partridges are also common here; and three large snakes, a puff-adder, yellow snake, and cobra, were killed.

On Saturday we halted for half an hour to trade with old Jan Jonker, bartering, in this short space, upwards of thirty head of cattle and about a hundred sheep and goats. The old man himself stood by to see that justice was done to us as well as to his people. They were obliged to be content with what he thought sufficient in exchange for an ox, and they seemed to respect his authority more than I have ever seen Namaqua subjects do before. But there is a reason for this, he being well known as a stern and unyielding disciplinarian—so uncompromising, indeed, that it is said of him that, when his own daughter had disgraced herself by becoming pregnant before marriage, he, after lecturing her severely in a most fatherly way, became the executioner himself, and shot her. In like manner, also, he shot his favourite son, to whom he had given an order to fetch in some refractory subject, in which the youth failed. But old Jan Jonker, determined that his orders should be obeyed, sent him again, warning him at the same time that if he had himself to fetch this man he would punish his son with death. The son again failed, through lack of spirit: Jan Jonker brought in the refractory subject, and punished his own son with death.

I regretted very much the not being able to photograph these people, for Jan Jonker's wife, who had been conveyed here in a bullock-wagon, is a woman of enormous proportions. Her arms were fully equal in size to my waist, and her body proportionately large; her breasts were of most unseemly dimensions, something wonderful to behold. She was so fat as to be almost incapable of voluntary locomotion, and totally blind, having been so from childhood; and her daughter, who was with her, has likewise been blind from infancy. Her case confirms the common observation, that the loss of one faculty or organ improves the acuteness of others. This girl is remarkably clever at working with beads, and judging of their respective colours by their touch alone. She can select any particular colour from amongst a dozen others, and sew them into all sorts of agreeable and fanciful patterns—generally on a black sheep-skin garment. This article, the bridegroom's present to his bride on her wedding-day, is called the "brock-kaross;" and, in order to put her in possession of such a skin, the bridegroom is expected always to give her, before marriage, a present of a black sheep. It is amusing to hear old Mrs. Jonker and her daughter discussing the colours of different things, and beads in particular.

On Saturday evening we trekked down the Quaiep, or Elephant river, passing the nests of two families of social grosbeaks, which I have known for many years. The old gnarled thorn trees (*Acacia giraffe*), on which the nests were built, have been completely killed by their enormous weight, nothing but the trunks and supporting branches remaining erect. These nests, so completely packed, must sometimes exceed a ton in weight. They are generally inhabited likewise by snakes (tree-climbers), which live from year to year on the birds and their eggs, and increase also in proportion to the birds.

Having found some rain-water in a pool of the river, we stayed over the Sabbath, to give our men and cattle rest. Hottentots, and young believers generally, are very particular also on this point of observance, but they spend the day according to their individual tastes. A pharisaic sort of person in my service, who was one of the principal men on a mission station, and an elder, could not think of travelling on a Sunday, but was, nevertheless, out the greater part of the day hunting, although there was no lack of meat in our larder; nor did I ever see him at his Sunday devotions.

Our next start was for Droogevley, about 20 miles west of Wittvley. We found a great many springboks at this place. We shot as many as was necessary for our people and dogs, and stayed over a day to trade with some outlying natives. At Droogevley (dry vley) were stationed a troop of 700 very fine cattle, and several hundred sheep and goats, which, together with about 4000 lbs. of ivory, had been bartered from the natives in the short space of two months. The goods given in exchange consist of general merchandise, but principally gaudy-patterned clothing, both for men and women; but guns and ammunition generally are the most in demand, and realise a good profit. They are, however, very particular in their selection of guns, of which they are good judges, and have in their possession several very valuable pieces. A few years ago a long flint rifle, or smooth-bore, with a piece of white ivory on the end of the stock, near the muzzle, used to be the fashionable article; but since the introduction of percussion rifles, the former have gradually been discarded by those who could afford it for the more handy and handsome little weapons in use with us, and the musket and starloops fall into the hands of the servants or the poorer classes. Fustians, ready-made clothing, prints, shawls, calicoes, handkerchiefs (silk and cotton), knives, tinder-boxes, large-brimmed felt hats, shoes, tools, &c., are the chief articles of trade.

Leaving Droogevley on Tuesday for Wittvley, otherwise Adullam, we slept half way between the two. Next morning, with a Damara to carry my gun, I walked in advance of the wagons to the village, on nearing which I was met by a party of Namaquas, vociferating at the utmost pitch of their voices, and menacing me with their long sticks in a very threatening way. I was at a loss to comprehend the meaning of this strange movement. After some delay, an interpreter was brought, by whose aid they gave me to understand that I could not be allowed to enter their village, either in person or with my wagon, as it was the chief Amraal's positive orders, and applied to anybody coming from the lung-sick country. Now, I had everywhere heard that lung-sickness had already broken out here, having been introduced by a party of these same vagabonds, who had been cattle-lifting in Ovambo Land, and I was most anxious not to delay here, and endanger the safety of Jaager's healthy cattle. I knew Amraal to be a respectable and reasonable old man; and although I have no doubt I might have forced my way through without much danger, for these men are as great cowards as they are rascals, I preferred a present delay, as I wished to conciliate old Amraal, explain matters to himself, and, if not permitted to pass through his town, to obtain guides to take me through the veldt to the north, in which case no responsibility would rest with me in the event of lung-sickness breaking out among the bartered cattle.

But now came a messenger from the old chief himself, forbidding the advance of myself or any of the party. This I looked upon as the mere invention of these rebel inhabitants of Adullam, who I very well knew had no respect for the authority of Amraal, nor cared so much for his interest as to exert themselves on his behalf. Indeed, I believe, on the contrary, if they could have sent the contagion there they would have done so, in order to spite the missionaries and the church-goers at Gobabies, whose influence was beginning to be strongly felt in these parts, and whom they heartily hated for trying to put a stop to murder and cattle-lifting, they themselves being the chief culprits, and expecting that some among them would soon feel the weight of the new law instituted by Amraal.

Messenger followed messenger, all of whom I affected to disbelieve, but at length a man from near Amraal's person, and whom I knew could be trusted, brought the same discouraging news, and orders to return, both verbal and written. On inquiring further into this strange conduct, I ascertained that the same evil spirit who had before tried to mar our progress had been again at work. The question now was, what was to be done. The Adullam people, drunk and uproarious, laughed and jeered at us; but their sneers only had the effect of inspiring me with a fresh determination, and, in answer to their taunts, I vowed that death alone should now prevent me from getting through to the lake. As these people were never sober, and consequently always troublesome, insisting upon trade whether we would or not, I determined to send for my oxen, take the wagons back a few miles, so as to have my people and property beyond their reach, and then, taking a boy to carry my gun, proceed to Amraal's on foot, and there meet my accuser face to face. But the Wittvley people, discovering my intention, determined to thwart it, and one portly individual, who professed, as many others did, to be chief, and quite independent of Amraal, imperatively forbade the removal of the wagons, threatening to attack and destroy every one of us if we disobeyed his orders. This fellow was near coming into a personal conflict with me, but, warned by my determined bearing, and by the proximity of my revolver, which I carried by my side, the old rascal at length took to his heels, covering his retreat with a furious threat, which I felt was never to be performed, of going to fetch all his "manschappen" (all his men) to stop us. Thus the old villain left us, gaining courage at every step as he shuffled off, and increased his distance, half screaming, half crying with rage and fear, and uttering the vilest curses he could call to mind.

Meanwhile my people were not idle; the wagons were inspanned and started, while I lagged behind to see that no attempt was made by the persons in our rear. Having seen the wagons safely to their outspan, I left, under cover of night, for Amraal's, and slept under a mogonono bush, a little beyond Adullam, on a neighbouring height. After a bitter cold night, we started at daybreak, and marched briskly till noon, when I lay down under the shade of a tree, while my boy grubbed up and roasted some bulbs and roots. The latter, round, like a small potato, but harder, is found at the root of a small, red-flowered, upright plant. I collected some of the seeds and roots of these bulbs, and several other edible plants, which are more plentiful on this plain leading to Amraal's than anywhere else I know of, and several kinds of edible gourds (*Cucumberi*) hang from the bushes, or lay ripening on the ground. Those climbing up the bushes are generally of a scarlet and green colour, with white freckles; the commonest of those on the ground have rows of soft prickles, longitudinally arranged, and are of a yellowish colour, with brown freckles or blotches. I collected, also, several kinds of seeds of convolvuli, and the gropler plant.

I see the country under very favourable circumstances this year. The grass stands very dense, just in its prime, and fully three feet high. A large quantity might easily be mowed, and stowed away as provision for hard times, when the fields get burnt up. Wild flowers of various kinds and colours are abundantly scattered throughout the plain, intermixed with the rich and flowing grasses, which look like miles of ripening corn.

After half an hour's rest, and a good feast of roots and water, we started again, and, walking briskly and without another halt, reached Elephant's Fountain, called by the Namaquas Gobabies, and by the Damaras Epako, having walked 35 miles since daylight. Tired and exhausted as I was, I was gratified at the kind welcome, good cheer, and warm bed, provided by Mrs. Krapohl at the mission station, and retired to rest soon after.

After a good night's rest I visited, on Sunday, the missionary church, a well-ventilated reed shed, in which were assembled about 300 men, women, and children. Mr. Weber officiated, and preached a very sensible, homely sermon in Dutch, quite suited to the understanding of his hearers, illustrating his subject throughout with appropriate similes, and calculated to make an impression upon the minds of the heathen groping for truth in the dark.

The mission house at Gobabies is a long, flat-roofed building divided into two sections, one half occupied by

Mr. Weber's family, and the other half by Mr. Krapohl's. Near the house, on the south side, a pretty fountain bubbles out between a clump of reeds, and the water, flowing westward through a limestone bed for a quarter of a mile, falls into the Swart Noosop river, which is densely bordered with fine acacias. The Swart (or Black) Noosop is joined by the White Noosop about 45 miles south of this station, and, after a long southwardly course, ultimately joins the Orange river.

At the back of the dwelling-houses, which have some detached offices, stand the walls of a substantial new brick church. This building, which is intended to accommodate 500 persons, is built in the form of a cross, seventy-five feet by twenty-five, the transepts being fifty feet by twenty-five. At the head of the cross there is to be a vestry. The church, when finished, will have a small steeple, with a bell. The windows and doors are all arched. It is being built by the two missionaries, Weber and Krapohl, assisted by all the Namaquas—men, women, children, and servants. The good old chief, Amraal himself, though ninety years of age, is never idle, being, indeed, one of the most industrious. It is to his consistent and exemplary conduct and evident sincerity of feeling that the missionaries here owe a great deal of their success, and travellers and traders are indebted for a very fair amount of peace and good order.

Next morning I had an interview with Amraal, and, after contradicting the false statements which had been made concerning my cattle, I requested the chief to obtain the opinion of his councillors, stating that I did not wish to force my way through their villages or herds, but, on the contrary, would rather be guided through the veldt past their villages, to prevent any possible after-blame that may arise from such an unforeseen accident as lung-sickness really breaking out, although we had no reason to anticipate it. At the same time, I assured him that I was determined to accomplish my design, even if I had to make a circuit of 500 miles.

A consultation accordingly was held, and it was settled that we were to come straight to this place; but, in order to make matters doubly sure, it was arranged that the oxen and harness of the chief's son, Lamert, were to be sent to fetch my wagons, and that Jaager's oxen were to be immediately sent back. This suited me exactly, as I wished to write to my brother to bring up Baines and the boats, if they had arrived, and also a load of goods which I was expecting, to replace those lost in the Canute.

Having gained my point I returned to my wagons, after walking about 77 or 80 miles in two days. We had no end of trouble with the drunken rabble at Wittvley, who, enraged at my having succeeded with the chief, swore that I should, nevertheless, not proceed. It would, however, be tedious to relate in detail the annoyances we experienced in the accomplishment of our purpose.

We reached Kobi-Kobis on Saturday evening. In our anxiety to get away from these people we should not have scrupled to travel on the Sabbath, but a message came from Amraal forbidding it, for fear of provoking the wrath of God to afflict them with lung-sickness. It was now insisted that we should produce every particle of hide or leather about the wagons, to be destroyed by fire, so that they might supply us with others of their own making; but this I would not agree to. Being pretty well knocked up with travel, excitement, and anxiety, and feeling, besides, very unwell, I should have been very glad to get a day's rest in the wagon; but we were now annoyed by a visit from a most turbulent fellow, named Adam Flerkmuis, who picked up a quarrel with my servant, John, which ended in a challenge to fight a duel with the ruffian. But when the arrangements were being made as to time, place, and distance, it appeared that the duel was to consist in their firing at a target at 100 yards, and that the worst marksman should be considered as shot and vanquished. John, however, assured his antagonist that in his country the practice was to fire at each other; whereon the affair was suffered to drop.

On Monday we left Kobi-Kobis, on the White Noosop, and travelled about twelve hours to Elephant's Fountain; on nearing which place we were met by a mounted party forbidding the approach of our loose cattle, now numbering about fifty, and I was obliged to ride on to expostulate, or make some arrangement concerning them. The chief said he had promised to let them pass, but he could not let them drink for some days, for fear of infecting his fountains. This I did not care so much about, as he promised to let me drive them on some days in advance, keeping them from drinking at his waters. Stauffir volunteered to go with them, but he returned the next day; and Barry had gone with the missionary wagon and oxen, which I had engaged to fetch up Baines and the boats from Walvisch Bay. It was arranged that this wagon was to remain within the borders of the healthy country, while Baines and the boats were being brought up by my brother's wagons to the Quarantine Camp.

In parting with my cattle I requested Amraal to permit me to send them on to the extreme eastern boundary of his country at Elephant's Kloof, but he dissuaded me from this project, assuring me that the lions had of late become so daring that no human being could live there; the Damaras and Bushmen who had escaped their ferocity had been obliged to remove to a district north-east of this place. The cowardice shown by these poor people had of late made the lions so bold that nothing but human flesh seemed to satisfy them; nor did their huts, fires, or fences afford them the slightest protection. Some of Amraal's people, who were returning the other day from a giraffe-hunt, were assailed by a troop of these daring animals in open daylight. The lions sprang upon the pack-oxen, who ran wildly about under the weight of their rough jockeys, plunging madly, until, fortunately, they had disencumbered themselves of their bundles of meat as well as their rude riders, the lions contenting themselves, after having a few shots fired at them, with the meat they had seized. Another party of these hunters, the same day, came upon the carcase of a Damara recently killed and partially eaten; and every night this same party were kept awake, or had to make circular fires round them, leaving their dogs to fight off the brutes until daylight. So changeable and uncertain is the character of the lion that, in some districts, by daylight he is timid as a mouse, and will scarce venture to attack man even by stealth, or by night; but when he comes upon a famished or mean-spirited race, he keeps near a village, and treats its inhabitants as though they were his flock of cattle, killing them as hunger urges. A hungry lion is a most daring animal; there is nothing that he will not dare in broad daylight, and in the most impudent manner, driving you off from your own game, or following you up in open ground, under every disadvantage to himself. But such cases are rare, and they are generally either driven to it by hunger, past success, or a keen relish for human above all other flesh. The general disposition of a lion—like that of all other animals—is to avoid man, and the districts which he haunts in South Africa being as yet abundantly stocked with game, man seldom becomes his victim.*

* I have seen several instances amongst the Boers and natives of wounds inflicted by the lion and the leopard, or panther, breaking out periodically with a discharge of matter. The sufferers from such wounds assert that they feel very acute pains previous to any change in the weather. Mr. Stewardson, whom I knew in Damara Land, informed me of his being troubled with an irruption twice during the year ever since he was bitten, many years ago. But however venomous the tooth of the lion may be, I

I had a long and interesting conversation with old Amraal. It is astonishing what a retentive memory he has of things that occurred very long ago. He remembers the taking of the Cape the first time by the English, and was there when it was captured by them the second time. He has been a very temperate man, and that must account for his vigour at this advanced age. From his account his grandfather must have been a chief of great importance. He lived where the town of Worcester in the colony now stands, that district belonging to him. Amraal tells me that when his grandfather came into power a great meeting was called of all the Hottentot tribes for the purpose of deciding upon the choice of a right royal wife from amongst the most noble families of all the land, and this meeting, he assures me, lasted a whole year. The derivation of the names of some of the most notable families, Namaqua and Griqua, does not say much for their descent. Thus Adam Kok's father or grandfather was a cook, and so of others. Amraal tells me that his eldest son, Lamert, was two years old at the final taking of the Cape by the English.

The chief has issued an ordinance forbidding his people from making or indulging in the use of intoxicating drinks. This might be considered, according to the ideas prevalent in more civilized states, as a very arbitrary measure; but these people are like children, unfit to have unlimited liberty, and therefore, I think, they are very properly treated in some respects as such. It is pleasing to see that Amraal, notwith-

---

believe there is much more virulence in the danger to be apprehended from the bite of the panther or leopard—a belief arising from several instances which have come under my own observation in Natal. One occurred to a respectable farmer named Meyer, who was bitten by a wounded panther in the wrist, and, though he had every attention paid him, he died after a lingering illness of more than a year. I have known other cases of death ensuing from similar wounds inflicted by the Natal panther.

standing his great age, possesses much influence over his people, though yet not quite as much as one could wish. Were he a younger man, his rule might, perhaps, be more effective.

Amraal was at present concerting measures with Zwartboi and other chiefs to put a stop to the stealing of cattle from the Damaras or Ovambo. These more enlightened men, who have generally missionaries settled among their people, are anxious to form an union for the suppression of all kinds of aggressions by the lawless and abandoned men who are gradually deserting their respective tribes, forming a combination of their own at Adullam, or with Piet Nanip on the Visch river, and at other places, where they can practise their villanies without restraint. The ruffians are, consequently, much incensed against the more civilized chiefs, and their advisers, the missionaries. If the latter, however, could have the countenance of the British government in their good designs, or could get any sort of encouragement in the matter, I feel confident that they would feel so flattered at the honour thus done them, that they would effectually carry out their just and humane undertaking; and their tasks would be all the easier were it known by these vagabonds that the proceedings of the respectable tribes of Namaquas are honoured with the approbation of so great a power. A complimentary notification of this kind, addressed by the colonial authorities to the principal independent chiefs, would have the happiest effects, not only as regards the state of affairs among the natives, but as tending very materially to the general safety of commercial enterprise, and to securing a greater degree of respect and civility to travellers in these lands.

The trade of this country ought now, I should think, to be held of sufficient importance to the colony to warrant the appointment of a British resident somewhere in a central position among the tribes, to whom all manner of disputes between either natives or white men might be referred for arbitration. Such an authority would, I am sure, be respected by all classes, and I believe that the security thus given to traders and others would indirectly increase the revenue of the colony sufficiently to pay the expenses attending its maintenance. Besides the ten or twelve thousand head of cattle, and as many or more sheep and goats, which, in dear times, yearly find their way overland into the colony, and the quantity of ivory, now upwards of 10,000 lbs. per annum, and always on the increase, there are hides, ostrich feathers, &c.; copper ore is likely soon to become a very important article of export from this country; to say nothing of the imports of sugar, coffee, and all descriptions of manufactured articles, all which trade is in constant danger of being interrupted by the least disturbance among the tribes. I feel convinced that even a letter addressed to one or more of the most respectable chiefs having missionaries with them, authorizing them jointly to act as arbitrators in questions which might arise, would have a very beneficial effect, and would be all that is required to give security, not alone to the poorer pastoral tribes, but also to the missionary, the trader, and the traveller. In any case of difficulty or importance reference might be made to the government at the Cape, as there is a regular postal communication between this colony and the mission stations across the Orange river.

The merchandise finding its way into these parts is chiefly brought from Walvisch Bay, Angra Pequena, Hondiklip, or from the colony directly overland. The presence even of missionaries, where the natives respect them, is generally found a security for traders at a station.

The Namaquas buy very freely; indeed, so fond are they of dress and show, that many among them are ready to impoverish themselves by the prodigal manner in which they sometimes dispose of their cattle, and even of their breeding stock. The ivory trade, in which we have been the pioneers, has opened to them a new source of profit, and in this the native chiefs and hunters have already made a very fair beginning; but at present their guns are hardly fit for the work, and they require to be at peace with their neighbours at home, in order to carry out the long hunting expeditions, which would occupy them from six to nine months of every year. The success they have had already, although not what would be thought considerable by us, is sufficiently encouraging to stimulate their perseverance; and as hunting is congenial to their habits, and the only toil they are disposed to undertake, it may be taken for granted they will soon improve on their present imperfect mode of pursuing the game.

To say that the Namaquas are fond of strong drink conveys but a very poor idea of their passion for intoxicating liquors. Nothing in the world, I believe, would stop them from stealing it if they knew where to get it, and when they do get it they believe the only object for which Providence intended it was to make people drunk; and this they are not long in accomplishing in their own behalf, whenever possible. Although all descriptions of intoxicating liquors are strictly forbidden by the chief, I find that the people here have as great a craving for them as ever. The smell of the collodion in my chemical box is quite irresistible, and several of their principal men, who had signed the pledge, betrayed their weakness by a sly request for a dram. In the absence of spirituous liquors their inquiry is always for tea or coffee. I have been amused to see them sitting round a hot spring, sipping the water as a substitute for tea. They will even make earnest demands for pepper, and indulge in the luxury of pepper-tea as a substitute for brandy.

Owing partly to their successful hunt last year, the Namaquas are very much better dressed than formerly, especially the women, who, instead of their native skin or leather garments, now wear European clothing; and, owing no doubt to missionary example and influence, they are much less ragged and dirty than before. They are, however, as great beggars as ever, the demand for tea and coffee being incessant. We had been without bread, rice, or any other vegetable food for some time, and the kind missionary's wife having sent us a very welcome loaf, it was seen by some of the people, who instantly ran off to inform Lamert, the chief's son and heir, and the latter immediately sent to demand his share of the present.

Ophthalmia is a great curse in this place, as it is everywhere within the tropics; the missionary families suffer severely from it. I cured a great many cases among the natives, but some are constitutional and stubborn. Leprosy, cancer, and other loathsome diseases, consequent upon a life of debauchery and intemperance, are very prevalent here; and the demand for vitriol, mercury, and alum is very great.

It is becoming painfully visible to those interested in the progress and propriety of the Namaquas that their cattle are fast diminishing, and many are already impoverished before they have taken to agricultural habits, even to the extent of planting of their own tobacco. They are seriously meditating a hunt some hundreds of miles north to the sources, as they say, of some river flowing south-east into Lake Ngami; and if they find the country suited to agriculture, they will all remove thither. The old chief Amraal has, however, an objection to this movement, as he fears that if they thus come into contiguity with tribes possessing cattle, some of his disaffected or impoverished subjects will take to plunder. Other tribes from the south, he fears, will in such an event also join them, but for no other object than cattle-lifting. Amraal, therefore, means to accompany the hunt, and see if he cannot find a suitable place for settlement far away from any such temptation.

A few of the Namaquas, at the instance of the missionaries, are building brick houses. These are very indifferent specimens of masonry, and most of the people still inhabit their portable mat houses, ranged, like those of the Korannas, all round the fountain, and these are generally surrounded by the huts of their Damaras and servants, which are of similar shape, but plastered with cow-dung, and are somewhat smaller. As the cattle are permitted to lie about all night amongst these huts, the ground in time comes to be covered with a very thick crust of dung, for which they have no use.

The church bell here being cracked, a koodoo's horn is blown to call the inhabitants to prayers or meeting. They seem rather fond of going to church, and soon after the signal are to be seen coming from all directions, bearing little cross-legged veldt-stoeljies over their shoulders, the women wearing prints and shawls of the gaudiest of colours. In church a good many behave with becoming decorum; others are in the habit of going to sleep during the service: and in order to check this propensity a *bode* or messenger, otherwise called the *corporaal*, moves about from one end of the building to the other, and, on finding a delinquent, arouses him or her, without distinction of person, by a most unmerciful box on the ears. The women, generally, have their heads bound up with three handkerchiefs of different colours, so adjusted as to show the colour of each, and one of these kerchiefs is generally of black silk. The men have a great predilection for long-tailed dress coats, " bell-toppers," and cast-off scarlet uniforms. Their usual head-dress is a broad-brimmed coarse felt hat of colonial manufacture; but forage caps, shakoes, and even helmets, are to be found.

THE NAMAQUA KRAAL.

Although many of the older members of the congregation at Gobabies are far from being steadfast in their professions, the missionaries have great hopes of the rising generation, who are much more consistent Christians than the older converts, from many of whom it seems to be a task of great difficulty thoroughly to eradicate their heathenish propensities. As an instance of this, Mr. Weber assures me that at Hoachanas, an older station than this, some distance to the south, the young people stand firm in their principles, and have already quite the upper hand in managing all the affairs of the tribe, so that the missionary work there has passed the Rubicon. Here also the young exhibit a strong desire to learn. If lung-sickness, however, should unfortunately break out, the tribe will be dispersed and impoverished; and, having no other means of existence than their flocks and herds, it is to be feared that their Christianity will suffer much from their dispersion: and I cannot understand why the missionary does not rather encourage a removal to some more fertile region, where they may plough and sow.

As there seems but little chance of the Namaquas ever becoming a great or powerful nation, or that their language will be found to exist beyond another century or two, it is satisfactory to see that there are men to be found who are writing and printing books in this very singular language, with its awkward clicks. When the Hottentot races shall have become absorbed, or otherwise lost sight of, in the great influx of Europeans into South Africa, or by the rapid progress of other tribes in comparison with their own fast-thinning ranks, these books will be looked upon with peculiar interest, as curious memorials of the language of an extinct and barbarous race.

The Namaquas are in many respects a strange people, and one hardly knows what to make of their character and feelings. The missionaries told me, as a fact, that when once a party was going out on a cattle-lifting expedition, they very innocently asked them to pray for their success. Even yesterday old Waterboer, a respectable elder, remonstrating with a Namaqua who, he heard, was going on a lifting expedition, asked him if he was not afraid. "Yes," said he, "I am afraid," but he "hoped that God would help!"

I questioned old Amraal about the history of his tribe, as I know him to be a truthful and trustworthy old man. He tells me that Gam Naka, or Leeuw-wol-haar, as he terms it in Dutch, is the title of his particular tribe. Wol-haar (or woolly hair) was the original name, but one of his female ancestors, a chieftainess, had, in some extraordinary manner, conquered and killed a lion, since which time they have added "lion," in order to perpetuate the memory of this occurrence. Amraal's own family name is Goedan, corrupted by the old Dutch colonists into "Goodman." His grandfather, the paramount chief of the Gam Naka nation, possessed, as before mentioned, what is now called Worcester, in the colony. This once powerful tribe is now dispersed among all the Namaqua, Griqua, and Bastaard tribes.

The Namaqua people, both men and women, have remarkably small and neat hands and feet. The lower limbs of the women, however, are very thick and ungainly, especially as they advance in years, when they assume a dropsical appearance. Their breasts are also very early developed, and attain an unusually long and pendulous form.

The Namaquas have some strange customs and strange superstitions, in common with all the other heathen tribes of South Africa. If a chief happens to have any hornless cattle (*koes-kop*) amnog his herd, he is obliged to give them to his nephews whenever they are demanded; and therefore whenever they wish to beg of one another they say, "Cannot I make a *koes-kop* of this" (cannot I claim it)? They believe also, among other follies, that a baboon will protect a female from lions and other ferocious animals. Once upon a time, they say, a woman ran away, and slept in the mountains upon a ledge of rock. At night the lions approached her, and would have devoured her but for the baboons.

CHAPTER XIX.

Start in search of Baines—An amicable Exchange of Wives—Meet Baines —Cattle-sickness—Peculiarities of Hottentot Language—Water-tubers —Riet Fontein—Gert, the Hottentot—Eland-shooting—Gnathais— Difficulties of watering the Cattle—Our Horses stolen—Ghanze—Whirl- winds—Thounce—Advance to Koobie—Poison-Grubs—Native Plants— Living on Ostrich Eggs—Plan of future Proceedings—Leave Koobie.

I MADE preparations to start in search of Baines, for whose arrival I had waited three weeks in the greatest suspense, without receiving any news either of him or of my brother. Mr. Krapohl kindly assisted me with the loan of a donkey-cart, which we gradually reduced in bulk till it was no heavier than a child's perambulator, retaining merely the axle with wheels, on which a board was placed, and on that a box mounted as a seat. During our preparations the missionary wagon arrived from Otjimbengue, but the party in charge of mine were laid up with fever, and it was now the general opinion that Jonker would stop my wagon. He had made two traders, Messrs. Cator and Smitz, pay dearly for passing, and report spoke now of his fetching me and Green back. He says that the white men had written their names on the trees in Ovambo Land, by which he judged they claimed the land which he had lately conquered. Baines also writing to the same effect, I determined to push on with all speed, to prevent any danger of extortion on the part of Jonker. Having bought two more horses, in addition to two screws my man had brought from the south, we started from Gobabies on the 16th of July, 1861, our untrained team dragging us along at a furious rate, at the imminent risk of breaking our necks. On the first night we slept at Wittvley; a rhinoceros had been shot there, and a man blew his hand off by the bursting of his gun. The rhinoceros had run so far (probably from Elephant's Kloof, as one that had been there had recently disappeared, and was traced in this direction) that the soles of his feet were said to have become quite loose. It was bitter cold sleeping on the open plain.

Passing Springbok and Droogevley on the 18th, we crossed Gemsbok Flat to the Qwaiep river, and, finding no water, proceeded to See-ace. Five lions had been killed here by the Hottentots since we passed here before. Lions do not like the trouble of catching game by day, excepting when they have good shelter. They are also very chary of the soles of their feet, and do not walk about in the heat of the sun, for fear of hurting them. Here I bought a sheep from one Klass Rymveld, who, with a friend of his, paid me a visit. The two had recently exchanged wives very amicably, but under very peculiar circumstances. One of the husbands was an irritable man, who had a very patient and submissive wife; the other, a patient man, had, on the contrary, a very irritable wife: so that they thought an exchange desirable. The irritable man found he had caught a Tartar, and wished to cancel the bargain, but the other found he had secured a prize, and determined to abide by it.

We saw many ostriches, and I noticed here, what I have often done before, that these birds have not the power of smelling at any distance like quadrupeds. The Bushmen say their ears and noses are in their eyes. They cannot smell far, but have a very keen sight, which makes up for this deficiency.

Continuing our course by way of the Qwaiep river and Eikhams, on the 20th of July I fell in with Baines, who was coming up with his wagon, and, leaving him to push on to

Gobabies, I proceeded to Barmen, where I expected to meet Jonker, intending to give him an opportunity of making the demand he was said to be contemplating, but determined not to submit to his impositions. I stayed at Barmen over the Sunday, and sent my respects to Jonker, who had heard that my wagon had passed, but made no further opposition. I then returned on Baines's track, and overtaking his wagon, we reached Eikhams on the 2nd of August. Here we delayed a few days to take the usual precautions of cleaning the harness, &c.

We found, while among Jonker's people, that there were strong rumours of a serious division in the tribe. It appeared that young Jan Jonker, some years ago, had killed a man wilfully during a lion-hunt, alleging that his victim's brother had been the accidental cause of the man's death. The brother, knowing it would be useless for him to dispute with the chief's son, said nothing, although several persons saw Jan Jonker kill the man. But now that old Jonker has been guilty of other arbitrary proceedings, besides taking the property of this murdered man and dividing it amongst his own children, the people are taking courage and withdrawing from him. All his brothers and their people are against him, so that Jonker has now a very small party of adherents, and a strong feeling exists everywhere against them.

Having remained at Eikhams until Thursday, the 4th of August, we trekked on, but only made half a mile, and then stuck fast, and in about five days, I think, we made no more than three miles, exhausting ourselves, and severely punishing the cattle. I had hired oxen and bought some: at length I hired a wagon and oxen to divide the load; but nothing would do; still we stuck fast. Happily Jonker did not, as most people feared, add to our difficulties by any attempt at extortion, or take any measure to prevent our passage through the country. He had probably thought better of it.

One night a great firing was heard, which greatly alarmed our Hottentot followers. It turned out, however, that a large meteor had fallen, and they fired blank powder to ward off the evil effects which this phenomenon is supposed to bring. On a previous night firing was also heard, but that turned out to be a rejoicing at the arrival of a trader (probably, as they surmised, bringing with him spirituous liquors) at the mines, 35 miles to the southward, and thither they all prepared to flock.

At last we got clear of the hills, and next day (Wednesday) I went ahead with my light cart, and, sleeping at the wagon of a trader (Oekenaar), waited till my wagon came up. We trekked on to See-ace, or Little Cheek (Kleine Bakjie), as Baines says, in contradistinction to the last place, which is called Groote Bakjie, or Great Cheek. Thence I again started ahead with the light cart, the wagon overtaking us at night. On the second night, shortly before reaching Wittvley, after dark I saw in the bed of the river, under the large mimosa trees, several fires, with people standing round examining something. I left the cart, and, creeping close up, found that they were busily engaged over the carcases of several head of cattle, newly dead of lung-sickness. This was a thunderbolt. Alarmed at the intelligence thus unexpectedly conveyed, I went up to the people to make some inquiry about the matter. They vouchsafed no reply, but glared savagely upon me as the author of the misfortune. Proceeding to Wittvley, I sent back instructions for the wagons to avoid this place, passing it through the fields to the south. Amraal had told me to bring the wagon on at once, and not to remain at Wittvley, as that village was supposed to have the contagion. I had felt delighted to think that my troubles were at an end, that I was so near my healthy cattle, and could now make a fair start, leaving behind all the cares connected with lung-sickness. Another week, I thought, and all would be right; but I now received the overwhelming intelligence that my cattle at Elephant's Kloof were dying of the dreaded disorder. It is impossible to describe my distress. I had been labouring hard for eight months, and had taken every possible precaution. Besides, I should now have to allay, if possible, the fears of the Hottentots. They already fancied all their cattle posts were infected, and already upbraided me as the cause of their ruin in the bitterest terms. I had, indeed, taken the precaution of insisting that my cattle should be sent a long way round to their station, and if they had not sent them far enough round it was their own fault. But Jonker's people had deceived me.

The missionaries, also, are in great alarm. They say, what are the Hottentots to do if their cattle all die? Their people must rob the Damaras and the Ovambos of their cattle; therefore great fears are entertained of the dispersion of the tribe if lung-sickness extends.

During my absence one of my wagon-drivers, a Berg-Damara named April, had stolen some of my cattle. He had just taken another wife (his third), and slaughtered my cattle for the occasion, without any thought of compensation. The Hottentots, though they knew he was doing wrong, made no opposition: they helped him to devour the flesh, and came to tell me of it immediately after. About the same time a murder was committed here by one of Amraal's Hottentots, who put to death his servant. The old chief was inclined to hang the offender, but commuted the extreme penalty to 250 severe lashes, the infliction of which nearly killed the man. Amraal, however, has proclaimed that from the beginning of the new year he will hang any of his people who commit murder. This new and stern administration of justice has set a great many of the people against him; but I hope he will succeed in his humanizing and civilizing efforts. What will become of his people when he dies no one can say: his son and successor has no ability.

The Hottentots have no word for "both;" they say, "this one and that one," or "two." They have also no word for thanks." If you do a Hottentot a favour and he makes any acknowledgment at all, which is not very frequent, he says simply, "Yes."

I bought all the missionaries' oxen which brought up my wagon, in order to get the doubtful ones out of the country and start the wagon off, and, remaining behind myself to look for the missing horses, followed at night. I slept half way, alone in the bushes, and next day overtook Baines and the wagon at Apollo's werft. There I heard that some of my men had shot a Bushman (a report which my servant, John, believed to be without foundation), and that Lamert, who is in the neighbourhood hunting, had sent to ask the chief, his father, for orders what to do with us. This promises fresh trouble. The cattle are now dying at the rate of four to six per day, and a great many more are falling sick. After a few days' rest, starting ahead with the sick cattle, I rode down and shot a fine eland, losing my powder-flask in the chase. Wolves and vultures follow us, scenting their prey, and we were surrounded by lions. During the night I was charged by a lung-sick ox in a state of delirium. I slept without any blanket, food, or water, and it was bitter cold. Next day our cattle came up, but most of the Damaras were missing. The horses looked so miserable that I took them on to the water at Riet Fontein (57 miles), which I reached late in the afternoon, tired, hungry, and thirsty. I was visited at night by a party of wild-looking Bushmen, who, fortunately, had some dried eland flesh killed by their poisoned arrows. This they gave me, and I appeased my hunger; next day, as the cattle did not arrive early, I went back to look for them, and at noon brought them up, with the sheep and goats. The cattle smelt the water a long way off, and came galloping on in a long line at a great rate; the weaker ones, calves and sick cattle, bringing up the rear, but miles behind. The Damara herdsmen, however, lost the race by a long way, as some of them did not arrive for several hours, and some did not reach it for many days, having knocked up and subsisted on the water-tubers, of which they fortunately found a field. These tubers (lerush, or chabba) grow in strong rocky places, and twine amongst scrubby bush. The juice of the plant is milky; the leaves long, narrow, lanceolate, and wavy; the flower small, white, and insignificant; the seed-vessel a long green pod, marked transversely with reddish-brown spots. The tuber, which is soft and watery as a melon, is large and slightly milky-white inside. A system of wavy white fibres runs through the tuber, giving it, when cut, an appearance like damask. It is brown on the outside. One which I measured was three and a half feet by two and a half feet in circumference, but I have seen them double that size. The juice is of a sweetish taste. This root constitutes both food and drink to Bushmen at some seasons, and to Damaras also.

At midnight Baines arrived with one wagon, and the other, with the oxen, reached the fountain on the following night. Before the arrival of the latter, one of the Damaras, and I both fancied we had heard lions about, where the horses were tethered. We scrambled after them in the dark, through thorn bushes which scratched my face and nearly tore my eyes out, but to no purpose. During the night I was seized with severe rheumatic pains all over, no doubt brought on by sleeping in the frost without a blanket after such a fatiguing ride.

At Riet Fontein I made further inquiries into the story of two of my Damaras having killed a Bushman, and it turned out that the report was true; it was alleged, however, that the Bushman had fired upon them first, and had refused to hold a parley. The Damaras had concealed the matter from me, fearing I should be angry with them, and I punished them now with from 75 to 100 lashes each for bringing their bows and arrows to my wagon, and for concealing the fact, and resolved to have further inquiries made into the matter.

On Sunday, the 1st of September, 1861, the Hottentot Gert, our evil genius, arrived, having followed us on the pretence that my refusing his services arose from a misunderstanding. He brought with him the other wretch, Jan, who stole away from us at Wittvley. Hoping to send letters back by him from Koobie, I take him on as far as that, on the promise that he will behave better in future. I despatched them, with some Damaras, to the fountain at Gnathais, directing them to open it, and on the 2nd of September started both wagons for that place. At night I sent the cattle back for another drink, and next day Baines and I went forward with the horses and overtook Gert and his company on the road. I got an attack of cholic, from eating a small quantity of bitter melons to quench my thirst; and I shot another eland, which the Bushmen were to have conveyed to the passing wagons, but they carried it all off for themselves: we fed the thirsty horses with melons.

On reaching Gnathais (Wahlberg's Pit), along with the Damaras, I kept them digging all night and part of next day. We have no food, so eat roots. Riding out, I fell in with a thirsty Bushman lying under the shade of a bush, and gave him a drink out of my canteen. The wagons were long in coming up, the oxen having run away. I had great trouble in watering only a portion of them here; and as the well was rather deep and steep, it required three men to dip it up and hand it out in buckets to the topmost man, who poured it into the small hollow scooped out for the purpose.

This was, however, the least part of the labour, for it was as much as all the other hands, myself included, could do to keep back the rush of cattle which stormed up to the water's edge, determined to take it by main force. The yelling and running about of men and women, and the flying about of stones and knob-kerries in order to drive off the cattle, was something terrible. I proceed to G'nuegga, with cattle and horses, to open the water there, as the quantity here was insufficient.

Reaching G'nuegga (or Fort Funk, as I christened it), I had a long and fruitless hunt after elands. When at last I fell in with some, I had no gun with me. Dig, dig, dig, was now the word for two days and a half. The savages round us seemed to glory in the misfortunes that leave me without cattle, because it provides them a continual feast. Five or six large oxen a day would suffice for a large number of white men, but these hungry wretches, who, perhaps, never had such a windfall in their lives, are always complaining of hunger, and seem to grudge the very dogs their share. In spite of my injunctions, they will eat even the rotten lungs.

I receive disastrous accounts of the wagons, which I hear are stuck fast near Riet Fontein, and, owing to bad oxen and drivers, cannot get along. I also heard from Gert that Lamert had sent after us to catch my Damaras, and take them back prisoners to Elephant's Fountain.

I sent a message to the Bushman chief, about the man who had been killed by my Damaras, inviting him to come and hear the case inquired into. Discover a pit in Limestone Rock, one mile south-east of G'nuegga. There is one also about two miles west of G'nuegga, which I found formerly.

Finding Gert and his companion, in spite of their promises, so utterly useless and lazy, I paid and discharged them, after which my own people, the Damaras, mutinied. I gave all who were unwilling to proceed farther full permission to leave me. One or two only volunteered to remain; the rest had put themselves in an offensive attitude, arming themselves with their kerries, in which manner they came to demand payment, which I peremptorily refused. It did not take them long, however, to come to their senses, on their knees, and the women, having an eye to the flesh-pots, did not like bearing their husbands' luggage back to Damara Land, feeding on roots and lunches. They had threatened to return to Damara Land in a body, because I objected to their women accompanying my caravan any farther, as they have all along been deceiving me with the promise that the women were only going a little farther, and would return, which I now saw to have been from the first a mere piece of deception.

Some of my dogs have died in consequence of an inflammation of the lungs and a swelling in the throat. I think the lung-sick beef must have something to do with it. My pack is now diminished from upwards of fifty to under thirty.*

Misfortunes never come singly, and troubles seem never to cease, for this morning (the 12th of September), intending to go out hunting for elands, I sent for the horses, which we had been in the habit of letting run all night, as there were no lions about. They were gone, and the spoors of Gert and his company, with those of three Bushmen, are on them; they are driving them to the south. John was very eager to follow them, and Baines volunteered also on that service; so I sent six Damaras with them, four being armed, to overtake

* I have since, as will be seen at a later period, had a great deal more experience of this disease amongst the dogs, which I find to have commenced simultaneously in an epidemic form throughout the whole length and breadth of the colony—indeed, from Cape Town as far north as the Zambesi at least. One out of a hundred only who took the disease have escaped, and that one is generally blemished for life with some local malady, usually confined to the back of the loins, which seems to be partially paralysed for ever after.

them, if possible. I would follow them myself, but I am
afraid that if I saw a Hottentot mounted on one of my horses
I should not wait for him to fire at me first, and Baines's
opinion seems to be that we should not fire the first shot, but
try to take the thieves alive. For my part, I think that
a case of this kind would justify the most summary measures
of reprisal; but I will not trust myself to go, John seeming
determined to bring back the horses, and as Baines volunteers
to go with him, I think it may be for the best. He is cool and
discreet, and will see that my Damaras perpetrate no cruelties.

I find that one old and useless horse has been left, by
which alone I could have identified Gert as the thief, even
if we had not found their spoors. It is only on fuller reflection that I feel the extent of my loss if we do not recover
the horses; but we *must*, even if I have to follow them for a
month. I begin to reproach myself for not going personally
in pursuit—not that I believe I could really do more than
John or Baines; on the contrary, I have become so enfeebled
by illness and a succession of difficulties and strong excitement, that I do not feel half the man I was; yet I believe
I can shoot perhaps a little better than Baines or John,
and, having travelled more in it, know more of the shifts
employed by the rogues of this country.

I am now left in charge of the cattle, sheep, and wagons;
I sleep on one side of them and a couple of sick Damaras on
the other—a very formidable guard certainly, in a country
where we are often visited by troops of upwards of a hundred
armed Bushmen at a time. I had sent several of my best
men on to Ghanze to open the water there, and, with one or
two useless exceptions, all the remainder are with Baines and
John.

14th September.—No signs of Baines or John, yet their long
absence shows that they are going the right way to work for
catching the thieves, by pursuing them up to some resting-

place or other. Temperature to-day, max. 96° in the shade, min. 70°. Water boils $205\tfrac{4}{10}$°. At night, about 10 o'clock, the Damaras which I had previously sent on returned from Ghanze, having opened the water there.

On Sunday evening three Damaras return from the pursuit; Baines and John are not with them. They report that the first night they came to a Bushman village, and learnt that Gert and his companion had just passed with the horses. They say that Baines and John, not being able to keep up with the Damaras, had sent them on in advance, and were themselves to follow, but by the moonlight they must have lost the spoor, and evidently struck out for Riet Fontein. The Damaras say they came up with the thieves, but unfortunately were discovered. Gert had left spies behind, to run and tell him if they saw anyone in pursuit, which was done—the Bushmen seeming to be all in league with him and inclined to screen him. He had shot three of the horses, for fear they should fall into our hands, and made use of to overtake them; then, each mounting one of the best horses and leading another with the hand, he and his companions galloped off when our people made their appearance. It was useless, they say, to try and overtake them, and the distance was too great for shooting at them. I afterwards discovered that all this was a very ingeniously got-up story of the Damaras to deceive me. They brought three tails, probably those of zebras or gnus, caught and killed by Gert, in proof of what they had seen, and we believed them. It appears they left John and Baines to go on the spoor alone, stopping behind with all the water, food, guns, &c., and probably went to sleep; and, as they returned minus a good riding-ox, it is very probable that they slaughtered and ate it, while poor Baines and John suffered hunger and thirst on a fruitless chase.

Immediately after the arrival of the Damaras I packed off

three other men to look for Baines and John, and carry them water and flesh. Three of the Damaras, on returning from the pursuit, took up their spoor and followed them in the direction of Riet Fontein.

I hardly know now what is best to be done, but wait until Baines comes, to receive his report. It is evident that we cannot go on to the lake; we cannot go back, and I shall soon have no food for my people. I wrote a letter, and sent three Damaras to Amraal's with it. Gert has displayed such good generalship already, that one needs to be clever to catch him, and yet it would be too humiliating to let him escape altogether. Another attempt must be made, but how?

At length Baines and John arrived, bringing a letter from Amraal, which they found at Riet Fontein, requesting me to come back and bring the two Damara homicides with me. Baines had been 50 miles back, but had finally lost the spoor after a long, toilsome, and fruitless search. As it seems, therefore, impossible to overtake the thieves on ox-back or on foot at present, since they are evidently in league with the Bushmen, who were stationed at different points to warn them of danger, we decide upon going on as if we were giving up the pursuit altogether, hoping that after awhile their confidence will be restored, and that we shall surprise and recapture them in some unguarded moment.

This being settled, I despatched letters to Amraal, and sent three Damaras with them. They are to go as far as Otjimbengue. We moved forward on the following day, and on Thursday morning reached Ghanze, where I prepared for a week's stay. From here I sent some of the Bushmen to make a further effort for the recovery of the horses. I placed one of my men, Kapenyoka, at Baines's disposal as a sort of body-servant. He is one of my best men, and more intelligent than the generality, so that I have no doubt, under Baines's tuition, he will turn out a smart and useful fellow.

A kind of horse-fly is very troublesome here, and its bite painful. It makes its attacks only in the shade. These flies live in trees, near water, and are not found on the plains. Their wings are closed, and spotted with black. They bite through one's clothes. Animals can brush them away, more or less, with their tails, but the poor Damaras and Bushmen, having no such useful appendage, are constantly assailed on their backs and shoulders. They carry in the hand the tail of an eland or gnu, which they have to employ continually. The oxen whose tails have been cut off suffer terribly.

At Ghanze we were, as usual, visited daily by parties of Bushmen, in larger or smaller numbers. Water boils here at $205 \tfrac{4}{10}°$: temperature 93°. Last week the thermometer was only once as low as 56° at sunrise. Generally 62°, from which it ranges throughout the day to 94°, 96°, 98° Fahrenheit in the shade. Latitude of Ghanze, 21° 33′ 39″.

The large parties of Bushmen not returning, they have very likely gone to make friends with Gert for the sake of a feed. These children of the wilds think nothing of walking 100 miles and back for the sake of a few meals of flesh.

I made some further collections for Dr. Harvey of seeds and specimens. I pointed out to Baines the spoors of numerous elephants in the limestone rocks near the fountain. The mud in which they had trodden has evidently since become solidified, bearing a deep and obvious impression; I wonder that Andersson and Green have never remarked it.

The wind blows every day, but if it were not so one does not know how great and destructive fevers and other diseases would be, owing to the great heat. Whirlwinds are more frequent and stupendous here than I have found elsewhere. They seem more prevalent even on calm days. Their approach is announced by a startling sound in the air like thunder, and can be traced by the occasional loud gusts that sweep and rush through the air with a fearful sound, or, tearing through the forest trees, carry up a cloud of dry leaves, grass, sticks, and dust. The traveller hearing it approach secures his hat in good time, and shuts his eyes and nostrils until the gust is over. Huge lurid columns are seen chasing each other in the most lively manner. These whirlwinds generally commence after the sun has passed the meridian.

Our Bushmen returned without the horses, but they heard that Gert is still amongst the Bushmen, keeping in the desert by day and sending the horses to drink at different pits by night. They advise us to catch some Bushmen in that neighbourhood, and compel them to bring us to the horses. As the cattle are showing renewed symptoms of the lung-disease, we resolved to take the wagons on at once as far as Koobie, while we have oxen left to draw them.

Tuesday, 24th.—We started at 1 o'clock, and, having travelled 12 miles, we intended inspanning again at midnight, when the moon would be up, but the oxen had run away, and we did not start till 7.30. A dissel-boom, or pole, of one of the wagons breaking, it stuck till 10.30 A.M., from which time till 5 P.M. we kept on, and reached Thounce with one wagon. I saw a troop of elands, and had a long shot. In the forenoon there was a strong glare peculiar to the country, especially in the autumn. Next morning we had frost and ice. After sunrise, thermometer 40°; after 10 o'clock, 86°; 1 o'clock, 96°; sunset, 90°; 8 o'clock A.M., 80°. The second wagon joined us at 8 o'clock, the oxen having been in yoke since 7.30 A.M.

26th September.—We started early, but the oxen were so tired and done up, and the heat was so great, 100° in the shade, that it took us till dark to make 12 miles. We were then still 11 miles from Koobie, to which place I sent the oxen to drink, as they could pull us no farther. The Damaras, who went in advance yesterday with picks and spades, will, probably, have procured sufficient water for them. All the pits are dried up. I never saw them so empty before.

Lat. 21° 14′ 36″ south. Therm. 62° at sunrise; increasing gradually till 1 o'clock, when it stood at 100°; sunset, 92°. Distance from Thounce, 12½ miles.

While waiting here, I strolled out to collect some seeds and botanical specimens. The level surface of the limestone looks everywhere like the former muddy bed of a great inland sea or lake.*

No cattle returning, I became very anxious. In the evening two Bushmen came from the water at Koobie, having been sent to inform me that hitherto the Damaras had not been able to get sufficient water for the oxen, but that they were digging another well, and hoped by to-night to get enough for all our wants.

Sunday, 29th September.—The wagons reached Koobie last night very late. The heat at noon was terrific. Friday there was not sufficient water. I had the pit enlarged, and placed the wagons in a good position for a lengthened encampment. It has not rained here during the last year, consequently water is very scarce. The trees are mostly dead, and the grass very scanty, and crumbling to powder. It was some days before all our cattle got water, though we dug at several places, and through the rock. Therm. noon 102°; 8 P.M. 85°; 9 P.M. 75°. Three oxen died; next day ten more—the heat being intense, and the effluvia from the carcases so offensive as to produce nausea and disease.

2nd October.—I sent John to the lake with a message to Lechulatèbe, informing him that it would be impossible for

---

* These limestone crusts are very extensive. They extend, to my knowledge, from Riet Fontein (or Tounobis) on the west, to the Shua on the east, and generally occur only in the lowest land. The crust of these limestone plains has usually an uniform thickness of about one foot.

me to visit him now, as my cattle were diseased and would infect his. I now re-inoculated about forty head of cattle, and many of my people were attacked by fever. Max. therm. 74°.

5th October.— The weather changed, becoming cloudy, with rain, the first notice we have had of approaching summer. On the day following (Sunday), max. therm. 90°, I got some of the poison-grubs mentioned in my journal of 1852. I had sent some Bushmen after them, promising them a handsome reward, as I wanted Baines to make me a sketch of the insect and cocoons. They were absent three days, but returned at length successful. They are only found in certain localities, where the maruric papierie (a trefoliate bush) grows. The beetle is a yellow spotted one. The grub is incased in a small earthy cocoon, in size and appearance much like that of sheep's dung. The Bushmen take one by the tail and squeeze it gradually until a drop of yellow fluid comes out of the mouth, with which they dot the blades of their spears and arrows all over.*

All our Bushmen mysteriously disappeared during the night, probably alarmed at our star-gazing. Water boils here at $208 \frac{8}{10}°$; therm. 98°. Two gnus were observed approaching their wagons on their way to the water. We were duly informed of it, and, on looking up, I observed they were eyeing us at a distance of 100 yards. My gun standing alongside of me, I fired at and struck one, which did not go far before the dogs pulled him down. I took a photograph of the creature.

8th October.—Last night a leopard (perhaps one of the hunting-leopards)† caught one of our dogs and carried him off, but could not kill him. This is the second time this has

---

* Some further particulars respecting the poison-grub will be found in the Appendix.

† Mr. Layard tells me this must be a new and undescribed variety of leopard. He has not met with it before. It is not the Indian hunting-leopard, nor the one of the Cape colony.

been done, and probably by the same animal. I called from my bed to the men to rush out and send the dogs to the attack, which they did. I joined them immediately, and the dogs soon had him at bay in a tree. The confusion from the noise made by the Damaras, all at cross purposes, was marvellous; but I levelled at the beast in the darkness as well as I could. Before, however, I could pull the trigger, he sprang from the tree, when I urged the dogs on, shouting as loud as I could. The pack of dogs, thirty in number, fell on him as he reached the ground, and gave him an awful mauling, but were so crowded round him that it was impossible to distinguish him. It was as much as I could do to prevent the Damaras, in their state of excitement, from shooting amongst the dogs and men. One old Damara woman seized the leopard by the tail almost simultaneously with the dogs, nor would she relinquish her hold until the animal had succumbed. These women often do most desperate things, both in war or in the chase, in order to encourage, shame, or stimulate the men.

Wednesday, 9th October. — Some Bushmen brought us ostrich eggs. They said that the other Bushmen fled in the night for fear of my camera. Therm. from 72° to 102°, and 98° near sunset. Next day I received a note from John at the lake, asking me to send pack-oxen to meet him half way, which I did. He tells me that Snyman (the Bastard or Hottentot) is there. Therm. 80°. Every day lately has been cloudy and stormy.

On the 11th John returned from the lake, with a message from the chief, who says he will leave it to me to come with my cattle when I think it safe. He would have asked me to come at once, but he has more confidence in my prudence than his own, anxious as he is to see me; and he thanks me for my consideration and the precautions I have taken.

One of the men, in the course of conversation, told me he

had killed a crocodile. I asked how this was, as in former days they would not do it. He said, "They had seen white men do it, and no harm come of it, so they had taken courage."

Last night there was rain and thunder. The mornings were cool and cloudy, the evenings clear, but rain and thunder generally ensued during the night. On the morning of the 13th it was still raining when the Bechuanas left for the lake, but afterwards cleared up. Therm. 80°. It grew warmer during the two ensuing days, and we took some lunar observations.

On the 16th of September we commenced preparations for a move forwards. The grass is beginning to spring. The sandal-tree blossoms perfume the air with their dense, drooping clusters of pale yellowish flowers. The flower is five-petaled; the stamens vary from twenty to thirty. We had some difficulty in finding the pistil, as the flowers on several racemes were perfectly devoid of it. At length some branches were discovered, on which two or three flowers amongst the lot bearing stamens only were found, but divided or spread in six parts. The motseara is the only tree that seems clothed, but it is with dense clusters of brown winged seeds of the last season. I noticed that on the sandal-wood trees, where a branch has broken off, the young sprouts grow so plentifully at the extremity that it forms a regular bushy cluster, and looks like a parasite threatening to kill the tree.

I have heard a story told of some tree which produces insects instead of seeds, and that these, after feeding on the tree and growing to maturity, burrowed underground, and from them the young trees or plants shoot up. This story has been very much ridiculed; but it may, perhaps, have some foundation in truth. It may be in the order of nature that the seed of that particular plant may need to undergo some preparation in the stomach of an insect, and then to be deposited by it a certain depth underground, to enable it to germinate, just as the seeds of many gourds require to be deposited in the manure of certain animals, such as the elephant, the buffalo, rhinoceros, &c.

Thursday, 17th September.—Therm. 96°. Some Bushmen arrived from Ghanze. They had seen Gert at Obekokolleleko. Lamert had sent people after him, but he had killed the remaining horses and hidden himself in the desert, and they returned unsuccessful, but got his broken gun, which they found with the Bushmen by whom he was concealed. I learnt, further, that two wagons are coming through Amraal's country to the lake. If not the Polsons, they are, perhaps, my brother's, as I trust will prove to be the case.

The first showers having now fallen, I knew it was time to be on the look-out for the beautiful, fragile, sweet-scented, white bulbous flower I had taken such pains to have propagated at Cape Town. Not feeling well, for want of exercise, I strolled out in search of this plant, and was rewarded by finding one. The flower had not yet opened. I wanted Baines to make a sketch of it, he having formerly made one at Cape Town from my description only. This I had forwarded to Dr. Harvey, of Dublin University. I rejoiced in having now the opportunity of getting a sketch made from the plant itself, as I wished to send it to Dr. Harvey for comparison with the other. This flower, and another sweet-scented amaryllis—a cluster of large white lily-flowers—with others growing on bushes, are all ephemeral. They open and emit their fragrance during one short night only, and die away as soon as they are exposed to the sun's rays.

The ground had been so thoroughly parched here that even the late heavy showers have not as yet produced much effect upon vegetation. Only the slightest tinge of green is perceptible in the motseara trees, the deep colour of their brown clusters of seed-pods contrasting strongly with the faint efforts of reviving nature. A few flowers already appear, but they fade so soon, especially those of the bulbous tribe, that it seems as if we had never seen them. The thick, drooping clusters of yellow-blossomed sandal flowers lasted for only about three days.

Monday, 21st October.—We have now been busy for some days repairing and patching broken things and places, sighting guns, &c. Wind has been blowing for some days, but to-day it was quite still; and although yesterday there was scarcely a vestige of green on the trees, to-day the motseara is green as grass, and the clusters of seed-pods on the distant trees are now almost lost in the abundance of the fresh green foliage. The white bulbous flower also, about which I have been so anxious, has at length opened, and diffused its delightful odour on the evening air. Baines has made a successful sketch of it.

The corolla of this flower (*Choretis* [?]) is of a delicate transparent white, hexagonal, and like a cut wine-glass. The calyx is divided into six long, narrow sepals, spreading in a funnel shape, and about three and a half inches long, to which the corolla adheres for about three quarters of an inch. The stem is about six inches long and about one-eighth of an inch thick, but bulging out a little over the ovary. It has two bracts, spirally turned. The leaves are of a dark glossy green, narrow, and also spirally turned. The margin of the corolla is scollopped out, leaving at each angle a pair of tongues, between every pair of which grows out a stamen inclining inwards. The filaments of these stamens grow spirally up in the wall, or side of the corolla, but is visible only in dried specimens. I have shown specimens of this flower to Dr. Pappe, our colonial botanist, and Mr. M'Gibbon, of the Botanic Garden. It was quite new to them.

Wednesday, 23rd.—Therm. 96° at 12 o'clock; in the sun, 123°. In the sun, at 3 P.M., 136°; in the shade, 100°.

We remained stationary at Koobie till the end of the month, receiving occasionally visits from parties of Bushmen. One of these brought intelligence of Gert and the stolen horses, which induced me to despatch eight of my Damaras, well armed, to make a fresh attempt at regaining them. But the Damaras, faithless and cowardly as almost all these savages invariably prove, returned without making any real effort to accomplish the object of their expedition; having also plundered their Bushmen guides, for which I made an example of two of them.

Our cattle were still dying off. My health and spirits failed under the anxiety and disappointments I was suffering. There was no hunting, and we were reduced to such extremities for food that we subsisted almost exclusively on ostrich eggs, brought by the Bushmen, of which we got very tired. I have heard it said, but have never seen proof of the assertion, that an ostrich egg is not more than a sufficient meal for a hearty man, and some one has even asserted that he could eat three. If this be so, both appetite and digestive powers must be very great. Baines and I, who have very good appetites, could manage about half an egg between us. We prepared it in the simplest possible manner—frying it with water instead of fat.

On Friday, the 1st of November, I followed the spoor of a rhinoceros for five miles, without success. The note of the *khali*, or rhinoceros-bird, is a thrilling cry to the hunter, whose pulse beats quicker when he sees this bird on the track of a rhinoceros, as we did now. It is a rule among sportsmen here never to speak when on the spoor of game. One should learn to do without talking, except when on a large open plain, where you can see all before and around you. Any sound above a whisper is carried a long way in the stillness of the wilds. Even when I am hunting with Bushmen we never speak when tracking; we indicate by the way we carry our hands, a sort of dumb language, whether we have lost the spoor, where we have lost it, where we last saw it, whether other spoors have joined or crossed it, whether the game has fled; whether we have seen anything, and what seen, &c., &c. The horseman waits patiently behind, or silently assists in recovering the lost spoor, while others look out ahead.

When the Damaras bring grass to thatch their huts and protect themselves from the coming rain, we may consider it a good sign that they really expect it to rain, as they generally put it off until the very last moment.

At night I lay by the water for rhinoceroses, and just when we were giving it up, an elephant came towering over us and approaching head-on. I fired over Baines's head into the elephant's breast, and he went off, bleeding very much from the trunk. Next day I sent the Damaras with dogs in search of the beast, but they returned at night unsuccessful. He had gone straight off and joined a troop, and led them away. These animals are sagacious enough to know exactly where to find their companions; we could not do the same unless we had fresh spoor before us.

3rd November.—Snyman, who had been trading for the Polsons last year, arrived at our camp. I had just made known to Baines that as we could not, in fairness, leave this place before giving my brother ample time to come to us, I intended going in a northerly direction hence for the purpose of hunting, and also to try and connect Andersson's trip to the Ovambo river with Lebèbè, leaving one wagon at the Kopjies, and returning in March to start for the Zambesi in the following month. Snyman's arrival proved of service, as he agrees to stay with me now, and afterwards guide us to Shapatani's and the Zambesi, at certain rapids which he knows, and about where the Kaloma river, according to Dr. Livingstone, runs from the north. Snyman engages to do all he can in the way of hunting and trading for me, and if he succeeds in getting 8000 lbs. of ivory by trade and shooting, I shall give him the two wagons and the remaining cattle when we are fairly launched in our boat on the Zambesi. My idea is to embark at the junction of the Gwai, which is probably the river called by Livingstone, Longwè, as, I hear, there is no "fly" there. On the way we will visit the falls, to sketch and photograph, and then proceed to Tètè down the stream. I am glad to find that the position in which I placed Shapatani and the Gwai, from native reports many years ago, turns out very correct. Snyman compares the rapids which he has seen below the Victoria Falls to the leaping of springboks.

The men whom I had sent out while here to train the young dogs, and endeavour to supply the larder, brought in two civet cats (muskiate kat). The odour, however pleasant to some people, is to me disagreeable in the extreme: dogs and Damaras, however, like it. The head of this animal resembles that of a fox, the tip of the nose is black, with a dark streak extending from there to the eyes; another streak diverges from it at right angles to the edge of the upper lip, half-way to the jaw, leaving a spot of white from which the whiskers spring; the ears, which are nearly bare, are also large. The general colour of the animal is a cold ashy-grey, paler behind; a dark sepia-brown stripe extends from between the eyes, over the forehead and the back, to the tail. Two other brown dots, indicative of stripes, are placed on either side. The tail is ringed alternately black and white. It has five black claws, partially retractile, on each foot, which is naked underneath up to the heel.

Sunday, 17th November.—My insect catchers brought me a lemur (*Galaga Moholi*, or *Mogoèli*, of the Bechuanas), but they had half-killed the poor animal, and I had to nurse it tenderly to save its life. We have now been at this place since the 29th of September, have lost between thirty and forty head of oxen, and have shot nothing fit for food during all that time. We have burst two guns, and had other accidents. As the cattle are no longer dying, and our people have finished eating the hides, and there are scarcely any gum or berries to be found, we must perforce make a move either to the lake, or to some place where game is to be got.

Yesterday was excessively hot, though the thermometer stood only ninety-six in the shade; we felt it the more as, since the cooling influence of a few showers, the temperature has been pleasant lately, generally about ninety in the shade at the hottest. To day the wind blows from eastward, and brings with it an agreeable cool breeze from where the rains must have fallen heavily last night. The air is very much charged with vapour, and though there be not a cloud in the sky, one can feel and smell its dampness and its freshness, and see it in a hazy, misty appearance of the air. We trek on four miles to the small pit called Mahalapyè, which we only reach at night. Bechuanas arrive from the lake.

END OF VOL. I.

www.ingramcontent.com/pod-product-compliance
Lightning Source LLC
Chambersburg PA
CBHW051851300426
44117CB00006B/349